MW00760947

Paramagnetic Resonance
of Metallobiomolecules

ACS SYMPOSIUM SERIES **858**

Paramagnetic Resonance of Metallobiomolecules

Joshua Telser, Editor

Roosevelt University

Sponsored by the
ACS Division of Inorganic Chemistry, Inc.

American Chemical Society, Washington, DC

Library of Congress Cataloging-in-Publication Data

Paramagnetic resonance of metallobiomolecules / Joshua Telser, editor.

    p. cm.—(ACS symposium series ; 858)

    Includes bibliographical references and index.

    ISBN 0–8412–3832–4

    1. Metalloproteins—Magnetic properties—Congresses. 2. Electron paramagnetic resonance—Congresses.

    I. Telser, Joshua A., 1958- II. II. American Chemical Society. Division of Inorganic Chemistry. III. American Chemical Society. Meeting (223rd : 2002 : Orlando, Fla.) IV. Series.

QP552.M46P37   2003
572'.6—dc21   2003051856

The paper used in this publication meets the minimum requirements of American National Standard for Information Sciences—Permanence of Paper for Printed Library Materials, ANSI Z39.48–1984.

PRINTED IN THE UNITED STATES OF AMERICA

Foreword

The ACS Symposium Series was first published in 1974 to provide a mechanism for publishing symposia quickly in book form. The purpose of the series is to publish timely, comprehensive books developed from ACS sponsored symposia based on current scientific research. Occasionally, books are developed from symposia sponsored by other organizations when the topic is of keen interest to the chemistry audience.

Before agreeing to publish a book, the proposed table of contents is reviewed for appropriate and comprehensive coverage and for interest to the audience. Some papers may be excluded to better focus the book; others may be added to provide comprehensiveness. When appropriate, overview or introductory chapters are added. Drafts of chapters are peer-reviewed prior to final acceptance or rejection, and manuscripts are prepared in camera-ready format.

As a rule, only original research papers and original review papers are included in the volumes. Verbatim reproductions of previously published papers are not accepted.

ACS Books Department

Contents

Nuclear Paramagnetic Resonance

Magnetic Circular Dichroism and Related Techniques

Indexes

ix

Preface

This ACS Symposium Series volume is based on a Symposium sponsored by the ACS Division of Inorganic Chemistry, Inc. at the American Chemical Society (ACS) National Meeting in Orlando, Florida, April 7–11, 2002. The symposium covered a wide variety of metallo-protein systems, as well as metal nucleic acid systems, and inorganic complexes of biological relevance. The common theme was to employ spectroscopic methods that take advantage of, rather than fearing, the presence of unpaired electrons in these systems. Tutorials on paramagnetic resonance techniques, including electron paramagnetic resonance, para-magnetic nuclear magnetic resonance, magnetic circular dichroism, and Mössbauer effect spectroscopy were provided. These tutorials were quite well attended and were a valuable service to the inorganic and biochem-ical communities.

I thank the ACS Division of Inorganic Chemistry, Inc. for spon-soring this symposium, which is somewhat far afield from the efforts in synthetic inorganic, organometallic, and solid-state chemistry that often dominate the presentations in this Division. I also thank the ACS Petrol-eum Research Fund for a special education grant that supported the attendance of two speakers from Europe and I thank Marc A. Walters (New York University) and William B. Tolman (University of Minnesota) for their help in this area. Financial support for the symposium from Bruker BioSpin, from Cambridge Isotope Laboratories, and from Wilmad Glass is very much appreciated. Lastly, I thank everyone who had a role in this symposium and/or this volume for their hard work and insight into this interesting area. I hope that the reader will gain a renewed appreci--ation for the contributions of paramagnetic resonance techniques toward understanding complex bioinorganic systems.

Joshua Telser
Chemistry Program
Roosevelt University
430 South Michigan Avenue
Chicago, IL 60605

Paramagnetic Resonance
of Metallobiomolecules

Chapter 1

Paramagnetic Resonance of Metallobiomolecules: Introduction and Overview

Joshua Telser

Chemistry Program, Roosevelt University, 430 South Michigan Avenue, Chicago, IL 60605

The historical background and goals of this ACS Symposium Series volume is given. An overview of the general contents of the volume is provided along with some aspects of paramagnetic resonance techniques, particularly electron paramagnetic resonance (EPR) and its related techniques electron spin echo envelope modulation (ESEEM) and electron nuclear double resonance (ENDOR). Paramagnetic nuclear magnetic resonance (NMR) and magnetic circular dichroism (MCD) are briefly mentioned as is the scope of the volume in terms of the metallo-biomolecule systems studied.

At the American Chemical Society National Meeting in San Francisco in April, 1997, Prof. Edward I. Solomon, Stanford University, organized a very successful symposium within the Inorganic Division on "Spectroscopic Techniques in Bioinorganic Chemistry". In the following years, the Inorganic Division sponsored a number of interesting symposia, but none was devoted specifically to the application of spectroscopic methods to bioinorganic chemistry.

I was therefore inspired to organize such a symposium within my particular area of interest, namely, paramagnetic resonance techniques. There was an additional inspiration, however, which cannot go unmentioned. The ACS National Meeting in August, 2001 was scheduled to take place in Chicago, my birthplace and current home. Chicago was also the birthplace in August, 1941 of Brian M. Hoffman, currently Professor of Chemistry and Biochemistry and Molecular Biology at Northwestern University, in adjacent Evanston, Illinois. The opportunity to have an ACS Symposium devoted to paramagnetic resonance studies in bioinorganic chemistry located in Brian Hoffman's birthplace and timed on the occasion of his sixtieth birthday anniversary was too good to pass up.

Sadly, this was not to be the case. The very same week that the ACS meeting would take place in Chicago, the tenth anniversary International Conference on Bio-Inorganic Chemistry (ICBIC-10) was scheduled to take place in Florence, Italy. Florence was the site of the first ICBIC, organized by Profs. Ivano Bertini, Claudio Luchinat, and others, and the tenth anniversary ICBIC was expected to be a very special event, which Brian Hoffman and much of the bioinorganic community in the USA (as well as Europe) would be expected to attend. As a result, the Inorganic Division suggested that the proposed symposium on paramagnetic resonance be postponed. ICBIC-10 was indeed a great success and highlighted a lecture by Brian Hoffman on the developments in bioinorganic chemistry since the first ICBIC in 1983.

The ACS Symposium on "Paramagnetic Resonance of Metallobiomolecules" took place at the next ACS National Meeting, in Orlando, Florida in April, 2001. Although the location was quite distant from Brian Hoffman's birthplace, the timing was within roughly 1% of the 60th birthday anniversary date; an acceptable error.

The goal of the Symposium was to highlight two aspects of paramagnetic resonance. The first was the application of a specific technique to multiple problems of interest in bioinorganic chemistry. The second was the application of multiple techniques to a specific problem in bioinorganic chemistry. An overview of certain aspects of both of these approaches will be presented here.

Paramagnetic resonance is not a widely used term, but perhaps should be more so. The most obvious forms of paramagnetic resonance, those in which Brian Hoffman has made many contributions, are electron paramagnetic resonance (EPR) and its 'double resonance' extensions, electron-nuclear double resonace (ENDOR) and electron spin-echo envelope modulation (ESEEM), of which more below. That EPR often is referred to as electron spin resonance (ESR) and, more recently, electron magnetic resonance (EMR), obscures the necessity of a paramagnetic system for this technique. This confusion in nomenclature is reflected in the titles of two classic texts in the field, that by Poole (*1*) and that by Abragam and Bleaney (*2*). The latter text is the best source of information on the fundamentals of much, if not all, of the EPR spectroscopy in this volume.

EPR was discovered by E. Zavoisky in 1944 in Kazan, then USSR (now Tatarstan Republic, Russian Federation). Its origin was in the wartime development of radio- and microwave-based detection of aircraft, ships, and so forth. A legacy of this wartime project is the use of code names for the various bands of the microwave spectrum: X-band for 8.2 - 12.4 GHz (with the corresponding EPR spectrometer operating at roughly 9 - 9.5 GHz); W-band for 75 - 110 GHz (spectrometer at 95 GHz); and, as anyone associated with Mr. Clark E. Davoust of Northwestern (formerly of the University of Chicago) knows, K_a-band for 26.5 - 40.0 GHz, with the corresponding spectrometer operating at 35 GHz; not the incorrectly used Q-band (*3*). These code names have added an aura of mystery to EPR that is perhaps unnecessary, but does provide a certain camaraderie among its aficionados.

EPR in its early days was, as was the case for its more popular relative, NMR, a continuous wave (CW) technique. The application of time resolved (pulsed) methods to NMR, pioneered in the mid 1960's by Dr. Richard R. Ernst (then at Varian Associates, Palo Alto, California), revolutionized that technique. The importance of FT-NMR in synthetic organic chemistry and in structural biological chemistry (i.e., for diamagnetic proteins, nucleic acids, carbohydrates, etc.) has led to the assumption (perception) that *only* pulsed techniques are useful in magnetic resonance spectroscopy. This is simply not the case for EPR, due in part to the inherent sensitivity of the electron versus the proton (and the rest of the nuclei) magnetic moment. Furthermore, diamagnetic molecules can have tens (for organic molecules) or thousands of magnetically active nuclei, yet paramagnetic molecules generally have only a single paramagnetic site. Thus, as described in several chapters in this volume, CW EPR is still extremely important in characterizing both steady-state and transient forms of paramagnetic metalloproteins.

Pulsed EPR actually began roughly simultaneously with pulsed NMR; it being developed also in the 1960's, in this case by Dr. William B. Mims at AT&T Bell Laboratories, New Jersey (*4*), and its basis, the electron spin-echo, was first observed by Prof. Erwin L. Hahn in the late 1940's. With improvements in computers and electronics in general, and in microwave technology in particular, the latter driven by the civilian and military telecommunications industry, pulsed EPR methods have become more feasible and more widely applied. In the development and application of pulsed EPR techniques, the contributions of Prof. Arthur Schweiger and co-workers at the ETH, Zürich, Switzerland are especially noted. For details and very helpful background, the reader is directed to an earlier review article (*5*) and to the recent book by Schweiger and Dr. Gunnar Jeschke (*6*).

Also very useful are the older books by Dr. Michael K. Bowman and the late Prof. Larry Kevan (7,8), both of whose groups have contributed greatly to pulsed EPR, particularly in applications to materials science.

An important application of pulsed EPR, particularly in the area of bioinorganic chemistry is in the technique of electron spin echo envelope modulation (ESEEM). In this technique, which is featured in several chapters in this volume, nuclear hyperfine and quadrupole interactions appear as time-domain modulation of the amplitude of an electron spin echo. ESEEM is in many ways analogous to pulsed NMR and Fourier transform of the echo decay provides quantitative (NMR) information on nuclei coupled, often very weakly, to the paramagnetic center. Prof. Jack Peisach, Albert Einstein College of Medicine, New York, has made tremendous advances in this area, initially through a fruitful collaboration with Mims (9,10). Recently, multi-dimensional ESEEM, known as HYSCORE (hyperfine sublevel correlation spectroscopy), analogous to 2D NMR methods, has been very successfully used, particularly by the Schweiger laboratory (6). A recent review of ESEEM as applied to inorganic complexes is also a very useful resource (11).

Another offspring of EPR is electron nuclear double resonance (ENDOR), first devised by Dr. George Feher, in the late 1950's, also at Bell Laboratories, with applications originally to solid state physics (12-14). Considering chemical applications, ENDOR spectroscopy has been applied to organic radical systems (15), allowing the unraveling of complex hyperfine patterns in EPR spectra, however this area is of no direct interest here. ENDOR has also been very profitably applied to understanding the electronic structure of transition metal complexes, and for this the reader is referred to a review by Schweiger (16). Of the utmost relevance here, however, is the application of ENDOR to metalloproteins, which has proven to be the most fruitful field for this technique. Pioneering ENDOR studies of iron-sulfur proteins were performed in the early 1970's by Sands and co-workers (17). The explosive success of ENDOR of metalloproteins occurred in the 1980's, with Hoffman's laboratory making a number of the major contributions in this area. The reader is directed to several relevant review articles (18-21). As with EPR itself, the original ENDOR experiment was a CW technique that involved monitoring an EPR signal at a fixed magnetic field, while a radio-frequency was swept. By analogy with ESEEM, magnetically active nuclei that are hyperfine coupled to the paramagnetic center affect the EPR signal intensity and yield specific transitions from which hyperfine couplings can be directly extracted. ENDOR is thus "EPR-detected NMR" and as such, provides the best connection between the "camps" of EPR and NMR practitioners, as well as a link between the two major sections of this volume. Several chapters in this volume indeed describe CW ENDOR studies of specific metallobiomolecules, such as nitrogenase. For additional studies, the reader is directed to a somewhat outdated, but still very helpful, review by Lowe (22).

Time-resolved ENDOR techniques, in which both the microwave (electron spin) excitation and the radio-frequency (nuclear spin) excitation are pulsed, combine some of the advantages of both ESEEM, a pure pulsed EPR technique, and CW ENDOR, essentially a CW NMR technique. Pulsed ENDOR using primarily the sequences developed by Mims (23) and by Davies (24) allow direct determination of hyperfine couplings over a very wide range of magnitude and often with very high precision, rivaling that obtained from x-ray crystallography or NMR methods. The effectiveness of pulsed ENDOR is demonstrated in several of the chapters in this volume.

The contribution by Hoffman and co-workers to the area of ENDOR of metalloproteins has been referred to above in general terms, and specific contributions are described in several chapters in this volume. However, I wish to mention several specific contributions from the Evanston group that are not otherwise noted herein. The names of Mims, Schweiger, and others have been highlighted for their seminal work in pulsed EPR and related techniques (ENDOR/ESEEM). However, Hoffman and co-workers, specifically Mr. Clark Davoust and Dr. Peter E. Doan, have made significant advances in the development of pulsed EPR/ENDOR/ESEEM instrumentation, such as the development of a versatile spectrometer operating at 35 GHz (25). Pulsed techniques have also been refined and developed in the Hoffman laboratory, such in hyperfine selective Mims ENDOR (26), quantitiative Davies ENDOR (27), and implicit TRIPLE effects in Mims ENDOR (28). Even in the more prosaic area of CW ENDOR, early developments were made by the Hoffman group in the application of the loop-gap resonator (29,30). This type of resonator was first developed (31) by Prof. James S. Hyde, Medical College of Wisconsin, Milwaukee (who, among other achievements, is chiefly responsible for the development of the beloved "E-line" of EPR spectrometers while at Varian in the 1960's). Also useful is the more recent application by the Hoffman group of radio frequency bandwidth broadening to enhance not only CW (32), but also pulsed ENDOR signals. In the even more prosaic area of CW EPR, the use in the Hoffman laboratory of dispersion mode detection adiabatic rapid-passage EPR (yielding an absorption lineshape) (33,34) has allowed observation of EPR signals that are essentially undetectable using conventional, absorption mode, slow-passage (first derivative lineshape) EPR, even at liquid He temperatures, as shown in a recent study of a Mn(II) protein (35).

Significant developments in the theory of ENDOR and ESEEM have also been made by Hoffman and coworkers. Rist and Hyde realized that directional information contained in an EPR envelope can be used to obtain orientation selective ENDOR information (36). Hoffman and co-workers provided the conceptual, mathematical, and algorithmic framework within which 2D, field-frequency plots of ENDOR/ESEEM spectra collected across substantial portions of the EPR spectrum of a paramagnetic center can be used to determine full nuclear hyperfine and quadrupole tensors (37-39), and they have shown how such orientation selected ENDOR (and ESEEM) form the basis for 3D structure determination, applying this to many metalloprotein systems (21).Prof. David J. Singel and co-workers have made seminal studies on understanding the quantitative aspects of ESEEM spectroscopy (40,41) but, here too, Lee, Doan and Hoffman expanded this work into a general theory for describing ^{14}N ($I = 1$) ESEEM (42).

In recent years, a key accomplishment of the Hoffman laboratory has been to extend the use of ENDOR/ESEEM to the study of reactive enzymatic intermediates. In their earlier work, they had studied intermediates with sufficient stability for 'hand freezing', such as compound ES of cytochrome c peroxidase (43). One advance on this front has been the introduction of rapid freeze-quench (RFQ) techniques, through a collaboration with Prof. JoAnne Stubbe, MIT, in the study of dioxygen activation by the diiron oxo protein, ribonucleotide reductase (RNR) (44) A second method is cryoreduction, namely, the γ-irradiation of a frozen solution of a diamagnetic metalloprotein. Irradiation generates electrons from the solvent that reduce a redox active site in the protein, while retaining the structure of the parent molecule. The frozen (77 K) solution can be studied to give structural information about the diamagnetic parent, and then stepwise annealed through intermediate temperatures during which process reactive intermediates can be observed (e.g., when cryoreduction is performed in the presence of substrate) and characterized, by EPR and ENDOR/ESEEM. This technique was first developed by Dr. Roman Davydov and co-workers in Moscow (then USSR) in the 1970's (45,46) Over the past few years, Davydov has worked with Hoffman in a very fruitful collaboration that has led to application of the cryoreduction method to a wide variety of metalloprotein systems, including heme proteins (47-49) iron-sulfur proteins (50), diiron oxo proteins (51-53) and nickel enzymes (54) Certain chapters in this volume touch on some of these results as well.

A final area of contributions from the Hoffman laboratory to note here is in unraveling the thorny problem of non-Kramers (integer-spin) systems. Pioneering work on EPR spectroscopy of biological non-Kramers systems has been made by Profs. Peter G. Debrunner and Michael P. Hendrich (55). The Hoffman laboratory, spearheaded by the solo theoretical analysis by Hoffman (56) has made significant advances in the observation and analysis of ENDOR and ESEEM signals from integer-spin model compounds (57,58) and metalloproteins (59,60). Prior to this work, the conventional wisdom was that ENDOR/ESEEM of e.g., high-spin mononuclear Fe(II), or coupled diferrous oxo proteins, would in the first case be impossible to record and in the second be impossible to interpret. Both assumptions have been shown to be fallacious by Hoffman and co-workers.

The above ENDOR/ESEEM studies of non-Kramers systems were performed in the Hoffman laboratory at X-band and 35 GHz. The laboratory is currently expanding its armamentarium to a pulsed/CW spectrometer operating at 95 GHz (W-band). Pulsed EPR and related techniques at this frequency have been very successfully exploited first by Prof. Klaus Möbius and co-workers in Berlin,

Germany, with applications to understanding radicals in photosynthesis (15,61) and by the group in Leiden, The Netherlands, who have done very beautiful work on metalloproteins (62) and more recently by Prof. Daniela Goldfarb and co-workers at the Weizmann Institute, Rehovot, also on metalloproteins (63,64), as well as materials science applications. EPR at much higher frequencies (HFEPR, high frequency/field EPR, with frequencies that can exceed 500 GHz and fields up to ~25 T) can be performed at selected laboratories around the world, such at that of Prof. Jack H. Freed at Cornell University, Ithaca, New York (65) (where I spent my undergraduate years in near total ignorance of EPR) and the Laboratory in Grenoble, France. An excellent review of HFEPR with applications to metallobiomolecules is that by Prof. Wilfred R. Hagen (66). Another HFEPR facility with which I am quite familiar is that led by Dr. Louis-Claude Brunel at the National High Magnetic Field Laboratory (NHMFL) in Tallahassee, Florida. In this area once again, it was Hoffman who played a crucial role, in initiating an early HFEPR study of an integer-spin "EPR-silent" system, Mn(III) ($S = 2$) in porphryrinic complexes (67). Since then, we have extended HFEPR to other related Mn(III) complexes (68) and I, with Brunel and Dr. J. Krzystek, have been involved in studies of other integer-spin systems, such as complexes of Ni(II) ($S = 1$) (69). In particular, a HFEPR study at the NHMFL of a model for reduced rubredoxin ($[Fe(SC_6H_5)_4]^{2-}$, Fe(II), $S = 2$) (70) shows the potential of this technique in bioinorganic chemistry. HFEPR has also been very useful in studying organic radicals in metalloproteins., such as a tyrosyl radical in RNR (71) and a flavin radical in DNA photolyase (72). HFEPR has been applied to high-spin Kramers (half-integer spin, here $S = 5/2$) metalloprotein systems, such as met-(i.e., Fe(III)) myoglobin (73) and Mn(II) in the p21 ras protein (74), but the promise of this technique for non-Kramers ions in metallobiomolecules has yet to be realized. Further advances in HFEPR instrumentation will likely achieve this goal, for example, pulsed ENDOR at high-field/frequency has recently been accomplished (75). However it is unlikely that HFEPR will become a routinely applied technique nor will it replace conventional (X-band or even 35 GHz) EPR (or ENDOR). This raises an important difference between EPR and paramagnetic NMR on the one hand, and diamagnetic NMR, on the other. Lacking the large internal magnetic field supplied by the electron magnetic moment, ever higher fields are the Holy Grail in diamagnetic NMR, so that lower fields (^1H frequencies) are successively used *and then discarded* (60 MHz, then 90-100 MHz, then 300 MHz, so that now 600 MHz is routine and 900 MHz is preferred). In EPR (and ENDOR) spectroscopy, in contrast, multiple frequencies are needed; e.g., for Cu(II)-containing systems, spectrometers operating at lower frequencies, L-band (~1 GHz) or S-band (~3 GHz), can often be as helpful (76,77) as those at higher frequencies, W-band (95 GHz) and above (62).

Shifting at last to NMR, this technique has been referred to several times above, albeit merely as a foil for EPR. Concerning NMR in its own right, note that in the very early days of the technique, NMR of paramagnetic systems was thought to be unfeasible, however this was proven false and it has been very successfully exploited since the 1960's, initially for coordination complexes. The reader is referred to the classic text edited by three of the pioneers in this area (*78*). I must also mention that my thesis advisor, the late Prof. Russell S. Drago, was also involved in such early studies (*79*). Starting in the 1970's, paramagnetic NMR of metalloproteins came onto the scene, and the reader is referred to the text by two of its pioneers, Bertini and Luchinat, for a comprehensive overview of the field through the mid 1980's (*80*). An appreciation for the current status of this area is given by several of the chapters in this volume.

In contrast to diamagnetic NMR, but a commonality with EPR (see above), is the continuing utility of relatively low magnetic fields (radio frequencies, e.g., 300 MHz). In fact, in the days of ^1H NMR at ≤ 100 MHz, a common application of paramagnetic NMR to organic chemistry was the use of shift reagents (*79*). In bioinorganic chemistry today, such hyperfine shifted NMR signals provide metric information that is often complimentary to that obtained from EPR, including ESEEM and (pulsed) ENDOR. These electron paramagnetic techniques interrogate the electron spin (including orbital effects) and nuclei that are within roughly 5 Å of the paramagnetic center (this depends greatly on the nucleus involved, the coupling mechanism, etc.) whilst NMR interrogates nuclei that are more distant from the paramagnetic center so that the combination can provide a complete picture of, for example, the active site of a paramagnetic metalloenzyme. Paramagnetic NMR of transition metal ion systems is most commonly applied to those with suitable relaxation properties, such as high-spin Co(II). Such ions that are suitable for NMR are often those for which EPR is more difficult (i.e., those with fast electron spin relaxation). Those ions that are suitable for EPR, such as Cu(II), are often less appropriate for paramagnetic NMR. However, as shown in this volume, even metalloproteins containing Cu(II) can be successfully studied by NMR.

EPR (and its relatives, ESEEM and ENDOR) and paramagnetic NMR are the two main branches of paramagnetic resonance and together comprise the bulk of this volume. However, there are other, important techniques that can also be considered as forms of paramagnetic resonance. These are Mössbauer effect spectroscopy and magnetic circular dichroism (MCD) spectroscopy. In both cases, the application of multiple, external magnetic fields is crucial for recording and/or analysis of complicated spectra, and thus these techniques are part of the "superfamily" (to steal a term from the biochemists) of magnetic resonance spectroscopies.

The Mössbauer effect is observable for a wide variety of nuclei (*79*), but paramount among them is ^{57}Fe, because of the importance of iron in biology. Among the many reviews in this area, we note one by Prof. Eckard Münck and co-workers (*81*), whose group has made the greatest contributions in this area. The

Mössbauer effect provides detailed information directly at the nucleus, such as Fe oxidation state and coordination geometry, and is thus complimentary to both EPR, which focuses slightly further away, and to NMR, effective at still greater a distance. Another complementarity among the techniques, which is both a blessing and a curse for each, is that Mössbauer spectroscopy detects all ^{57}Fe in the sample (including, e.g., isolated, low-spin Fe(II), $S = 0$), while EPR/ENDOR and paramagnetic NMR detect only those ^{57}Fe nuclei that are coupled to (or indeed make up) the paramagnetic site. The ACS Symposium on Paramagnetic Resonance of Metallobiomolecules featured two excellent talks on Mössbauer spectroscopy, however it was not possible to obtain contributions to this volume on Mössbauer studies of metalloproteins, so no specific discussion of this area is present herein. Nevertheless, Mössbauer spectroscopy is extremely important in bioinorganic chemistry and several of the chapters in this volume refer to the results of Mössbauer studies.

MCD is also a very powerful technique and is particularly important in metalloproteins since these systems, as do many transition metal complexes, exhibit optical absorption bands. MCD thus provides a link between magnetic resonance and optical spectroscopy. This crucial connection between higher and lower energy electronic states is often neglected in pure EPR studies. The volume contains a third section comprising several chapters on MCD, including the very recent and exciting extension of this technique from its traditional energy domain (UV and visible radiation) into the high energy regime (x-rays).

The second characteristic of this symposium is that it provides a feel for the breadth of biological systems amenable to study by one or more paramagnetic resonance techniques. The coverage is by no means comprehensive; even a large conference devoted to bioinorganic chemistry, such as ICBIC, does not cover every class of metalloprotein nor the role of every biologically relevant metal ion. Nevertheless, a number of important systems are represented in this volume. These include heme proteins, such as hemoglobin and heme oxygenase, iron-sulfur proteins, such as putidaredoxin (Pdx) and pyruvate formate lyase activating enzyme (PFL-AE), and non-heme iron proteins, such as protocatechuate 3,4-dioxygenase (3,4-PCD). In addition to these numerous iron-containing proteins, the other biologically relevant metal ions have not been neglected. Other chapters feature copper proteins, true molybdenum proteins (i.e., pyranopterin Mo proteins, as opposed to nitrogenase, a MoFe protein, which is also featured), nickel hydrogenase (a NiFe protein), and manganese proteins. The arrangement of chapters within the two majors sections (EPR and NMR) is as follows: theoretically inclined and/or general technique oriented chapters first, succeeded by heme proteins, then other iron proteins, then other metal proteins, and lastly non-protein systems.

The symposium title advertised "metallobiomolecules". Naturally, the bulk of paramagnetic resonance studies are on proteins that contain one or more transition metal ion units, whether mononuclear, as in hemoglobin, or polynuclear, as in nitrogenase. Such metalloproteins almost invariably can exist in a paramagnetic state, whether part of their normal activity, or induced externally. However, there are other important biological molecules besides metalloproteins (metalloenzymes). These include lipids and carbohydrates, for which the application of inorganic compounds is essentially non-existent (although this may change!), but there are many important small molecules, such as those in sugar metabolism, in which metals can play a role. A final class of biological molecules is the nucleic acids, DNA and RNA. Metal ions such as Mg(II), which is spectroscopically "silent", have been shown to play an important role in the function *in vivo* of DNA and RNA. Use of paramagnetic dications such as VO^{2+} (V(IV), $S = 1/2$) (*82*) and Mn(II) (*83*) in conjunction with EPR, and particularly ENDOR/ESEEM (*83*) has provided structural information on metal-nucleotide interactions in general. This method has also provided information on enzymes that interact with nucleotides (*74,84,85*). Lastly, as described in this volume, the recent discovery of catalytic RNA, ribozymes, has opened up a new role for metals, and hence paramagnetic resonance, in particular ENDOR/ESEEM in studying this new type of non-protein enzyme (*86,87*).

It is thus hoped that this volume will provide both a greater understanding of the principles and practice of paramagnetic resonance techniques such as EPR, NMR, and MCD, and a greater appreciation of the complexity, variety, and importance of metallobiomolecules.

References

1. Poole, C. P. *Electron Spin Resonance*; Dover Publications, Inc.: Mineola, NY, 1983.
2. Abragam, A.; Bleaney, B. *Electron Paramagnetic Resonance of Transition Ions*; Dover Publications, Inc.: New York, 1986.
3. EPR spectrometers operating at 35 GHz generally employ K_a-band (also known as R-band or WR-28) waveguide (7.11 mm inside width, 3.56 mm inside height) and fittings because the K_a-band is centered near 35 GHz. Q-band waveguide is a different, smaller type (5.69 mm inside width, 2.84 mm inside height) that propagates a different, higher frequency range, 33.0 - 50.0 GHz, and thus is centered at 41.5 GHz, well above the "Q-band" spectrometer operating frequency.
4. Mims, W. B. *Physical Review B* **1972**, *5*, 2409-2419.
5. Gemperle, C.; Schweiger, A. *Chem. Rev.* **1991**, *91*, 1481-1505.
6. Schweiger, A.; Jeschke, G. *Principles of Pulse Electron Paramagnetic Resonance*; Oxford University Press: Oxford, UK, 2001.

7. Bowman, M. K. In *Modern Pulsed and Continuous-wave Electron Spin Resonance*; Kevan, L., Bowman, M. K., Eds.; John Wiley & Sons: New York, Chichester, Brisbane, Toronto, Singapore, 1990.

8. Kevan, L.; Bowman, M. *Modern Pulsed and Continuous-Wave Electron Spin Resonance*; Wiley: New York, 1990.

9. Mims, W. B.; Peisach, J. In *Biological Magnetic Resonance*; Berliner, L. J., Reuben, J., Eds.; Plenum Press: New York, London, 1981; Vol. 3.

10. Mims, W. B.; Peisach, J. In *Advanced EPR. Applications in Biology and Biochemistry*; Hoff, A. J., Ed.; Elsevier: Amsterdam, 1989.

11. Deligiannakis, Y.; Louloudi, M.; Hadjiliadis, N. *Coord. Chem. Revs.* **2000**, *204*, 1-112.

12. Feher, G. *Phys. Rev.* **1959**, *114*, 1219-1244.

13. Feher, G.; Fuller, C. S.; Gere, E. A. *Phys. Rev.* **1957**, *107*, 1462-1464.

14. Feher, G.; Gere, E. A. *Phys. Rev.* **1959**, *114*, 1245-1256.

15. Möbius, K.; Lubitz, W.; Plato, M. In *Advanced EPR. Applications in Biology and Biochemistry*; Hoff, A. J., Ed.; Elsevier: Amsterdam, 1989, pp 441-499.

16. Schweiger, A. In *Structure and Bonding*; Clarke, M. J., Goodenough, J. B., Hemmerich, P., Ibers, J. A., Jorgenson, C. K., Neilands, J. B., Reinen, D., Weiss, R., Williams, R. J. P., Eds.; Springer-Verlag: Berlin-Heidelberg-New York, 1982; Vol. 51, pp 1-121.

17. Anderson, R. E.; Dunham, W. R.; Sands, R. H.; Bearden, A. J.; Crespi, H. L. *Biochim. Biophys. Acta* **1975**, *408*, 306-318.

18. Hoffman, B. M.; Gurbiel, R. J.; Werst, M. M.; Sivaraja, M. In *Advanced EPR. Applications in Biology and Biochemistry*; Hoff, A. J., Ed.; Elsevier: Amsterdam, 1989, pp 541-591.

19. Hoffman, B. M. *Acc. Chem. Res.* **1991**, *24*, 164-170.

20. Hoffman, B. M.; DeRose, V. J.; Doan, P. E.; Gurbiel, R. J.; Houseman, A. L. P.; Telser, J. *Biol. Magn. Reson.* **1993**, *13*, 151-218.

21. Hoffman, B. M. *Acc. Chem. Res*; In press.

22. Lowe, D. J. *Prog. Biophys. Molec. Biol.* **1992**, *57*, 1-22.

23. Mims, W. B. *Proc. Roy. Soc. Lond.* **1965**, *283*, 452-457.

24. Davies, E. R. *Phys. Lett.* **1974**, *47A*, 1-2.

25. Davoust, C. E.; Doan, P. E.; Hoffman, B. M. *J. Magn. Reson.* **1996**, *119*, 38-44.

26. Doan, P. E.; Hoffman, B. M. *Chem. Phys. Lett.* **1997**, *269*, 208-214.

27. Fan, C.; Doan, P. E.; Davoust, C. E.; Hoffman, B. M. *J. Magn. Reson.* **1992**, *98*, 62-72.

28. Doan, P. E.; Nelson, M. J.; Jin, H.; Hoffman, B. M. *J. Am. Chem. Soc.* **1996**, *118*, 7014-7015.

29. Venters, R. A.; Anderson, J. R.; Cline, J. F.; Hoffman, B. M. *J. Magn. Reson.* **1984**, *58*, 507-510.

30. Anderson, J. R.; Venters, R. A.; Bowman, M. K.; True, A. E.; Hoffman, B. M. *J. Magn. Reson.* **1985**, *65*, 165-168.
31. Froncisz, W.; Hyde, J. S. *J. Magn. Reson.* **1982**, *47*, 515-521.
32. Hoffman, B. M.; DeRose, V. J.; Ong, J. L.; Davoust, C. E. *J. Magn. Reson.* **1994**, *110*, 52-57.
33. Mailer, C.; Taylor, C. P. S. *Biochim. Biophys. Acta* **1973**, *322*, 195-203.
34. Mailer, C.; Hoffman, B. M. *J. Phys. Chem.* **1976**, *80*, 842-846.
35. Smoukov, S. K.; Telser, J.; Bernat, B. A.; Rife, C. L.; Armstrong, R. N.; Hoffman, B. M. *J. Am. Chem. Soc.* **2002**, *124*, 2318-2326.
36. Rist, G. H.; Hyde, J. S. *J. Chem. Phys.* **1970**, *52*, 4633-4643.
37. Hoffman, B. M.; Martinsen, J.; Venters, R. A. *J. Magn. Reson.* **1984**, *59*, 110-123.
38. Hoffman, B. M.; Venters, R. A.; Martinsen, J. *J. Magn. Reson.* **1985**, *62*, 537-542.
39. Hoffman, B. M.; Gurbiel, R. J. *J. Magn. Reson.* **1989**, *82*, 309-317.
40. Flanagan, H. L.; Singel, D. J. *J. Chem. Phys.* **1987**, *87*, 5606-5616.
41. Flanagan, H. L.; Gerfen, G. J.; Lai, A.; Singel, D. L. *J. Chem. Phys.* **1988**, *88*, 2161-2168.
42. Lee, H.-I.; Doan, P. E.; Hoffman, B. M. *J. Magn. Reson.* **1999**, *140*, 91-107.
43. Huyett, J. E.; Doan, P. E.; Gurbiel, R.; Houseman, A. L. P.; Sivaraja, M.; Goodin, D. B.; Hoffman, B. M. *J. Am. Chem. Soc.* **1995**, *117*, 9033-9041.
44. Burdi, D.; Sturgeon, B. E.; Tong, W. H.; Stubbe, J.; Hoffman, B. M. *J. Am. Chem. Soc.* **1996**, *118*, 281-282.
45. Davydov, R.; Blumenfeld, L. A.; Magonov, S. N.; Vilu, R. O. *FEBS Lett.* **1974**, *45 & 49*, 256-259 & 246-249.
46. Blumenfeld, L. A.; Burbaev, D. S.; Davydov, R. M.; Vilu, R. O.; Kubrina, L. N.; Vanin, A. F. *Biochim. Biophys. Acta* **1975**, *379*, 512-519.
47. Davydov, R.; Macdonald, I. D. G.; Makris, T. M.; Sligar, S. G.; Hoffman, B. M. *J. Am. Chem. Soc.* **1999**, *121*, 10654-10655.
48. Davydov, R.; Makris, T. M.; Kofman, V.; Werst, D. W.; Sligar, S. G.; Hoffman, B. M. *J. Am. Chem. Soc.* **2001**, *123*, 1403-1415.
49. Davydov, R.; Ledbetter-Rogers, A.; Martasek, P.; Larukhin, M.; Sono, M.; Dawson, J. H.; Siler Masters, B. S.; Hoffman, B. M. *Biochemistry* **2002**, *41*, 10375-10381.
50. Telser, J.; Davydov, R.; Kim, C. H.; Adams, M. W. W.; Hoffman, B. M. *Inorg. Chem.* **1999**, *38*, 3550-3553.
51. Valentine, A. M.; Tavares, P.; Davydov, R.; Krebs, C.; Hoffman, B. M.; Edmondson, D. E.; Huynh, B. H.; Lippard, S. J. *J. Am. Chem. Soc.* **1998**, *120*, 2190-2191.
52. Davydov, R.; Valentine, A. M.; Komar-Panicucci, S.; Hoffman, B. M.; Lippard, S. J. *Biochemistry* **1999**, *38*, 4188-4197.

53. Smoukov, S. K.; Kopp, D. A.; Valentine, A. M.; Davydov, R.; Lippard, S. J.; Hoffman, B. M. *J. Am. Chem. Soc.* **2002**, *124*, 2657-2663.

54. Telser, J.; Davydov, R.; Horng, Y. C.; Ragsdale, S. W.; Hoffman, B. M. *J. Am. Chem. Soc.* **2001**, *123*, 5853-5860.

55. Hendrich, M.; Debrunner, P. *Biophys. J.* **1989**, *56*, 489-506.

56. Hoffman, B. M. *J. Phys. Chem.* **1994**, *98*, 11657-11665.

57. Song, R.; Doan, P. E.; Gurbiel, R. J.; Sturgeon, B. E.; Hoffman, B. M. *J. Magn. Reson.* **1999**, *141*, 291-300.

58. Doan, P. E.; Hoffman, B. M. *Inorg. Chim. Acta* **2000**, *297*, 400-403.

59. Hoffman, B. M.; Sturgeon, B. E.; Doan, P. E.; DeRose, V. J.; Liu, K. E.; Lippard, S. J. *J. Am. Chem. Soc.* **1994**, *116*, 6023-6024.

60. Sturgeon, B. E.; Doan, P. E.; Liu, K. E.; Burdi, D.; Tong, W. H.; Nocek, J. M.; Gupta, N.; Stubbe, J.; Kurtz, D. M.; Lippard, S. J.; Hoffman, B. M. *J. Am. Chem. Soc.* **1997**, *119*, 375-386.

61. Möbius, K. *Chem. Soc. Revs.* **2000**, *29*, 129-139.

62. Coremans, J. W. A.; Poluektov, O. G.; Groenen, E. J. J.; Canters, G. W.; Nar, H.; Messerschmidt, A. *J. Am. Chem. Soc.* **1996**, *118*, 12141-12153.

63. Manikandan, P.; Carmieli, R.; Shane, T.; Kalb, A. J.; Goldfarb, D. *J. Am. Chem. Soc.* **2000**, *122*, 3488-3494.

64. Carmieli, R.; Manikandan, P.; Kalb, A. J.; Goldfarb, D. *J. Am. Chem. Soc.* **2001**, *123*, 8378-8386.

65. Freed, J. H. *Ann. Rev. Phys. Chem.* **2000**, *51*, 655-689.

66. Hagen, W. *Coord. Chem. Revs.* **1999**, *190*, 209-229.

67. Goldberg, D. P.; Telser, J.; Krzystek, J.; Montalban, A. G.; Brunel, L.-C.; Barrett, A. G. M.; Hoffman, B. M. *J. Am. Chem. Soc.* **1997**, *119*, 8722-8723.

68. Krzystek, J.; Pardi, L. A.; Brunel, L.-C.; Goldberg, D. P.; Hoffman, B. M.; Licoccia, S.; Telser, J. *Spectrochim. Acta, Part A* **2002**, *58*, 1113-1127.

69. Krzystek, J.; Park, J.-H.; Meisel, M. W.; Hitchman, M. A.; Stratemeier, H.; Brunel, L.-C.; Telser, J. *Inorg. Chem.* **2002**, *41*, 4478-4487.

70. Knapp, M. J.; Krzystek, J.; Brunel, L.-C.; Hendrickson, D. N. *Inorg. Chem.* **2000**, *39*, 281-288.

71. Gerfen, G. J.; Bellew, B. F.; Un, S.; Bollinger, J. M., Jr.; Stubbe, J.; Griffin, R. G.; Singel, D. J. *J. Am. Chem. Soc.* **1993**, *115*, 6420-6421.

72. Fuchs, M. R.; Schleicher, E.; Schnegg, A.; Kay, C. W. M.; Torring, J. T.; Bittl, R.; Bacher, A.; Richter, G.; Möbius, K.; Weber, S. *J. Phys. Chem. B* **2002**, *106*, 8885-8890.

73. van Kan, P. J. M.; van der Horst, E.; Reijerse, E. J.; van Bentum, P. J. M.; Hagen, W. R. *J. Chem. Soc., Faraday Trans.* **1998**, *94*, 2975-2978.

74. Bellew, B. F.; Halkides, C. J.; Gerfen, G. J.; Griffin, R. G.; Singel, D. J. *Biochemistry* **1996**, *35*, 12186-12193.

75. Bennati, M.; Farrar, C. T.; Bryant, J. A.; Inati, S. J.; Weis, V.; Gerfen, G. J.; Riggs-Gelasco, P.; Stubbe, J.; Griffin, R. G. *J. Magn. Reson.* **1999**, *138*, 232-243.
76. Froncisz, W.; Scholes, C. P.; Hyde, J. S.; Wei, Y.-H.; King, T. E.; Shaw, R. W.; Beinert, H. *J. Biol. Chem.* **1979**, *254*, 7482-7484.
77. Hyde, J. S.; Froncisz, W. *Ann. Rev. Biophys. Bioeng.* **1982**, *11*, 391-417.
78. *NMR of Paramagnetic Molecules*; LaMar, G. N., Horrocks, W. D., Jr., Holm, R. H., Eds.; Academic Press: New York, 1973.
79. Drago, R. S. *Physical Methods in Chemistry*; W. B. Saunders: Philadelphia, 1977; for paramagnetic NMR, see Chapter 12 and references therein; for Mössbauer spectroscopy, see Chapter 15 and references therein.
80. Bertini, I.; Luchinat, C. *NMR of Paramagnetic Molecules in Biological Systems*; Benjamin/Cummings: Menlo Park, CA, 1986.
81. Münck, E.; Surerus, K.K.; Hendrich, M. P. *Meth. Enzym., Part D* **1993**, *227*, 463-479.
82. Mustafi, D.; Telser, J.; Makinen, M. W. *J. Am. Chem. Soc.* **1992**, *114*, 6219-6226.
83. Hoogstraten, C. G.; Grant, C. V.; Horton, T. E.; DeRose, V. J.; Britt, R. D. *J. Am. Chem. Soc.* **2002**, *124*, 834-842.
84. Halkides, C. J.; Bellew, B. F.; Gerfen, G. J.; Farrar, C. T.; Carter, P. H.; Ruo, B.; Evans, D. A.; Griffin, R. G.; Singel, D. J. *Biochemistry* **1996**, *35*, 12194-12200.
85. Crampton, D. J.; LoBrutto, R.; Frasch, W. D. *Biochemistry* **2001**, *40*, 3710-3716.
86. Morrissey, S. R.; Horton, T. E.; Grant, C. V.; Hoogstraten, C. G.; Britt, R. D.; DeRose, V. J. *J. Am. Chem. Soc.* **1999**, *121*, 9215-9218.
87. Morrissey, S. R.; Horton, T. E.; DeRose, V. J. *J. Am. Chem. Soc.* **2000**, *122*, 3473-3481.

Electron Paramagnetic Resonance and Related Techniques

Chapter 2

Electron Spin Echo Methods: A Tutorial

R. David Britt

Department of Chemistry, University of California at Davis, Davis, CA 95616

The electron spin echo methods of *electron spin echo envelope modulation (ESEEM)* and *electron spin echo – double resonance (ESE-ENDOR)* provide for sensitive and selective detection of spin transitions of magnetic nuclei coupled to paramagnetic metals and radicals. In this tutorial, I describe the basic operation of these pulsed EPR methods, with illustrative examples from metal and radical centers in Photosystem II.

Introduction

This article is derived from a tutorial lecture on pulsed (electron spin echo) EPR techniques given at the April 2002 American Chemical Society meeting in a symposium dedicated to Brian M. Hoffman, whose defining work in EPR/ENDOR studies on metalloenzymes has inspired all of us. In this tutorial chapter, I focus on the basic techniques of electron spin echo – electron nuclear double resonance (ESE-ENDOR) and electron spin echo envelope modulation (ESEEM). Illustrative examples are provided from our work on Photosystem II, its oxygen evolving complex, and related enzymes and model systems.

Photosystem II (PSII) presents an extensive set of interesting EPR signals, originating from chlorophyll, pheophytin, quinone, and tyrosine radicals, as well as from the S-state intermediates (*1*) of the oxygen evolving complex (OEC), which contains an ensemble of four manganese atoms along with the essential cofactors calcium and chloride (2-4). For example, Figure 1 displays those S-state EPR signals that exhibit some degree of resolved ^{55}Mn hyperfine interaction (5). Much has been learned by conventional continuous-wave (CW) EPR spectroscopy of these paramagnetic intermediates of the OEC, but the information inherent to these signals is ultimately limited by the relatively broad linewidths that result from various inhomogeneous broadening mechanisms.

Figure 2 shows the structure of di-μ-oxo-tetrakis(2,2'-bipyridine) dimanganese(III,IV), a simple "spectroscopic model" for the S_2-state multiline EPR signal. The di-μ-oxo bridges provide for a large antiferromagnetic coupling between the electron spin $S = 2$ Mn(III) ion and the $S = 3/2$ Mn(IV) ion. The observed hyperfine-rich CW EPR signal (labelled "bipy" in the figure) arises from the well-isolated net spin $S = 1/2$ ground state interacting with the two nuclear spin $I = 5/2$ ^{55}Mn nuclei. The resulting ^{55}Mn nuclear spin state multiplicity is $6 \times 6 = 36$. However there is appreciable overlap among the 36 allowed ($\Delta m_S = \pm 1; \Delta m_I = 0$) EPR transitions, and only 16 major peaks are resolved. Similar 16-line spectra are observed with the analogous 1,10-phenanthroline complex ("phen") and the Mn(III)Mn(IV) state of the dinuclear Mn catalase enzyme ("MnCat") (6). The ability to resolve hyperfine features becomes far worse for magnetic nuclei of ligands, which are generally more weakly coupled than the metal nuclei. Figure 3 focuses on a section of the "bipy" EPR signal with the most observed substructure, and compares the spectrum of the natural abundance compound with that of the compound synthesized with ^{15}N-labelled bipyridine. In this case, eight directly coordinated $I = 1$ ^{14}N ligands are substituted by eight $I = 1/2$ ^{15}N ligands, yet there are no clear differences in the two EPR spectra (7). The weak ligand hyperfine interactions are buried within the inhomogeneously broadened lineshapes.

In general, inhomogeneously broadened lineshapes result from the overlap of resonances (shown as "spin packets" in Figure 4) resulting from a near continuum of magnetic environments. One origin of inhomogeneous broadening, alluded to in the example above, is the unresolved overlap of hyperfine lines from multiple coupled nuclei. This is an issue because the number of hyperfine lines increases *multiplicatively* with the number of classes of coupled nuclei. For example, for the case of a spin $S = 1/2$ complex coupled to k different classes of n_i equivalent spin I_i nuclei with coupling constants A_i, the number of EPR lines and the EPR spectral density are given by the following expressions (8):

$$\text{number of EPR transitions} = \prod_{i=1}^{k} (2n_i I_i + 1), \tag{1}$$

$$\text{EPR spectral density} = \frac{\prod_{i=1}^{k} (2n_i I_i + 1)}{\sum_{i=1}^{k} 2|A_i|n_i I_i}. \tag{2}$$

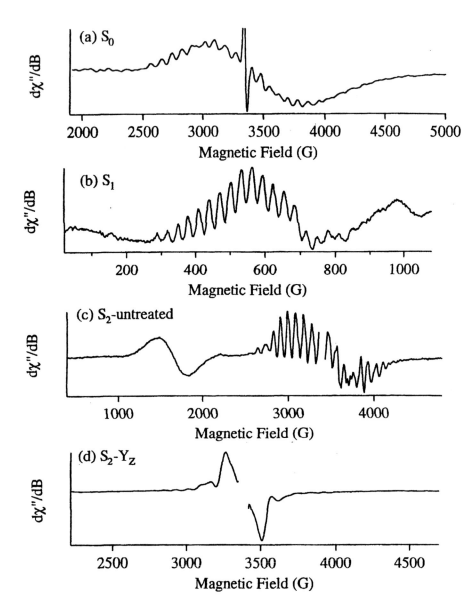

Figure 1. Manganese EPR signals from the S-states of the oxygen evolving complex
(Reproduced from reference 5. Copyright 2001 American Chemical Society.)

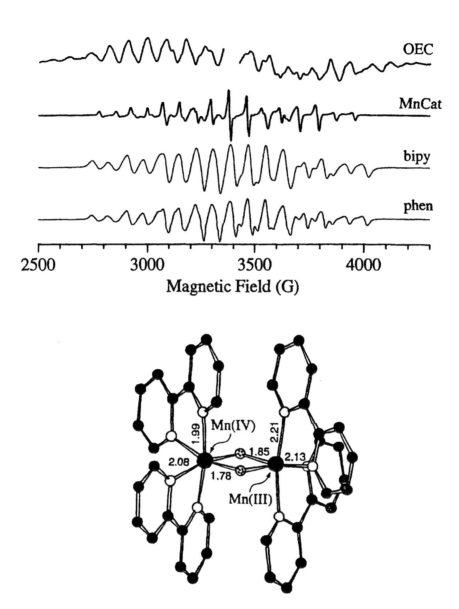

Figure 2. Upper panel: CW EPR spectra of the S_2-state of the OEC compared with that of the Mn(III)Mn(IV) state of Mn catalase and with Mn(III)Mn(IV) synthetic complexes with bipyridine and phenanthroline ligands (ref 6). Lower panel: the structure of di-μ-oxo-tetrakis(2,2′-bipyridine)dimanganese(III,IV).

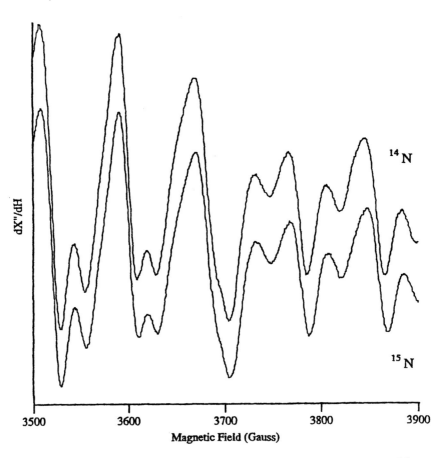

Figure 3. Comparisons of "bipy" EPR spectra with natural abundance ^{14}N vs. ^{15}N-labelled bipyridine ligands.

If the spectral density becomes sufficiently high, the spacing between individual hyperfine lines will become less than the instrinsic lifetime-broadened linewidth of each spin packet, resulting in the inhomogeneously broadened lineshape. Other sources of inhomogeneous broadening in frozen solution samples include hyperfine and g anisotropies, zero-field splittings (for $S > 1/2$ spins), and site-to-site "strain" of such parameters (9).

Fortunately, such inhomogeneous broadening can be negated by the use of multipulse magnetic resonance sequences to generate spin echoes (9). The sequence and underlying mechanism of the simplest spin echo sequence, that of the 2-pulse Hahn echo (10), is illustrated in Figure 4. At point (a) illustrated in the sequence and in the underlying rotating frame vector picture, a resonant H_1 field is applied along the x-axis to rotate magnetization from its initial equilibrium direction along the z-axis coincident with the static H_0 field. By point (b) the magnetization has been rotated by $\pi/2$ onto the rotating frame y-axis and the H_1 field is removed. In the subsequent interval, individual spin packets, of which a select set of five are displayed, will precess at different frequencies due to the effect of inhomogeneous broadening. Spin packet 0 has a resonance frequency equal to the rotating frame frequency relative to the laboratory frame, and therefore remains fixed along the y-axis. The remaining spin packets have higher or lower resonant frequencies, and therefore precess away from the y-axis during the free induction (c) following the $\pi/2$ pulse. At time τ (d) a π pulse is applied to the spin system. The spin packets are rotated into the positions displayed at point (e). Subsequent to this second pulse, each spin continues to precess with its original sense and rate relative to the y-axis, and at a time τ after this π pulse all spin packets refocus simultaneously unto the $-y$-axis. This evanescent magnetization coherence is referred to as a spin echo, and it manifests itself as a burst of radiation that can be detected with the spectrometer.

The detected echo can be used to generate a field swept EPR spectrum analogous to the conventional CW EPR spectrum, but since field modulation is not employed, the electron spin echo (ESE) spectrum corresponds to the direct absorption rather than its field derivative. This can be beneficial for the detection of signals with broad, relatively featureless lineshapes, but for most cases CW EPR still provides for more sensitive detection. Since the ESE experiment is carried out in the time-domain, it can readily be applied to study kinetics of electron transfer and spin polarization and to accurately measure relaxation times of electron spin systems. However the primary focus of this chapter is on applications of ESE spectroscopy to detect weak hyperfine interactions masked by inhomogeneous broadening. Specifically, the techniques of electron spin echo envelope modulation (ESEEM) and electron spin echo – electron nuclear double resonance (ESE-ENDOR) are utilized to detect the nuclear spin transitions of hyperfine-coupled nuclei. In these experiments, the spin echo is the carrier onto which nuclear spin

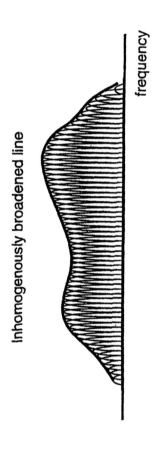

Inhomogenously broadened line

frequency

homogeneously broadened
"spin-packet"
linewidth ~ $1/T_2$

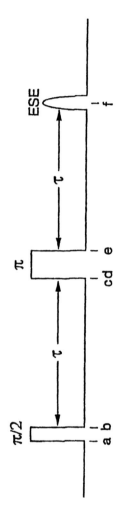

$\pi/2$

π

τ

τ

ESE

a b

cd e

f

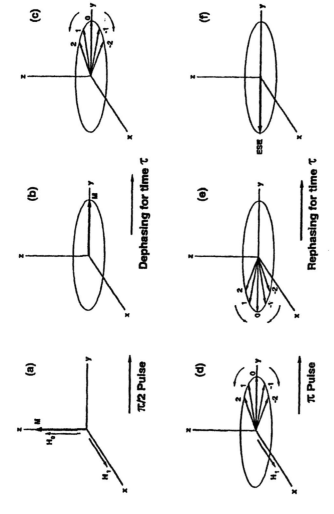

Figure 4. Upper panel: An inhomogeneously broadened line resulting from overlap of spin packets; Lower panel: The 2 pulse Hahn echo sequence and associated vector diagram.

information is encoded, either by time domain interference (ESEEM) or through RF-driven magnetization transfer (ESE-ENDOR).

ESE-ENDOR

Figure 5 shows the ESE-ENDOR sequences introduced by Davies (*11*) and Mims (*12*). In both cases, high power microwave and rf pulses are interleaved. Specifically, an alteration of the initial electron spin magnetization is created by one or more high-power resonant microwave pulses. Application of the radio frequency pulse further perturbs the electron magnetization if the rf pulse induces spin transitions of nuclei magnetically coupled to the electron spins. The nuclear spin transition frequencies are measured by varying the radio frequency while monitoring the effect of the rf pulse on a subsequent electron spin echo (*13, 14*).

Davies ENDOR

The Davies ESE-ENDOR sequence is particularly useful for nuclei with relatively strong hyperfine couplings, as illustrated in Figure 6. The inverting microwave π pulse inverts the electron magnetization in a narrow resonant bandwidth (proportional to the inverse of the π-pulse length) within the inhomogeneously broadened line (Figure 6, top panel). When the radio frequency is resonant with the spin transition frequency of a strongly hyperfine-coupled nucleus, the resulting nuclear spin flip transfers magnetization between the "hole" and an unperturbed region of the spectrum, decreasing the extent of magnetization inversion in the hole (Figure 6, bottom panel). This results in an alteration of the spin echo induced by the final two-pulse ESE sequence. The Davies sequence works best for relatively strongly coupled nuclei, because for weakly coupled nuclei the nuclear spin flips simply transfer magnetization within the hole, giving a negligible increase in echo amplitude (imagine Figure 6 with the hyperfine coupling A reduced to a fraction of the hole width). For nuclei with a modestly large couplings this effect can be mitigated by using long microwave pulses to form narrow holes in the inhomogeneously broadened line (*15*). However, this results in smaller, broader spin echoes, so this approach can in practice only be taken so far.

Because the ^{55}Mn couplings are quite large for the S_2-state multiline signal and analogous signals from dinuclear Mn complexes (Figure 2), the Davies ESE-ENDOR sequence works well for probing these systems. Figure 7 shows the field swept ESE-EPR and ^{55}Mn ENDOR spectra for the Mn(III)Mn(IV) phenanthroline complex along with simulations performed with identical spin Hamiltonian parameters (*6*, also see *16*). The field swept simulation shows that multiple EPR transitions contribute to the spectrum towards its center, and this must be taken into account, as well as the effects of hyperfine anisotropy, when simulating the ^{55}Mn ENDOR at a given field value. The ENDOR transitions for the ^{55}Mn of the Mn(IV) ion occur near 100 MHz, while the Mn(III) ion ^{55}Mn transitions occur in the 200 MHz range. The d^4 Mn(III) ion displays much greater hyperfine anisotropy than the d^3 Mn(IV) ion.

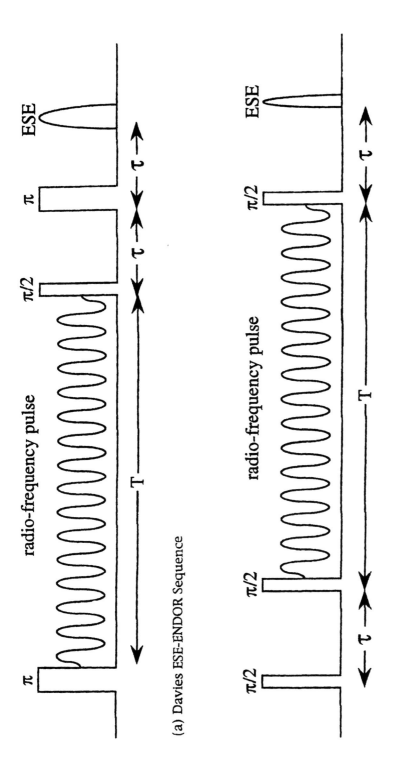

(a) Davies ESE-ENDOR Sequence

(b) Mims ESE-ENDOR Sequence

Figure 5. (a) Davies and (b) Mims ESE-ENDOR sequences

Magnetization

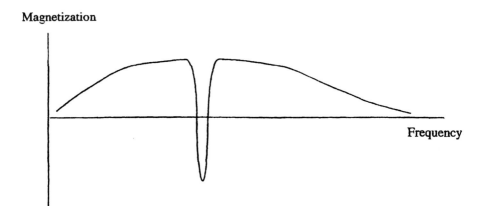

Frequency

Magnetization

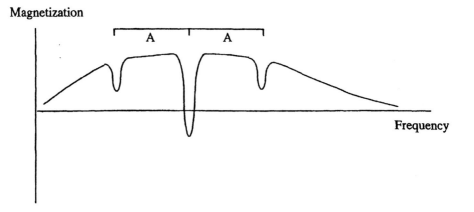

Frequency

Figure 6. Illustration of the Davies ENDOR mechanism.

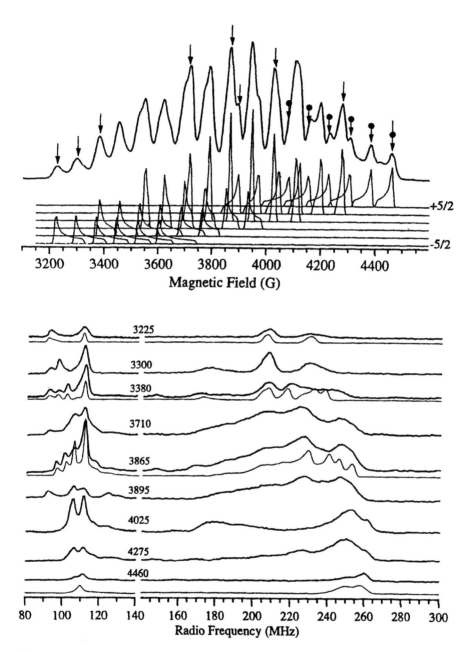

Figure 7. Mn(III)Mn(IV) phenanthroline compound data. Upper panel: ESE-EPR with powder pattern simulations; Lower panel: Davies ENDOR with simulations.

For such Mn(III)Mn(IV) clusters, the Davies ENDOR technique also provides resolution of ligand nuclear spin transitions, in contrast to the CW EPR spectra where the ligand hyperfine couplings are masked (Figure 3). Figure 8 shows the Davies ENDOR spectrum of the ^{15}N-labelled bipyridine complex in a lower frequency (0-25 MHz) region, where contributions are observed from both ligand protons and ^{15}N nuclei (6, 17).

As shown in eq. 1, the number of EPR transitions increases geometrically with the number of coupled nuclei. For multinuclear Mn clusters this rise in complexity is dramatic, from 36 transitions (counting only ^{55}Mn spin states) for a dinuclear cluster up to 1296 for a tetranuclear cluster. On the other hand, the number of ENDOR transitions scales only linearly with the number of coupled nuclei, and therefore the ENDOR experiment becomes even more useful for such complex systems. Figure 9 shows the CW EPR spectra of two S_2-state multiline forms, a native form (labelled MeOH because addition of a small amount of methanol increases resolution and intensity) and from a form altered by ammonia binding. Figure 10 displays the corresponding Davies ^{55}Mn ENDOR spectra (18, 19). The dashed lines show simulations, again with identical spin Hamiltonian parameters used for EPR and ENDOR of each spectral type. For both native and ammonia-treated cases, we can only achieve good simulations of EPR and ENDOR spectra using four ^{55}Mn hyperfine couplings. These results clearly demonstrate that the S_2-state of the OEC contains a tetranuclear manganese cluster at its core.

By incorporating Mn-Mn distances and Mn oxidation state assignments from X-ray spectroscopic experiments (4), we have interpreted these EPR/ENDOR results to favor a "3+1" model (Figure 11) for the S_2-state cluster, in which 3 Mn ions are strongly exchange coupled (corresponding to a relatively short 2.7Å Mn-Mn distance such as seen in the aforementioned Mn(III)Mn(IV) dimers (Figure 2)), and a fourth Mn ion is more weakly coupled to this trimer core (corresponding to a longer Mn-Mn distance). A recent PSII X-ray crystallographic report shows electron density assigned to the Mn cluster which appears consistant with such a 3+1 structure (20).

Figure 11 includes some details of substrate and cofactor ligation. We favor substrate water ligation by the S_2-state based on further pulsed EPR measurements. Protons of water or hydroxide ligands to paramagnetic metals are good targets for Davies ENDOR. Figure 12 shows Davies ENDOR in the proton region of the spectrum, with two different pulse widths. Note that the short pulse widths (lower panel) accentuate the more strongly coupled proton features, those farthest split from the proton Larmor frequency (≈ 15 MHz) (15). A powder pattern simulation is shown along with the experimental spectra. This will be further discussed in the ESEEM section.

Davies ^1H ENDOR also works well for detecting the strongly coupled protons associated with amino acid radicals, such as the relatively stable $Y_D^{\bullet}$ radical of PSII

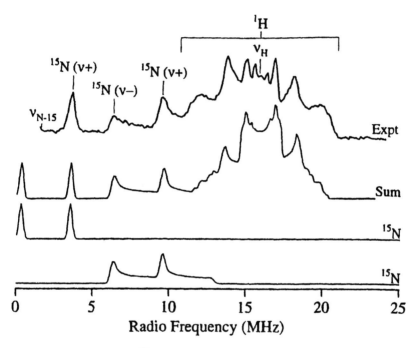

Figure 8. Mn(III)Mn(IV) ^{15}N-bipyridine Davies ENDOR data in the lower frequency range (0-25 MHz).

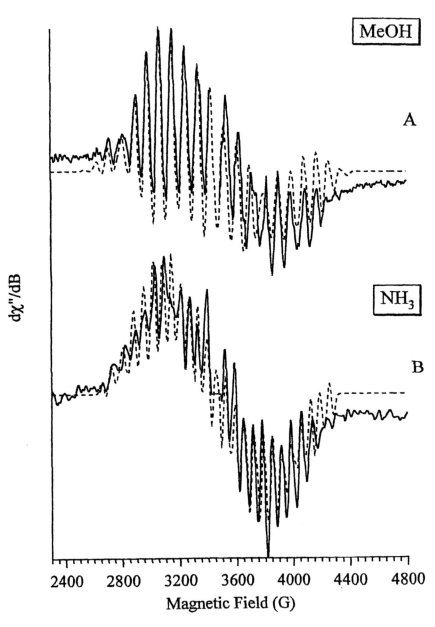

Figure 9. (A) the native, and (B) the ammonia-altered S_2-state multiline CW EPR spectra (solid lines) and simulations (dashed lines).

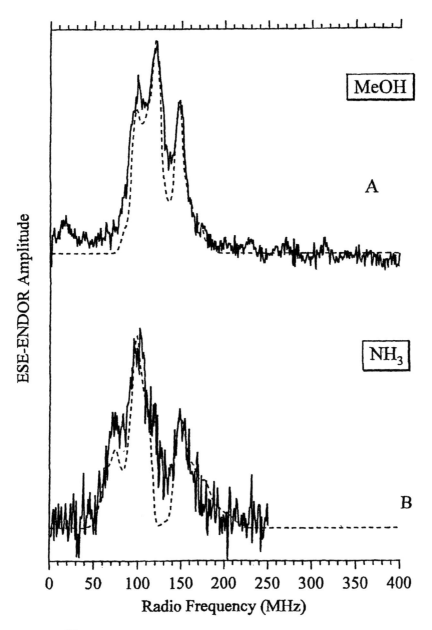

Figure 10. ^{55}Mn Davies ENDOR spectra (solid lines) and simulations (dashed lines) of (A) the native, and (B) the ammonia-altered S_2-state multiline EPR signals.

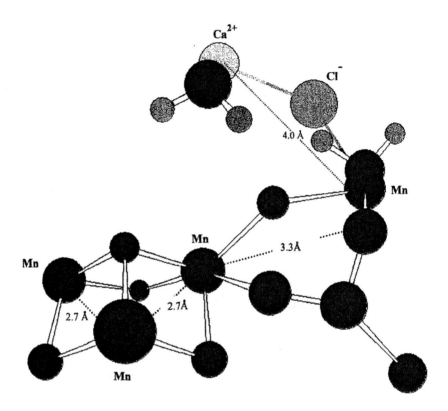

Figure 11. A "3+1" model for the S_2-state tetranuclear Mn cluster. Figure courtesy of Michelle Dicus and Constantino Aznar.

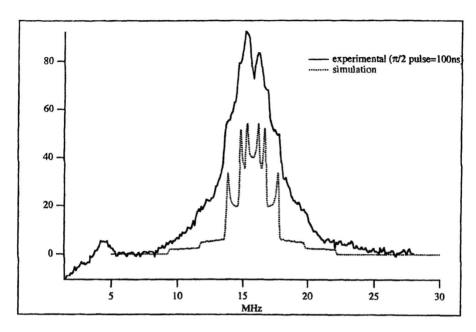

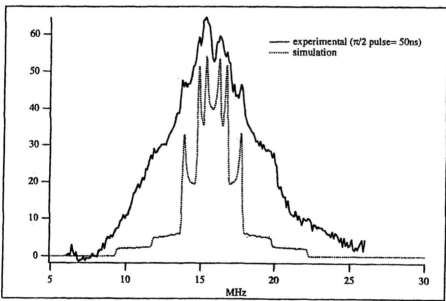

Figure 12. Davies ^{1}H ENDOR of the S_2-state multiline signals. Data are shown for two pulse conditions: Upper panel, $\pi/2$ pulse length 100 ns; Lower panel, $\pi/2$ pulse length 50 ns. The dashed line is a simulation.

(Figure 13) (21). In such cases, ENDOR-derived hyperfine couplings provide strong experimental contraints for associated electronic structure calculations (for example, 22).

Mims ENDOR

As discussed above, the use of the Davies ESE-ENDOR sequence becomes limited when targeting relatively weakly coupled nuclei, because the frequency difference between equilibrium electron magnetization and magnetization inverted by the initial preparation pulse tends to be relatively large. Fortunately, the Mims sequence (12) works well for weakly coupled nuclei. This is the case because the first two $\pi/2$ pulses in a "stimulated echo" sequence generate a sinusoidal M_z magnetization pattern, with the frequency spacing $2\pi/\tau$ under experimental control (Figure 14). This can greatly decrease the hyperfine coupling needed for an RF-induced nuclear spin flip to affect the echo signal (in this case, a stimulated echo). In fact, if the preparation phase microwave pulse widths were the same, the entire sinusoid pattern of the Mims experiment would reside in the inverted magnetization bandwidth of the Davies experiment.

In the Mims experiment, the τ value can be set to optimize sensitivity for a given targeted coupling. A disadvantage of the Mim sequence is that it contains "blind spots" resulting from insensitivity to nuclei with couplings an integral multiple of the sinusoidal spacing. In practice, the experiment may need repeating at multiple τ values to make sure no nuclei are missed. Doan and Hoffman have used an additional refocusing pulse to decouple the mimimum τ value from the spectrometer deadtime, and in many cases this procedure can be used to position the Mims blindspots outside the spectral region of interest (23). Increased sensitivity to distant, weakly-coupled nuclei results from using long τ values. In practice, the maximum τ that can be used is limited by spin-spin relaxation, with phase memories ($\approx T_2$) typically on the order of few hundred nanoseconds for biological paramagnetic centers.

We have found Mims ENDOR to be very useful for probing hydrogen bonding interactions of the PSII tyrosine radicals. Understanding these are key to understanding a possible role of the Y_Z tyrosine in proton-coupled electron transfer (21, 24). Figure 15 targets the exchangeable hydrogen(s) of the relevant hydrogen bond (25). Trace (a) shows the difference ^1H Davies ENDOR spectrum for spinach $Y_D^\bullet$ resulting from a data subtraction of spectra of PSII preparations incubated in normal natural abundance buffer and ^2H_2O-enriched buffer. The exchangeable hydrogen in the hydrogen bond is revealed. ^2H Mims ENDOR provides similar information in ^2H_2O-enriched buffer only, without any required data subtraction. Traces (b) and (c) show the ^2H Mims ENDOR for $Y_D^\bullet$ of spinach and Synechocystis preparations. These are very similar to the subtracted Davies ENDOR spectrum, but with the addition of resolved quadrupolar splittings for the $I = 1$ ^2H. In contrast, the ^2H Mims ENDOR for $Y_Z^\bullet$ of Synechocystis shows a broad unresolved line, which we

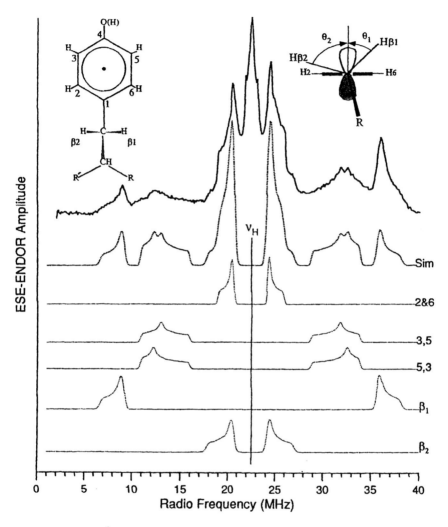

Figure 13. Davies ^{1}H ENDOR of the $Y_D^\bullet$ PSII tyrosine radical. Simulations show contributions from different classes of protons, along with the summed contribution.

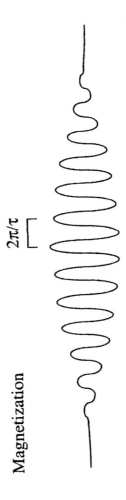

Magnetization

$2\pi/\tau$

Frequency

Figure 14. The electronic magnetization along the z axis following the first two $\pi/2$ pulses, separated by time τ, in a stimulated echo sequence.

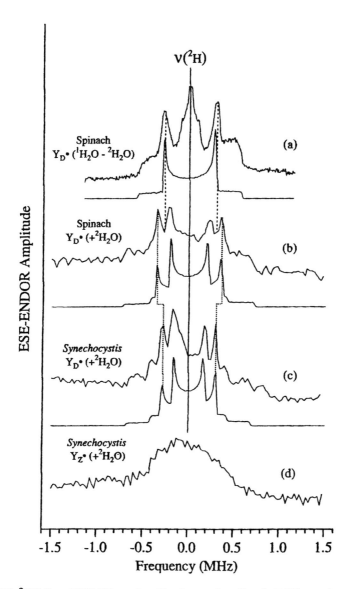

Figure 15. ^{2}H Mims ENDOR study of hydrogen bonding in PSII tyrosine radicals (ref 25). (a) Spinach $Y_D^\bullet$ ^{1}H Davies ENDOR, difference between ^{1}H$_2$O and ^{2}H$_2$O-enriched buffer, and scaled by the ^{2}H/^{1}H magnetic moment ratios; (b and c) $Y_D^\bullet$ ^{2}H Mims ENDOR for spinach and *Synechocystis* in ^{2}H$_2$O-enriched buffer; (d) $Y_Z^\bullet$ ^{2}H Mims ENDOR for *Synechocystis* in ^{2}H$_2$O-enriched buffer. Simulations provided for (a-c).

take as evidence of disorder in the hydrogen bonding of this tyrosine radical.

[15]N Mims ENDOR works very well in probing nitrogen interactions around the tyrosine radicals. Figure 16 illustrates for the case of the $Y_D^\bullet$ radical in *Synechocystis*, where mutants can be readily generated (26). The [15]N Mims ENDOR features of globally ^{15}N-labelled wildtype (trace a) resembles a pair of axial powder patterns, but by using preparations where only the histidine nitrogens are ^{15}N-labeled (trace b), we see that the wildtype spectrum consists of two overlapping components. A comparison with spectra of wildtype ^{15}N-labelled at only the π nitrogen of histidine (trace c) and of an ^{15}N globally labeled H189Q mutant (trace d) show that the τ nitrogen of the histidine189 (D2 protein) is the donor of the hydrogen bond studied above. The second set of ^{15}N transitions arises from the peptide nitrogen of the tyrosine (data not shown). Analogous studies of the $Y_Z^\bullet$ tyrosine show no histidine interactions, suggesting that no histidine hydrogen bonding is present for the $Y_Z^\bullet$ species (27).

ESEEM

Like ESE-ENDOR, the ESEEM experiment shares the use of 2 and 3 (or more) pulse electron spin echo sequences. However, no externally applied RF pulses are used. The ESEEM experiment works because, in addition to inducing fully allowed ($\Delta m_S = \pm 1; \Delta m_I = 0$) electron spin transitions, the microwave pulses may also induce semi-allowed ($\Delta m_S = \pm 1; \Delta m_I \neq 0$) transitions involving simultaneous electron and nuclear spin transitions, resulting in quantum mechanical coherences in the nuclear spin sublevels associated with the electron spin levels. These coherences create interference effects which can be measured by varying the electron spin echo pulse timing (Figure 17) (9, 14, 28, 29). Fourier analysis of the resulting time-domain electron spin echo envelope modulation pattern reveals the frequencies of the nuclear spin transitions. The frequencies and amplitudes of the Fourier peaks can be interpreted to determine hyperfine and electric quadrupolar interactions of the coupled nuclei. In a electron and nuclear Zeeman basis picture, off-diagonal spin Hamiltonian terms are needed to turn on the semi-allowed transitions needed for the modulation effect. Often, dipolar hyperfine couplings serve this purpose. For $I > 1/2$ nuclei, the electric quadrupolar interaction can make an appreciable contribution. Regardless, the relative transition probabilities can be calculated from the spin Hamiltonian, and ideally this leads to accurate quantitative analysis of the modulation depths in the ESEEM experiment.

Typical one-dimensional ESEEM experiments utilize two-pulse Hahn echo (Figure 17a) or three-pulse 'stimulated' echo (Figure 17b) sequences. Without the need for RF coils and high power RF pulses, ESEEM data are somewhat easier to obtain than ESE-ENDOR data, and currently there are more pulsed EPR laboratories using ESEEM than using the double resonance spin echo experiments.

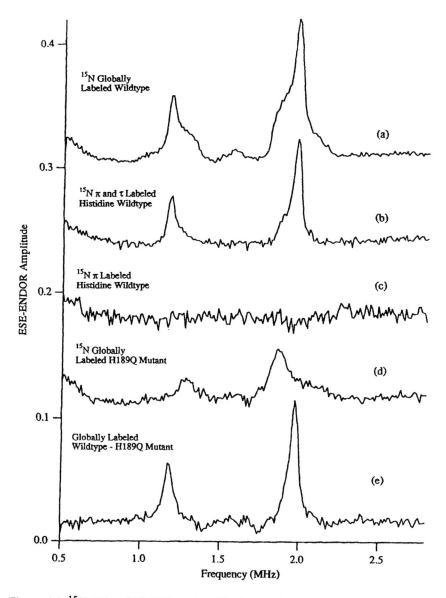

Figure 16. ^{15}N Mims ENDOR study of hydrogen bonding in *Synechocystis* Y$_D^{\bullet}$ (ref 26). (a) ^{15}N globally labeled wildtype; (b) ^{15}N histidine-labelled wildtype; (c) ^{15}N π-histidine-labelled wildtype; (d) ^{15}N globally labeled H189Q mutant; (e) (a-d) difference spectrum.

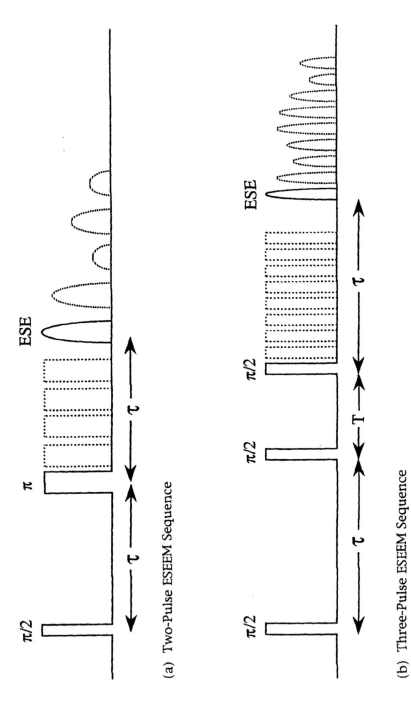

(a) Two-Pulse ESEEM Sequence

(b) Three-Pulse ESEEM Sequence

Figure 17. (a) Two and (b) Three pulse ESEEM sequences.

2-Pulse ESEEM

In the two-pulse ESEEM experiment, the electron spin echo amplitude is measured as a function of the interpulse spacing τ (Figure 17a). As discussed above, a quantum mechanical treatment such as the density matrix approached used by Mims (*9, 28*) is required to properly describe the ESEEM phenomenon, including the amplitude of the modulation and the effects of other nuclear spin Hamiltonian terms such as the nuclear Zeeman and electric quadrupolar interactions. One specific result of the density matrix analysis for the 2-pulse ESEEM experiment is that, in addition to the fundamental nuclear spin transition frequencies, their sum and difference frequencies also appear in the modulation (though 180° out of phase).

The first stable electron acceptor in PSII is a tightly bound plastoquinone designated Q_A. The reduced Q_A^- anion radical gives a strong EPR signal when decoupled from the proximal Fe(II) by one of several procedures (*30*), and this signal shows nice spin echo modulation, so we will use it as an example of the ESEEM method.

Figure 18 shows 2-pulse ESEEM results for the Q_A^- anion radical of PS II (*30, 31*). The time domain modulation pattern is shown in Figure 18a, and the frequency domain spectrum shown in 18b is generated as the Fourier Transform of this time domain pattern. The FT peak at 14.1 MHz arises from weakly coupled protons. The strongly coupled protons of the Q_A^- radical do not produce measurable modulation, and these are better studied with ESE-ENDOR. However there is much information in the lower frequency range of the spectrum. The pattern of three sharp low frequency peaks along with a broader peak at higher frequency is characteristic of ^{14}N ESEEM when the hyperfine and external magnetic fields are of approximately the same amplitude (*32, 33*). Under this "exact cancellation" condition the local magnetic field at the ^{14}N nucleus is essentially nulled for one electron spin orientation, and three sharp peaks arise because the electric quadrupole interaction splits the three levels of the $I = 1$ ^{14}N nucleus even in the absence of a magnetic field (*34*). The higher frequency "double-quantum" peak arises from the transition between the outer energy levels of the ^{14}N for the other electron spin orientation where the hyperfine and external magnetic fields add. The two corresponding transitions involving the inner ^{14}N level are typically too broad to observe in non-crystalline samples. Analysis of the frequencies and lineshapes of the quadrupolar peaks and the double quantum peak provides a full determination of the electric quadrupole and hyperfine interactions. This "exact cancellation" limit is one of the most fruitfully explored arenas for ESEEM spectroscopy, particularly for targeting nitrogen ligation (ref 32 for example).

As in other forms of magnetic resonance, stable magnetic isotopic substitution provides one of the most powerful assignment tools in pulsed EPR spectroscopy. Figure 18c shows the frequency domain ESEEM results for Q_A^- in globally ^{15}N-labelled PS II particles. The low frequency region of the spectrum is completely altered, confirming our assignment of the previous transitions to an ^{14}N nucleus

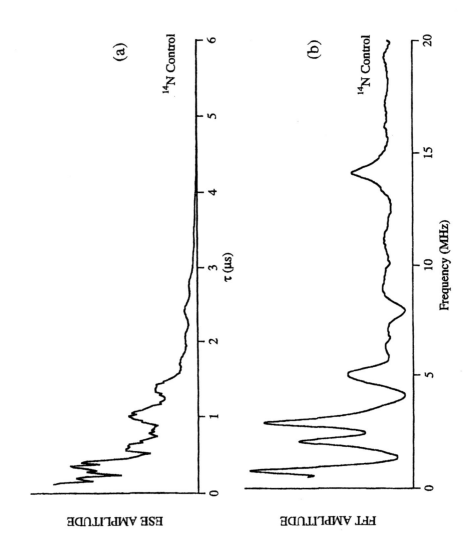

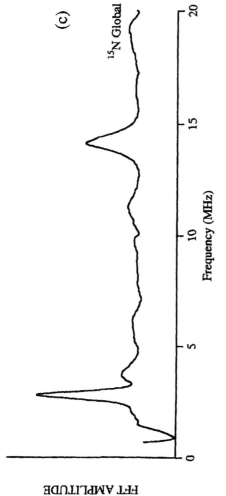

Figure 18. 2-pulse ESEEM results for the Q_A^- semiquinone anion radical of PS II (ref 30). (a) The time domain envelope modulation pattern. (b) The frequency domain ESEEM spectrum obtained as the Fourier Transform of (a). (c) The frequency domain spectrum of an ^{15}N-labelled PSII sample.

coupled to the Q_A^- radical. For the $I = 1/2$ ^{15}N nucleus we resolve a single transition at 2.8 MHz, twice the ^{15}N Larmor frequency of 1.4 MHz at the applied field of 3316 G. This confirms that we are indeed in the "exact cancellation" limit for this class of nitrogen. We do not observe a low frequency peak from the ^{15}N nucleus for the other electron spin orientation where the two fields cancel because the frequency is too low to observe with the 2-pulse ESEEM sequence. This illustrates an important drawback to the 2-pulse sequence. The 2-pulse echo amplitude decreases rapidly with increased τ due to spin-spin (T_2-type) relaxation processes. In the representation of Figure 4 this corresponds to the magnitude of the magnetization vector of each spin packet decreasing over time due to random spin-spin dephasing. The effect of this limited "phase-memory" is dramatically seen in Figure 18a, where the echo amplitude has essentially vanished after $\approx$3 μs. This loss of the echo carrier with time leads to a loss of resolution of low frequency modulation components. Fortunately, this problem can be circumventing by using the 3-pulse ESEEM method based on the stimulated echo sequence.

3-Pulse ESEEM

As discussed in the Mims ENDOR section, the first two $\pi/2$ pulses of a stimulated echo sequence give rise to a sinusoidal nonequilibrium magnetization pattern encoded along the z-axis (Figure 14). Because this magnetization pattern is imposed along the z-axis, it decays on the order of the spin-lattice relaxation time (T_1), which is typically much longer than the phase memory time. The stimulated echo, elicited by a final $\pi/2$ pulse, corresponds to the free induction decay of this sinusoidal magnetization pattern. In the 3-pulse ESEEM experiment (Figure 17b), the stimulated echo (which lags the last pulse by the time τ) is measured as a function of the time T. The great advantage is that the carrier echo now lasts a much longer time than in the 2-pulse case, allowing better spectral resolution of the modulation frequencies. A disadvantage is that the echo is only one half the initial intensity of the 2-pulse echo because magnetization components remaining along the $\pm x$ axes after the first two pulses are lost on the timescale of the phase memory. Another difference is that only fundamental nuclear spin transition frequencies appear in the 3-pulse ESEEM experiment, although their amplitudes depend on the specific value of the the first interpulse time τ used in the experiment, and certain frequency components may be completely suppressed for certain τ values. An investigator typically performs the 2-pulse ESEEM experiment along with a set of 3-pulse ESEEM experiments with different τ values to completely characterize the ESEEM effects.

Figure 19a illustrates the longer time range that can be exploited in the 3-pulse ESEEM experiment for the Q_A^- radical. The "carrier" stimulated echo shows negligible diminution out to the 16 μs maximal time of this data set. Over this time the modulation damps out due to the intrinsic linewidth of the nuclear transitions, rather than because of the lifetime of the carrier signal. One observes far superior frequency resolution in the Fourier Transform 3-pulse ESEEM spectrum (19b) when compared to the corresponding 2-pulse spectrum (18b). The weakly coupled proton modulation in this 3-pulse spectrum is suppressed by using a τ-value that is a multiple of the proton Larmor precession period, and the frequency domain spectrum is expanded to the 0-6 MHz range to emphasize the low frequency region, with baseline resolved peaks at 0.75, 2.06, and 2.85 MHz. The calculated ^{14}N electric quadrupole coupling parameters are $e^2qQ = 3.29$ MHz and $\eta = 0.50$, which are characteristic of an ^{14}N in a peptide bond (35). We postulate that this modulation arises from a peptide nitrogen hydrogen-bonded to the quinone (30, 31). From the 5.0 MHz frequency of the broader double quantum transition, we calculate a hyperfine coupling of 2.0 MHz to this nitrogen.

The 3-pulse ESEEM spectrum of the ^{15}N-labelled sample (Figure 19c) shows a very sharp low frequency ^{15}N-transition at 0.27 MHz. The origin of such exceedingly sharp ESEEM lines for $I = 1/2$ nuclei such as ^{15}N near the exact cancellation limit and their frequency dependence with respect to dipolar coupling has been discussed (35). As mentioned above, this feature is too low in frequency to be seen in the 2-pulse data. On the other hand, the 2.8 MHz transition seen in the ^{15}N 2-pulse spectrum (Figure 18c) is completely suppressed at the τ value of 213 ns used in this 3-pulse data set. We also observe a small peak at the ^{15}N Larmor frequency of 1.4 MHz arising from additional weakly coupled ^{15}N nuclei.

The multifrequency advantage

There is often a great advantage in terms of signal intensity and resolution in carrying out the ESEEM experiment as close as possible to the "exact cancellation" condition. For robust ESEEM detection for a variety of nuclei with varied coupling constants, this necessitates a "multifrequency" pulsed EPR approach, either with multiple instruments, broad bandwidth instruments, or a combination of both. Figure 20 provides an example for the Mn(III)Mn(IV) state of manganese catalase (Figure 2) using our 8-18 GHz pulsed instrument (7, 37). Trace (a) shows the X-band (near 11 GHz) 3-pulse FFT spectra of both the native and azide treated enzyme. Histidine ^{14}N modulation for the native enzyme is ideal at this frequency and field (3910 G), but the corresponding low frequency histidine ^{14}N peaks are much broader for the azide treated form. The ^{14}N double quantum peak is at higher frequency for the azide form, revealing that the nitrogen has a stronger hyperfine coupling following azide treatment (and no change upon using ^{15}N-azide). This suggests that a higher frequency/field would be beneficial for approaching "exact cancellation" for the azide sample, and Figure 20b illustrates this very well. The low frequency ^{14}N peaks sharpen greatly for the azide sample at 14 GHz/5300 G. On the other hand, the native sample data quality suffers at higher frequency/field values.

Dipolar coupled nuclei

In addition to the "exact cancellation" condition, ESEEM is also often applied

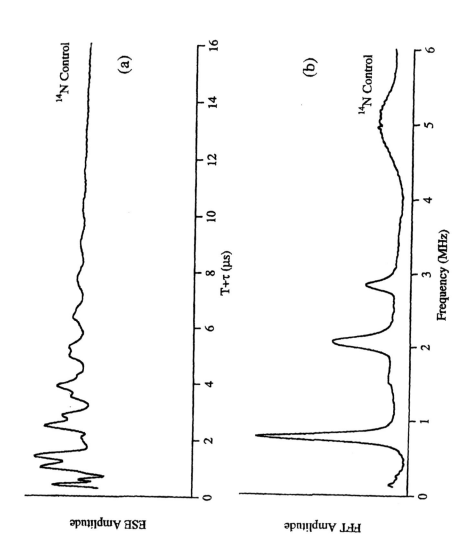

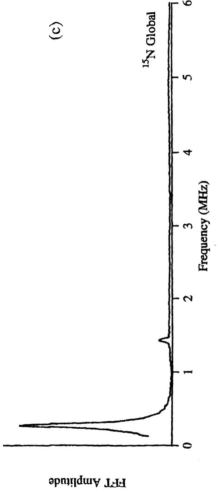

Figure 19. 3-pulse ESEEM results corresponding to Figure 17.

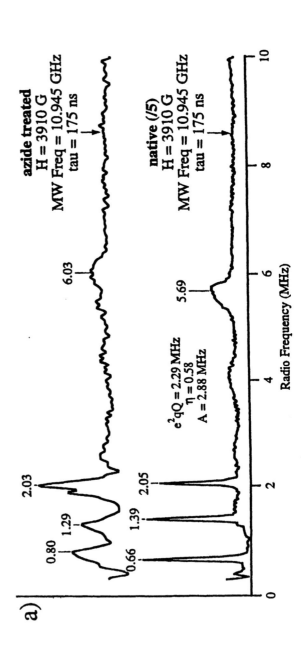

a)

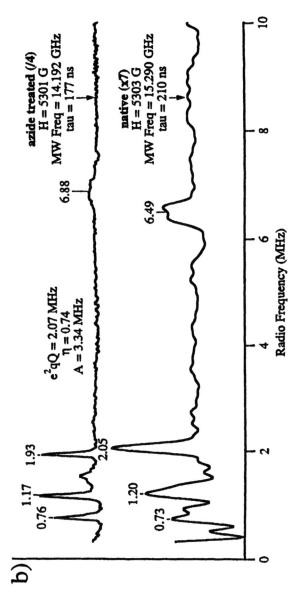

b)

azide treated (/4)
H = 5301 G
MW Freq = 14.192 GHz
tau = 177 ns

native (x7)
H = 5303 G
MW Freq = 15.290 GHz
tau = 210 ns

$e^2qQ = 2.07$ MHz
$\eta = 0.74$
A = 3.34 MHz

6.88

6.49

1.93

1.17

0.76

2.05

1.20

0.73

Radio Frequency (MHz)

Figure 20. FFT 3-pulse ESEEM spectra of Mn(III)Mn(IV) catalase, native and azide treated. (a) X-band; (b) P-band.

to weakly coupled nuclei where the anisotropic (often dipolar, through space) coupling dominates. In such cases the modulation depth is inversely proportional to the square of the magnetic field, so instrumentation at relatively low microwave frequencies is most useful.

An appreciable fraction of ESEEM studies target weakly-coupled deuterons exchanged or site-specifically introduced into sites near paramagnetic centers. Figure 21 provides an example related to tyrosine hydrogen bonding (*38*). Four different radical forms were studied (details in caption). For each case, Mims ^{2}H ENDOR derived hyperfine and quadrupolar couplings were used in constrained ESEEM simulations in which only the number of coupled deuterons were varied (one or two in the figure). In this case, the quantitative analysis of modulation depth is combined with the high resolution of ^{2}H Mims ENDOR (Figure 15) to provide a more complete analysis than would be obtained with either method alone. Analogous ^{2}H ESEEM has been performed on the S_2-state multiline signal following ^{2}H$_2$O exchange (Figure 22). In this case, the hyperfine couplings derived from the ^{1}H Davies ENDOR power pattern simulations (Figure 12) were used in the constrained ESEEM simulations. The good quality of the ^{2}H ESEEM simulation both supports the ^{1}H ENDOR assignment and provides information as to the number of exchangeable hydrogens in each class (*39*). This information is incorporated into the S_2-state structural model of Figure 11.

Final comments

This tutorial provides only a glimpse of the possibilties of pulsed EPR methods to fill in the many outstanding lacunae in interesting biochemical systems such as Photosystem II. The pulsed EPR field is rapidly advancing, with many new methods being introduced. Just a few examples include high frequency ($\geq$ 94 GHz) pulsed EPR, 2-D ESEEM methods, and pulsed EPR of protein single crystals. The number of pulsed EPR instruments is rapidly increasing, with both "laboratory built" and commercial instruments available. Thus it continues to be an exciting time to be a practitioner in this interesting field.

References

(1.) Kok, B.; Forbush, B.; McGloin, M. *Photochem. Photobiol.* **1970**, *11*, 457-475.
(2.) Debus, R. J. *Biochim. Biophys. Acta* **1992**, *1102*, 269-352.
(3.) Britt, R. D. In *Oxygenic Photosynthesis: The Light Reactions*; Ort, D., Yocum, C. F., Eds.; Kluwer Academic: Dordrecht, The Netherlands, 1996; pp 137-164.
(4.) Yachandra, V. K.; Sauer, K.; Klein, M. P. *Chem. Rev.* **1996**, *96*, 2927-2950.
(5.) Peloquin, J. M; Britt, R. D. *Biochim. Biophys. Acta* **2001**, *1503*, 96-111.
(6.) Randall, D. W.; Chan, M. K.; Armstrong, W. H.; Britt, R. D. *Mol. Phys.* **1998**, *95*, 1283-1294.

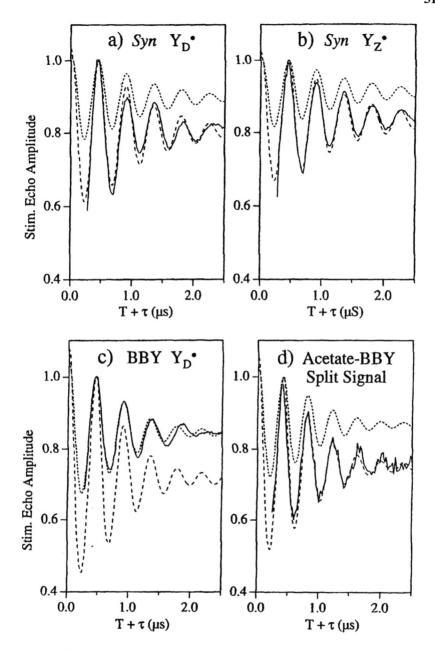

Figure 21. ^{2}H 3-pulse time domain ESEEM patterns for tyrosine radical signals from ^{2}H$_2$O-exchanged PSII preparations. (a) *Synechocystis* Y$_D^\bullet$; (b) *Synechocystis* Y$_Z^\bullet$; (c) spinach Y$_D^\bullet$; (d) spinach inhibited S_2-Y$_Z^\bullet$ interaction signal. Experimental data: Solid lines. Simulations with 1 coupled deuteron: dotted lines. Simulations with 2 coupled deuterons: dashed lines.

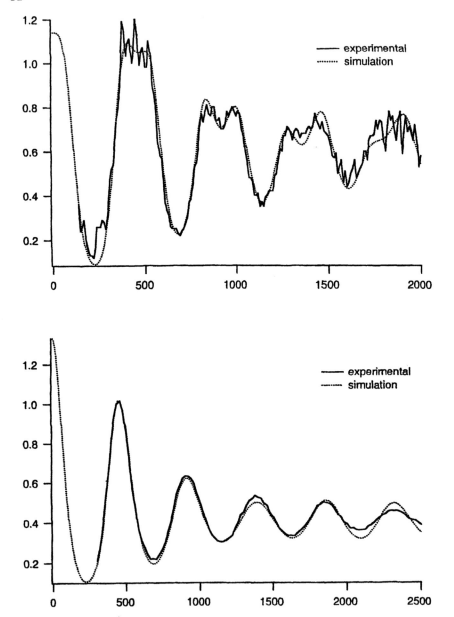

Figure 22. ^{2}H 2-pulse (upper panel) and 3-pulse (lower panel) time domain ESEEM patterns for the S_2-state multiline EPR signal from ^{2}H$_2$O-exchanged PSII preparations. These time domain spectra correspond to the ratio of time domain patterns obtained after and before ^{2}H$_2$O exchange. This procedure nulls contributions from other nuclei such as ^{14}N.

(7.) Sturgeon, B. E. Ph.D. thesis, University of California, Davis, CA, 1994.

(8.) Kurreck H.; Kirset, B; Lubitz, W. *Electron Nuclear Double Resonance Spectroscopy of Radicals In Solution*, VCH, Weinheim, 1988.

(9.) Mims, W. B. In *Electron Paramagnetic Resonance;* Geschwind. S. Ed.; Plenum Press, New York. 1972; pp. 263-351

(10.) Hahn, E. L. *Physical Review.* **1950**, *80*, 580-594.

(11.) Davies, E. R. *Physics Lett.* **1974**, *47A*, 1-2.

(12.) Mims, W. B. *Proc. Royal. Soc. London.* **1965**, *283*, 452-457.

(13.) Hoffman, B. M.; DeRose, V. J.; Doan, P. E.; Gurbiel, R. J.; Houseman, A. L. P.; Telser, J. In *Biological Magnetic Resonance;* Berliner, L. J., Reuben, J., Eds.; Plenum: New York, 1993; Vol. 13, pp 151-218.

(14.) Thomann, H.; Bernardo, M. In *Biological Magnetic Resonance;* Berliner, L. J., Reuben, J., Eds.; Plenum: New York, 1993; Vol. 13, pp 275-322.

(15.) Fan, C. L.; Doan, P. E.; Davoust, C. E.; Hoffman, B. M. *J. Magn. Reson.* **1992**, *98*, 62-72.

(16.) Randall, D. W.; Sturgeon, B. E.; Ball, J. A.; Lorigan, G. A.; Chan, M. K.; Klein, M. P.; Armstrong, W. H.; Britt, R. D. *J. Am. Chem. Soc.* **1995**, *117*, 11780-11789.

(17.) Randall, D. W. Ph.D. thesis, University of California, Davis, CA, 1997.

(18.) Britt, R. D.; Peloquin, J. M.; Campbell, K. A. *Annu. Rev. of Biophys. and Biomolecul. Struct.* **2000**, *29*, 463-495.

(19.) Peloquin, J. M.; Campbell, K. A.; Randall, D. W.; Evanchik, M. A.; Pecoraro, V. L.; Armstrong, W. H.; Britt, R. D. *J. Am. Chem. Soc.* **2000**, *122*, 10926-10942.

(20.) Zouni, A.; Witt, H. T.; Kern, J.; Fromme, P.; Krauss, N.; Saenger, W.; Orth, P. *Nature* **2001**, *409*, 739-743.

(21.) Gilchrist, M. L.; Ball, J. A.; Randall, D. W.; Britt, R. D. *Proc. Natl. Acad. Sci. U.S.A.* **1995**, *92*, 9545-9549.

(22.) O'Malley, P. J. *Biochim. Biophys. Acta* **2002**, *1553*, 212-217.

(23.) Doan, P. E.; Hoffman, B. M. *Chem. Phys. Lett.* **1997**, *269*, 3-4.

(24.) Hoganson, C. W.; Lydakis-Simantiris, N.; Tang, X.-S.; Tommos, C.; Warncke, K.; Babcock, G. T.; Diner, B. A.; McCracken, J. ; Styring, S. *Photosynth. Res.* **1995**, *46*, 177-184.

(25.) Force, D. A.; Randall, D. W.; Britt, R. D.; Tang, X. S.; Diner, B. A. *J. Am. Chem. Soc.* **1995**, *117*, 12643-12644

(26.) Campbell, K. A.; Peloquin, J. M.; Diner, B. A.; Tang, X.-S.; Chisholm, D. A. Britt, R. D. *J. Am. Chem. Soc.* **1997**, *119*, 4787-4788.

(27.) Campbell, K. A.; Force, D. A.; Diner, B. A.; Britt, R. D., manuscript in preparation.

(28.) Mims, W. B. *Phys. Rev. B* **1972** *5*, 2409-2419.

(29.) Mims, W. B.; Peisach, J. In *Biological Magnetic Resonance;* Berliner, L. J., Reuben, J., Eds.; Plenum: New York, 1981; Vol. 3, pp 213-263.

(30.) Peloquin, J. M.; Tang, X.-S.; Diner, B. A.; Britt, R. D. *Biochemistry* **1999**, *38*, 2057-2067.

(31.) Tang, X.-S.; Peloquin, J. M.; Lorigan, G. A.; Britt, R. D.; Diner, B. A. In *Photosynthesis: from Light to Biosphere;* Mathis, P., Ed.; Kluwer Academic Publishers, Amsterdam, 1995; Vol. 3, pp 775-778.

(32.) Mims, W. B; Peisach, J. *J. Chem. Phys.* **1978** *69*, 4921-4930.

(33.) Flanagan, H. L.; Singel, D. J. *J. Chem. Phys.* **1987** *87*, 5606-5616.

(34.) Das, T. P.; Hahn, E. L. *Nuclear Quadrupole Resonance Spectroscopy*, Academic Press, New York, NY; 1958.

(35.) Edmonds, D. T. *Physics Reports* **1977**, *4*, 233-290.

(36.) Lai, A.; Flanagan, H. L.; Singel, D. J. *J. Chem. Phys.* **1988** *89*, 7161- 7166.

(37.) Stemmler, T. L.; Sturgeon, B. E.; Randall, D. W.; Britt, R. D.; Penner-Hahn, J. E. *J. Am. Chem. Soc.* **1997**, *119*, 9215-9225.

(38.) Diner, B. A.; Force, D. A.; Randall, D. W.; Britt, R. D. *Biochemistry* **1998**, *37*, 17931-17943.

(39.) Aznar, C. P.; Britt R. D. *Phil. Trans. R. Soc. Lond. B.* **2002**; *357*, 1359-1366.

Chapter 3

The Past, Present, and Future of Orientation-Selected ENDOR Analysis: Solving the Challenges of Dipolar-Coupled Nuclei

Peter E. Doan

Department of Chemistry, Northwestern University, 2145 Sheridan Road, Evanston, IL 60208

Extracting structural information from Electron-Nuclear Double Resonance (ENDOR) spectra of metalloproteins requires the understanding of the mapping of hyperfine and quadrupole tensors onto the **g** tensor, a process known as orientation selection. The original work in this field focused predominantly on central metal and ligand ENDOR patterns with large hyperfine couplings that had substantial isotropic components. In this paper, I will update some of the mathematics and explore the differences between these more traditional types of ENDOR patterns and the spectra from systems with little or no isotropic hyperfine interaction.

Introduction

During the past decade, the work in the Hoffman group attempting to understand the many nuances of ENDOR patterns in metalloproteins has taken us in many different directions. In the earliest work, now nearly 20 years old, Brian Hoffman and his group laid out a simple algorithm for extracting the relevant hyperfine tensor values and Euler angles that related the **A** tensor to the **g** tensor (1,2). They were attempting to understand a specific problem, ^{57}Fe

ENDOR spectra (3-5) from the FeMo cofactor in the resting state of the nitrogenase enzyme, and were responding to the basic 'necessity is the mother of invention,' aspect of science. The program they developed, GENDOR (GENeral ENDOR simulation), therefore was strongly motivated in its design by aspects of this problem, which is an $S = 3/2$ spin system with the large hyperfine couplings that arise from metal-ion ENDOR spectra. These spectra contain both frequency and intensity information, but they (quite rightly) assumed that the frequency information was far more reliable than the intensity information, and therefore tended to ignore subtle intensity variations. In two papers, they investigated the underlying mathematics of what they called orientation-selection in ENDOR, namely the frequency envelopes that arise from a series of different **A** tensors and the Euler angles relating these **A** tensors to the **g** tensor. In a separate paper, Hoffman and Gurbiel (6) examined the frequency envelopes that arose from dipolar couplings of an $I = 1/2$ nucleus with a nitroxide spin label. In this work, orientation-selection was introduced by the ^{14}N hyperfine interaction using a slight variant of the GENDOR program. As in the original work on the [57]Fe ENDOR of nitrogenase, they considered the intensity functions to be of secondary importance to the frequency patterns, due to the inherent difficulties in measuring and understanding ENDOR intensities, especially from continuous-wave (CW) methods.

In the late 1980s, a number of groups, including ours, began using echo-detected (pulse) techniques to measure ENDOR of metalloproteins (7-10). These pulse methods quickly demonstrated their advantages in obtaining quantitative lineshape and intensity information, though these methods tend to suffer from much lower signal-to-noise than the CW ENDOR techniques. From this point on, when possible, CW-ENDOR spectra were checked against the data obtained from the pulsed ENDOR methods and we found that the intensity patterns for CW ENDOR in many systems were more reliable than was first thought. Concurrently, the original intensity formula used in GENDOR was replaced by a more appropriate one, and the simulation intensity patterns more closely resembled the experimental data than in the previous version of the program. The increases in computer power that took place during this time allowed for rapid inclusion of EPR linewidth effects that we had previously been forced to minimize due to time constraints. Given these advantages and nearly a decade of successfully analyzing metalloprotein ENDOR data, our group has formulated guidelines for understanding the basis of ENDOR frequency envelopes and lineshapes in a variety of systems. I will present the recent advances that apply to a specific problems of dipolar coupled nuclei in the limit of very small hyperfine couplings, those with $|A| < 0.5$ MHz; a type of system that was not thought to be amenable to ENDOR study just a few years ago.

The ultimate goal of this work is to progress to a point where a limited set of spectra collected in a short period of time can be used to predict accurately and quickly the specific field values at which principal hyperfine and quadrupole values will be observed. Achieving this goal could greatly reduce the number of spectra required and thereby increase the throughput of samples in ENDOR and the related techniques that fall under the broad category of Electron Spin Echo Envelope Modulation (ESEEM) spectroscopies (*11*).

EPR and ENDOR Powder Spectra

To first order, the energy levels for an $S = 1/2$ electron, coupled to an $I = 1/2$ nuclear spin for a single orientation in an applied magnetic field, B_0 are given by

$$E(M_S, M_I) \ = \ g_e \beta_e B_0 M_S + g_n \beta_n B_0 M_I + A M_S M_I \qquad (1)$$

where A is the orientation dependent hyperfine coupling, and $g_e \beta_e B_0/h$ (v_e), $g_n \beta_n B_0/h$ (v_n) are the electron and nuclear Zeeman interactions, respectively (*12*). This gives the four level energy diagram shown in **Figure 1**. For $I > 1/2$, nuclear quadrupole splittings would also have to be included, and there would be $2I + 1$ nuclear spin levels per electron spin manifold. Each of these interactions is anisotropic and would be diagonal within their own axis system that can be related to the molecular framework, although the chemical shift anisotropy of the nuclear Larmor term is too small to be detected in most ENDOR experiments.

Though the various tensor quantities and orientations can be extracted by using oriented systems such as magnetically dilute single crystals, this would not be practical for a majority of metalloprotein samples. The most common form for an EPR/ENDOR sample is an isotropically frozen solution in which it is assumed that all molecular orientations are equally probable and thereby contribute equally to the resulting spectrum. With a change of perspective, this is equivalent to stating that in a magnetic resonance experiment, the external field has equal probability of having any orientation relative to a given molecular axis system. For completeness, the definition of 'orientation' is a unit vector in 3-space axis system $l = (x, y, z)$. In a given axis system, the probability associated with a specific orientation of the magnetic field is proportional to the differential surface area element, $d\sigma$, of that particular orientation. Consider for example a rhombic EPR **g** tensor having principal values ($g_1 > g_2 > g_3$). An EPR powder pattern of such a center is comprised of contributions from all possible orientations, each of which has a g value, $g(l)$ or resonance field, $B(l) = h\nu/g(l)\beta_e$, that falls between g_1 and g_3. An orientation can be specified by two independent variables and is traditionally parameterized in terms of the spherical coordinates, (θ, ϕ) giving $l(\theta, \phi) = (\sin\theta\cos\phi, \sin\theta\sin\phi, \cos\theta)$, with the associated area element $d\sigma(\theta, \phi) = \sin\theta \, d\theta d\phi/4\pi$. The orientation dependent g factor then becomes

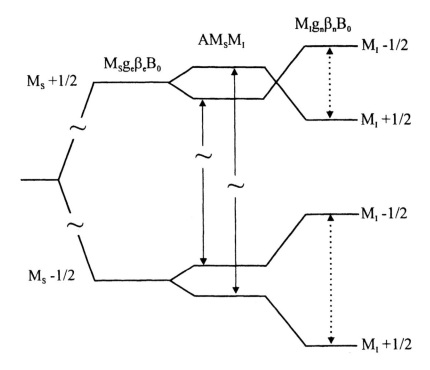

Figure 1. Four-level energy diagram for an S = 1/2 electron coupled to an I = 1/2 nucleus in an applied magnetic field of B_0. In this figure, the nuclear Larmor frequency is greater than the hyperfine interaction. The two allowed EPR transitions are shown in the solid lines; the NMR (ENDOR) transitions are shown in dashed lines.

$$g^2(\theta,\phi) \;=\; \sin^2\theta\cos^2\phi\, g_1^2 + \sin^2\theta\sin^2\phi\, g_2^2 + \cos^2\theta\, g_3^2 \qquad (2)$$

Kneubuhl (*13*) produced the first analytical solution to this problem in 1960, showing that the three principal *g* values can be calculated with a high degree of accuracy from such a randomly oriented sample. The resulting EPR-absorption lineshape for such a system is shown in **Figure 2**. This lineshape derived by Kneubuhl is a mathematical construct and is sometimes called a "statistical lineshape." It does not account for transition-dipole moments, relaxation effects, anisotropic unresolved hyperfine interactions or any of a myriad of other possible complications in EPR spectroscopy. As noted by Kneubuhl, similar lineshapes had been derived for nuclear quadrupole and NMR lineshapes. Not surprisingly, the same mathematics can hold for ENDOR lineshapes as well though there are multiple complications due to the relative magnitudes of *A* and v_n. From equation 1, the two ENDOR frequencies for a single orientation occur as the frequencies $|v_n + A/2|$ and $|v_n - A/2|$. When $|v_n| > |A/2|$, the peaks are centered at the nuclear Larmor frequency and split by the hyperfine value, $|A|$. When $|A/2| > |v_n|$, the peaks are centered at $|A/2|$ and split by $2|v_n|$. Each of the two transitions has its associated lineshape that add to produce the spectrum. The two transitions typically are denoted as v_+ and v_-, though this should not be taken as an assignment to a specific electron spin manifold as an ENDOR spectrum alone cannot determine the absolute sign of *A*. **Figure 3** shows one type of ENDOR powder pattern for an axial hyperfine tensor with $|v_n| \gg |A_\parallel/2| > |A_\perp|$.

Orientation-Selection in ENDOR

The Past

One way to view the Kneubuhl lineshape is to see it as a sorting mechanism. Though all orientations are equally probable, not all orientations contribute equally to the EPR spectrum at a specific *g* value. The resonance condition sorts the orientations of the molecules into mathematically well-defined subsets. An ENDOR spectrum is literally an EPR-detected NMR spectrum; at a constant field and microwave frequency setting, an EPR transition is excited and concurrently a radio frequency is swept. NMR transitions are detected as changes in the EPR absorption, therefore only the molecules whose electron spin system are in resonance can be detected via an ENDOR experiment. Therefore, in an ENDOR experiment, only a well-defined subset of orientations can contribute to the NMR spectrum, and that subset is defined by the g tensor values, the microwave quantum, and the applied magnetic field.

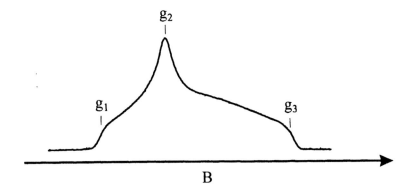

*Figure 2. Frozen-solution EPR spectrum with **g** tensor (g$_1$, g$_2$, g$_3$) shown in absorption mode.*

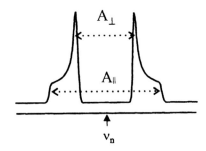

Figure 3. Simulated statistical ENDOR powder average lineshapes for a Larmor-centered doublet with $|v_n| > |A_{||}| > |A_\perp|$.

Rist and Hyde (*14*) were the first to exploit this in ENDOR studies of Cu(II) complexes doped into diamagnetic hosts. They observed that at the low field edge ($g_\parallel$) of the Cu(II) EPR spectrum, the ^{14}N ENDOR spectra tended to resemble the high resolved single-crystal ENDOR spectra. They were able to resolve both hyperfine and quadrupole splittings from coordinated N-donor atoms in these 'single-crystal-like' spectra. At other positions on the EPR spectrum, the ENDOR spectra resembled broad 'powder-type' spectra, and were therefore difficult to interpret. At the $g_\parallel$ turning point, the subset of molecules that contribute to the EPR intensity becomes extremely small, to the point where it begins to resemble a single orientation, as in an single-crystal ENDOR experiment.

They also attempted to understand the lineshapes of the 'powder-type' ENDOR spectra that were collected at other field positions across the EPR envelopes of the various complexes with some success. As Cu(II) generates an axial EPR spectrum with $g_\parallel > g_\perp$, instead of either a true powder average or a single-crystal-like set, they would generate a 'plane averaged' set of orientations, which reproduced the data to some degree of accuracy (*14*).

At this point, the literature on orientation-selection breaks into two parallel approaches: a mathematical construct more similar to the early lineshape papers of Kneubuhl (Hoffman), and the use of computer simulation software that attempts to replicate experimental spectra (*15*). As this paper is focused on the former method, I will briefly summarize the latter approach and its advantages and disadvantages. By the early 1970s, computer time was becoming more available and more reasonable in cost, providing EPR spectroscopists with the ability to simulate powder spectra with a fair degree of accuracy. These programs could not rely on the original lineshape functions as they were designed to account for precisely the factors that complicate the exact mathematical constructs. It was recognized by Sands and coworkers that one could simulate ENDOR spectra of FeS proteins using the same program that they used to simulate the EPR spectra of these proteins. From the point of view of reproducing ENDOR data, this is an incredibly robust method for orientation-selection, as it makes so few assumptions about spin systems. Extremely complicated systems with resolved fine structure, resolved hyperfine structure, overlapping signals, etc can be handled that would not be approachable using a more mathematically direct approach. The speed and low costs of today's computers makes this direct spectral simulation an even more useful technique.

The one major disadvantage is that any understanding of the process by which the lineshapes are produced is lost in the computational details. This reduces the field to an empirical approach, correlating observation (or simulation) with input parameters. The work done by Kneubuhl and others with the mathematical constructs for EPR, NMR and NQR provides us with a rich vocabulary of terms to describe the features in powder spectra, even if the details of the spectrum could not be faithfully reproduced by the mathematics used. For example, the concept of a 'turning point,' an orientation at which the derivative of the resonance condition with respect to the orientation parameters goes to zero. In an EPR spectrum, this could be stated as $dg/d(\theta,\phi) \to 0$. The turning point implies that the resonance condition tends to change slowly with orientation around that g value so that the spectrum builds up intensity at that point. Despite the fact that no one has used the analytical lineshape functions to simulate an EPR spectrum for at least 30 years, we still refer to the narrow features of these spectra as turning points.

The approach taken by Hoffman, Martinsen, and Venters (*1*) was stated specifically to be an attempt to understand the mathematics of orientation-selected ENDOR, rather than a direct attempt to simulate experimental spectra. They extended the approach of Kneubuhl by assuming both a δ-function EPR and a δ-function NMR linewidth. The first step, identical to that of Kneubuhl is to express the one of the two spherical parameters instead by the observable g. From equation 2 and the resonance condition, $g = h\nu_e/\beta_e B_0$, the parameter $\sin^2\theta$ can be solved for in terms of g and ϕ

$$\sin^2\theta \;=\; \frac{g^2 - g_3^2}{(g_2^2 - g_3^2) + (g_1^2 - g_2^2)\cos^2\phi} \tag{3}$$

The area element $d\sigma(\phi, g_{obs})$ associated with this orientation given in equation 4 is obtained by the standard change of variable technique.

$$d\sigma(\phi, g) \;=\; \left(\frac{g}{g - g_3^2}\right)\frac{\sin^2\theta(\phi, g)}{\cos\theta(\phi, g)}\,d\phi dg \tag{4}$$

(In their original formulation, they used an arc segment length $dl(\phi, g)$ rather than this area element.) If $d\sigma(\phi, g)$ is integrated over the allowed values of ϕ for a given g, the result is the lineshape function $S(g)$ reported by Kneubuhl. Along a given g value, equations 3 and 4 provide functions of a single variable (ϕ) that can completely define both the orientation of the magnetic field and the proper intensity associated with that orientation.

Assuming the hyperfine values are small compared to the microwave quantum (the strong field approximation) the ENDOR frequencies for a $S = 1/2$, $I = 1/2$ system with a hyperfine tensor **A** and the **g** tensor can be calculated using the matrix formulation of Thuomas and Lund (*16*) given in equation 5

$$v_{\pm}^2 = l \cdot \left[\pm \frac{1}{2g} \mathbf{g} \cdot {}^{\mathbf{g}}\mathbf{A} - v_n \mathbf{I} \right] \cdot \left[\pm \frac{1}{2g} {}^{\mathbf{g}}\mathbf{A}\mathbf{g} - v_n \mathbf{I} \right] \cdot l^T \tag{5}$$

where ${}^{\mathbf{g}}\mathbf{A}$ is the hyperfine tensor expressed in the $\mathbf{g}$ tensor frame, l is the row vector (x, y, z) of the magnetic field orientation in the $\mathbf{g}$ tensor frame, $\mathbf{I}$ is the identity tensor and v_n is the nuclear Larmor frequency. By using the orientation selected subset described by equation 3, at a given g value, the row vector l can be expressed in terms of the single parameter, ϕ. As all the other terms in equation 5 are constant, the ENDOR frequencies, $v_{\pm}$ are also then dependent upon a single variable (ϕ). This is the heart of their approach to orientation-selection, as now there is *a simple mathematical formula based on a single variable that gives the ENDOR frequencies at a given value of g*. Turning points can be calculated directly by taking the derivatives $dv_{\pm}/d\phi$ and setting them equal to 0, which can also be describe as divergences in ENDOR intensities by the standard change-of-variable technique. Simulations can be accomplished by creating a frequency histogram with the frequencies calculated by applying equation 5 and intensities calculated using equation 4, which led to the program GENDOR

Hoffman and coworkers examined a number of different symmetries both of the $\mathbf{g}$ and $\mathbf{A}$ tensors, and relative orientations between these two. To simplify matters (and because of their interest in [57]Fe ENDOR spectra), they looked at patterns in the limit of zero v_n, so that $v_+ = v_- = v = A(\phi,g)/2$. In the limit of coaxial, rhombic $\mathbf{g}$-$\mathbf{A}$ tensors, a 'triangle' diagram **(Figure 4)** appears in a plot of g versus v, with the legs of the triangle connecting the points $(g_1,v(g_1))$, $(g_2,v(g_2))$, and $(g_3,v(g_3))$. At any g value between g_1 and g_3, the two peaks observed in the ENDOR lie along these lines. There are multiple peaks in orientation-selected spectra for all g between g_1 and g_3 in the case of a single anisotropic hyperfine tensor. Only at the two extrema of the EPR envelope do all the peaks coalesce into the one resonance condition or the single-crystal-like pattern.

They also investigated the number of peaks and the field dependence of the ENDOR patterns that are expected for simple cases of non-coaxial tensors, where one of the three principal axes from each the $\mathbf{g}$ and $\mathbf{A}$ tensors are coaxial, say $g_3 \parallel A_3$, and the A_1, A_2 axes are rotated around g_3 by an angle α. In this case, the g_1 - g_2 leg of the triangle splits into two arcs as shown in on either side of the original line. The other legs of the triangle are unaffected so that in the region between g_1 and g_2, one would expect three peaks from a single $I = 1/2$ nucleus and two peaks between g_2 and g_3. Extending this to a system with arbitrary symmetry, they found that there could be up to six turning points at a single g value of observation. The effects of a non-zero EPR linewidth investigated numerically simulating a number of these spectra at closely spaced g values and

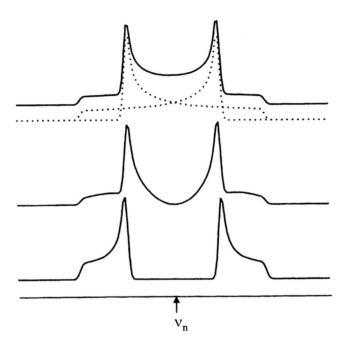

Figure 4. (Top) The statistical ENDOR powder spectrum of a dipolar-coupled nucleus (no isotropic component) showing the overlapping v_+ and v_- branches. (Middle) The observable lineshape of the same system with a Mims pulsed ENDOR experiment. (Bottom) A lineshape similar to the observable (middle) using an A tensor that has a large isotropic hyperfine interaction and small axial anisotropic component.

adding them with the proper relative weighting factors, as discussed in the 1985 paper (2).

An analysis procedure for obtaining principal hyperfine tensor values orientations from a suite of ENDOR spectra collected across the EPR envelope was suggested by Hoffman et al. in a review article (17). First, obtain the ENDOR spectra across the entire EPR envelope. Second, use the two single-crystal-like spectra at g_1 and g_3 to approximate A_1 and A_3; A_2 is estimated from the spread of frequencies (if available) at g_2. The nature of the relative orientations of the tensors is inferred from the development of the ENDOR patterns as the field increases from the low-field (g_1) of the EPR spectrum. Simulations of selected spectra typically are performed by varying the hyperfine interaction, principal values and relative orientations of the **g** tensor and nuclear coordinate frames.

The Present State of Orientation-Selection

The most exhaustive study on a single enzyme system using the basic approach laid out above is on the modified heme in allylbenzene-inactivated chloroperoxidase (CPO) (18). The biological role of CPO is the halogenation of organic substrates but the enzyme also is found to epoxidize alkenes with high facioselectivity (19). During the epoxidation of allylbenzene (AB), CPO eventually is converted to an inactive green species whose low-spin ($S = 1/2$) ferri-heme prosthetic group (AB-CPO) is modified by addition of the alkene plus an oxygen atom. Determination of the structure of the adduct gives insights into the catalytic mechanism and information about the AB binding geometry.

The details concerning this work (18) are far too intricate to summarize in the space available, so only an exceedingly brief synopsis is provided here. The rhombic **g** tensor of AB-CPO (g_1, g_2, g_3) = (2.32, 2.16, 1.95) provides for excellent orientation selection. This orientation-selection combined with the nitrogen ENDOR using natural abundance ^{14}N and ^{15}N-labeled heme allows the **g** tensor to be mapped relative to the $^{14,15}N$ ligand hyperfine and ^{14}N quadrupole tensors, and therefore onto the molecular framework of the enzyme using a process that is beyond the scope of this paper. The structure of the inactive enzyme then was placed onto this framework by the use of orientation-selected CW and pulsed ENDOR by including isotopic labels (^{13}C, ^2H) at specific sites of the substrate. The **A** and **P** tensors derived from each individual site are mapped onto the **g** tensor and thereby onto the molecular framework. The various possible geometries were tested with MM2 calculations.

It is clear that the ideas laid out in the two original papers on orientation-selected ENDOR analysis (1,2) provide a strong foundation for analyzing molecular geometries derived from ENDOR spectra. The authors achieved one of their goals, namely to categorize all the types of orientation-selected hyperfine ENDOR patterns. In one aspect, however, these papers fell short of their ultimate goal to provide an *analytical solution* to the lineshapes and frequency envelopes as was done by Kneubuhl. As such, the analysis procedure still relies too heavily on multiple simulations rather than a more analytical approach where, say, a turning point at given g value would suggest a specific possible set of Euler angles between the **g** and **A** tensors.

An example of one of the major difficulties involved in applying this procedure can be demonstrated using a point-dipole coupled nucleus. In general, these will be cases where $|v_n| >> |A/2|$ so the pattern will be centered at the nuclear Larmor frequency and split by the hyperfine interaction. For an isotropic g value, the orientation dependent hyperfine for a nucleus at distance r from an electron is given by the well-known relation, $A(\theta) = T(3cos^2\theta - 1)$, where θ is the angle between the electron-nuclear vector and the applied magnetic field and T is proportional to the magnetic moment of the nucleus divided by r^3. The resulting powder pattern gives the well-known 'Pake pattern' shown in **Figure 5** (top). The pattern is decomposed into the two separate branches, v_+ and v_- that overlap between $|v_n - T/2|$ and $|v_n + T/2|$. The major difficulty comes about from the fact that the ability to observe a specific ENDOR transition is dependent upon the magnitude of the A value associated with that orientation. As $\theta \rightarrow$ 54.7°, $A(\theta) \rightarrow 0$, and the experimental ENDOR intensity will vanish. For example, in a Mims pulsed ENDOR experiment *(20)*, the ENDOR intensity is determined both by the A value and the time interval between two microwave pulses (τ) according to the formula $E(\tau) = (1 - cos(2\pi A\tau))/2$. Local maxima in ENDOR intensity occur at $A(MHz)\tau(\mu s) = 0.5, 1.5, 2.5,...$ and local minima or 'blind spots' at $A\tau = 0, 1, 2,...$ Multiplying $E(\tau)$ by the statistical lineshape function produces a simulated "observable lineshape" given in Figure 5 (middle). This observable lineshape can easily be mistaken for a system with an axial hyperfine tensor but with an isotropic component that is larger than the anisotropic component, as shown in Figure 4 (bottom). The use of only the frequency envelopes of these patterns does not produce a unique solution, as the only major differences between the spectra in Figure 5 (middle) and Figure 5 (bottom) will be seen in the intensity patterns of the ENDOR responses (*vide infra*). In an orientation-selected experiment, there is a further complication that arises if the angle between one of the EPR envelope extrema, say g_1, and the electron-nuclear vector approaches ~55° as the single-crystal-like ENDOR pattern would then be predicted to have zero intensity.

A New Approach to Orientation-Selection

The use of the spherical coordinates (θ, ϕ) in describing a specific orientation (x, y, z) is traditional and advantageous in almost any mathematics involving a unit sphere parameterization. It is, however, not the only set of parameters that will work effectively. If instead, the explicit definition of a unit sphere is combined with equation 2, there are four variables and two equations relating these four variables: $1 = x^2 + y^2 + z^2$ and $g^2 = x^2 g_1^2 + y^2 g_2^2 + z^2 g_3^2$. These two equations can be manipulated to use one of the coordinates and the g^2 value to define the other two coordinates. Choosing x, and defining $U(g) = (g^2 - g_3^2)/(g_2^2 - g_3^2)$ and $V = U(g_1) = (g_1^2 - g_3^2)/(g_2^2 - g_3^2)$ gives the following parameterization of the first octant of the unit sphere.

$$x(x, g) = x$$
$$y(x, g) = \sqrt{U(g) - x^2 V} \tag{6}$$
$$z(x, g) = \sqrt{1 - U(g) + x^2 (V - 1)}$$

Other octants are related by symmetry. Because this is an unconventional approach, I will describe some of the characteristics in some detail. The region in the (x, g) plane that corresponds to the unit sphere is bounded by a piecewise continuous curve shown in **Figure 6**, $C = C_1 + C_2 + C_3$ where

$$C_1 : x = 0 \quad (g_3 \le g \le g_2)$$
$$C_2 : x = \sqrt{(g^2 - g_3^2)/(g_1^2 - g_3^2)} \quad (g_3 \le g \le g_1) \tag{7}$$
$$C_3 : x = \sqrt{(g^2 - g_2^2)/(g_1^2 - g_2^2)} \quad (g_2 \le g \le g_1)$$

In essence, C_1 lies in the yz plane, C_2 lies in the xz plane, and C_3 in the xy plane. The differential surface area element $d\sigma(x, g)$ of the unit sphere associated with the point (x, g) can be derived using the two tangents to the surface, T_x, and T_g

$$T_x = \left(\frac{dx}{dx}, \frac{dy}{dx}, \frac{dz}{dx} \right) = \left(1, \frac{-2xV}{y}, \frac{2x(V-1)}{z} \right)$$
$$T_g = \left(\frac{dx}{dg}, \frac{dy}{dg}, \frac{dz}{dg} \right) = \left(\frac{2g}{(g_2^2 - g_3^2)} \right) \left(0, \frac{1}{y}, \frac{-1}{z} \right) \tag{8}$$

giving

$$d\sigma(x, g) = \left\| T_x \times T_g \right\| dx\, dg = \frac{2g}{g_2^2 - g_3^2} \frac{dx\, dg}{y(x, g)z(x, g)} \tag{9}$$

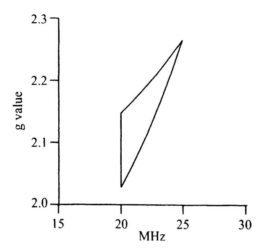

*Figure 5. The frequencies of ENDOR intensity divergences (peaks) for an axial A tensor (50, 40, 40) MHz, coaxial with the rhombic **g** tensor (2.27, 2.15, 2.03) in the limit of zero nuclear Larmor frequency. With the exception of the two single-crystal-like positions, g_1 and g_3, two peaks are expected in the ENDOR spectrum that arise from a single $I = 1/2$ nucleus.*

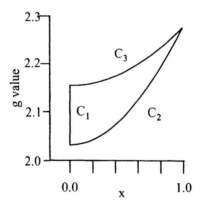

*Figure 6. The domain of the first octant of the unit sphere in the (x, g) parameter space for the rhombic **g** tensor (2.27, 2.15, 2.03). The curves C_1, C_2, and C_3 represent the points on x = 0, y = 0, and z = 0 planes of the unit sphere, respectively.*

Under this parameterization, the statistical lineshape function shows divergences both along C_2 and C_3 in equation 6, but not along C_1.

One advantage of this parameterization can be seen by the looking at one of the examples examined in depth in the 1984 paper, coaxial rhombic g-A tensors. When A is coaxial with g, then in equation 5, ^gA is diagonal, and the frequency equation can be readily expanded to a form

$$v_{\pm}^2(x, g) = x^2 M_{\pm,x} + y^2 M_{\pm,y} + z^2 M_{\pm,z} + v_n^2$$

$$where \tag{10}$$

$$M_{\pm,l} = \frac{g_i^2 A_i^2}{4g^2} \pm \frac{v_n g_i A_i}{g}$$

Substituting for y^2 and z^2 using equation 6, it is easy to rearrange equation 10 into the simple form

$$v_{\pm}^2(x, g) = x^2 N_{\pm}(g) + O_{\pm}(g)$$

$$where$$

$$N_{\pm}(g) = M_{\pm,x} - M_{\pm,z} - V(M_{\pm,y} - M_{\pm,z}) \tag{11}$$

$$O_{\pm}(g) = v_n^2 + M_{\pm,z} + U(g)(M_{\pm,y} - M_{\pm,z})$$

This shows that for coaxial g and A tensors there is a 1:1 relationship between the v_+ and v_- ENDOR frequencies at a given g value and the parameter x, without using any simplifying assumptions other than the initial strong field approximation. Therefore, it is possible to invert equation 11, substituting v_+ and v_- as separate calculations into the equation to give an analytical formula for the differential area element in terms of the two experimental parameters, (v, g).

$$d\sigma(g, v_{\pm}) = d\sigma(g, x) \left|\frac{dx}{dv_{\pm}}\right| |dv_{\pm}|$$

$$= \frac{1}{|N_+(g)|} \frac{1}{g_2^2 - g_3^2} \frac{g v_{\pm}}{x(g, v_{\pm})y(g, v_{\pm})z(g, v_{\pm})} dg \, dv_{\pm} \tag{12}$$

The regions of the (v, g) plane associated with these intensity functions, shown in **Figure 7**, are obtained by calculating $v_{\pm}(x, g)$ along the boundary curve $C = C_1 + C_2 + C_3$ described in equation 7, represent the triangle(s) that was discussed in the 1984 paper for coaxial g-A tensors. What we learn from equation 12 is that the orientations associated with the triangle diagram discussed in the 1984 paper (*1*) arise from divergences that occur along the three planes ($x = 0$, $y = 0$, $z = 0$). The relative area elements associated with these divergences contain additional information. The intensity associated with the peaks along the C_1 ($x = 0$) and C_3 ($z = 0$) curves is always greater than the intensity of the peak at the same g value along the C_2 curve ($y = 0$). The intensity function is well-defined at all points other than at the three principal g values, where two of the three terms (coordinates) in the denominator go to zero.

With this last step, we have begun to approach the ultimate goal that was originally set in 1984 (*1*), to define a true analytical lineshape for orientation-selected ENDOR. Much like the work of Kneubuhl (*13*), equation 12 serves little purpose in simulating 'real' orientation-selected ENDOR patterns. The second-generation software, now called ENDORSIM, uses the weighting factor of equation 9, and just as GENDOR, numerically generates multiple frequency histograms. What we have obtained from equation 12 is a greater appreciation for the subtle effects that can create changes in both lineshape and intensity.

Applications of the New Approach

Knocking out the 'Cornerstone' of Orientation-Selection

From all points of view, the 'cornerstone' of orientation-selection is that at the high-field and low-field edges of a rhombic EPR spectrum, the orientation-selection process narrows to a point, and this gives rise to 'single-crystal-like' ENDOR spectra (*14*). One of the first observations that arose from use of the newer weighting factors and equations is that in many cases, especially with dipolar-coupled nuclei, spectra taken at the 'single-crystal-like' positions of g_1 and g_3 do not bear much resemblance to single crystal ENDOR spectra. For example, consider the proton from a coordinated hydroxo ligand in the protein nitrile hydratase (*21*). The active site of this enzyme is a nonheme mononuclear low-spin Fe(III) center with a **g** tensor $(g_1, g_2, g_3) = (2.27, 2.14, 1.975)$ (*22*) that has exceedingly narrow EPR linewidths so that the orientation-selective process is well-defined. By all indications, the ^{1}H ENDOR spectra associated with the hydroxide obtained both at g_1 and g_3 should be sharp. This is certainly true for the spectrum at g_3, both in the native ^{1}H and ^{2}H-exchanged system, with measured ENDOR linewidths less than 50 kHz. Surprisingly, the single-crystal-like ENDOR spectrum measured at g_1 is so broad that it is actually difficult to detect, even in the ^{2}H spectrum where there are no overlapping peaks arising from nonexchangeable hydrogens (*23*). It was only a serendipitous finding of the 'Implicit-TRIPLE' effect in Mims ENDOR that provided clues to the proper assignments in this system (*24*).

Using simple geometric arguments, if the Fe-O bond of the hydroxide lies along the g_1 axis, then the Fe-H vector is between 15-20° off that axis, projecting somewhere into the g_2 - g_3 plane. If I choose to place the proton in the g_1 - g_2 plane, then the hyperfine matrix will have a nonzero element in the *xy* position. Expanding equation 5 and regrouping as was done in equations 9 and 10 produces the following expression similar to equation 11:

$$v^2 = x^2 N + O \pm xyP \qquad (13)$$

For clarity, I have dropped the $\pm$ notation describing the two different ENDOR transitions. The off-diagonal element in ^gA creates a mixed quadratic term, which requires that at least two octants be used to describe the frequency envelope, and is the origin of the splitting of the leg of the triangle into two curves described in the 1984 paper (1). Following the development above and taking the derivative dv/dx gives

$$\frac{dv}{dx} = \frac{1}{v}\left(xN \pm P(xy - \frac{x^2 V}{y}\right) \qquad (14)$$

As $x \rightarrow 1$ at the g_1 edge, $y \rightarrow 0$ and the third term in equation 14 diverges. So even though the frequency approaches a 'single-crystal-like' value at g_1, the area element associated with that specific frequency is vanishingly small, and the peak tends to smear out. This effect cannot be reproduced by simulation without the proper use of an EPR linewidth argument, so that a range of orientations in the vicinity of the g_1 axis can be interrogated.

To illustrate, I use the **A** matrix defined by a dipolar proton at 2.8 Å from a metal center with a rhombic **g** tensor (2.2, 2.0, 1.8) (25). Spectra are simulated by placing the proton at 5 different polar angles off the g_1 axis (0,5,10,15,20 degrees), defined by the directional cosines $(l_x, l_y, l_z) = (\cos\theta, \sin\theta, 0)$. I first examine spectra in the field region near g_1 and use a small EPR linewidth value of 150 MHz (fwhh) at a microwave frequency of 35 GHz. Two fields are simulated, one at $g_{obs} = 2.20$ and the second off the EPR envelope g value by approximately one linewidth, $g_{obs} = 2.21$. The range and relative contributions of g values for each of these two g_{obs} values is described by one-sided Gaussians centered at the g_{obs} value and overlapping the g_1 - g_3 region **(Figure 8)** The relative intensities of the simulated EPR spectra at these two g values using this linewidth is approximately 75:1 $(g = 2.20) / (g = 2.21)$. The simulated v_+ peaks for $g = 2.20$ (solid lines) and $g = 2.21$ (dashed lines) are shown in **Figure 9**. The spectra at both 0° and 5° are reasonably sharp at the g value, roughly 80 kHz and 120 kHz (fwhh) respectively, and would certainly be labelled as single-crystal-like ENDOR spectra. But by 10°, the breadth of the pattern is approximately 320 kHz, the two separate peaks that are associated with the two octants are observed, with the dip in the intensity envelope occurring precisely at the calculated single-crystal-like frequency. The frequency breadths at the final two angles are 500 kHz and 800 kHz for 15° and 20°, respectively. It is sometimes assumed that when one has not observed the single-crystal-like peak at the g_1 (or

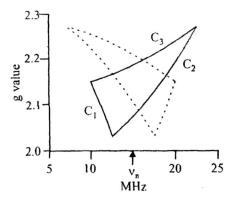

Figure 7. Domains for the parameterization of the unit sphere in the (g, v_+) (solid line) and (g, v_-) spaces for the rhombic g tensor (2.27, 2.15, 2.03) coaxial with the rhombic A tensor (15, -10, -5) MHz. Horizontal lines (constant g) represent the set of frequencies that would be observed in an ENDOR spectrum at that g value. The C_1, C_2, and C_3 curves represent the x = 0, y = 0, and z = 0 planes on the unit sphere and denote the turning points (peaks) in an ENDOR spectrum at a given g value.

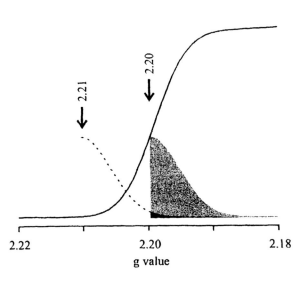

Figure 8. Simulated low-field (g_l) edge of a 35 GHz EPR-absorption envelope for the rhombic g tensor (2.20, 2.00, 1.80) using an isotropic EPR linewidth of 150 MHz. An ENDOR spectrum taken at g_{obs} = 2.20 contains contributions from the set of orientations described by the lighter shaded half-Gaussian curve. An ENDOR spectrum taken at g_{obs} = 2.21 contains much smaller set of orientations described the darker shaded region. This darker region is defined by the intersection of a half-Gaussian centered at g_{obs} = 2.21 (dashed line) and the set of g values g < g_l = 2.20.

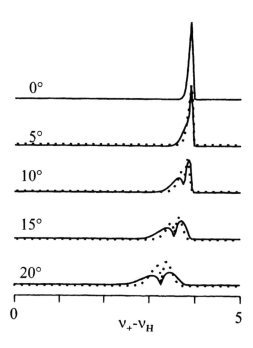

*Figure 9. Simulated ENDOR v_+ spectra of a proton at 2.8 Å from a metal center with rhombic **g** tensor (2.20, 2.00, 1.80) at five different polar angles with respect to the g_1 axis. The solid lines are simulations using $g_{obs} = g_1 = 2.20$, the dashed lines use $g_{obs} = 2.21$. Both sets of simulations use an EPR linewidth parameter of 150 MHz (see Figure 8). The simulated spectra for the angles 10°, 15°, and 20° do not coalesce into a 'single-crystal-like' peak even at $g_{obs} = 2.21$.*

g_3) edge of the EPR envelope it is because the experimenter has not pushed far enough into the region outside the g values. The dashed lines show that the peaks do not become substantially narrower when the field is set at $g_{obs} = 2.21$, while the 75-fold reduction in EPR intensity would likely make any such experiment a life-long project.

What is missing from the concepts engendered by the single-crystal-like spectrum ENDOR spectrum at extrema of an EPR envelope is that the *narrow features of an ENDOR spectrum are determined not by the turning points of the EPR spectrum, but are instead determined by the turning points of the ENDOR frequency equations.* Though as stated, this sounds both obvious and trivial, it is a point that has not been emphasized in the field of orientation-selection in ENDOR or ESEEM. In fact, when narrow features are observed at g_1 or g_3 edges of a rhombic EPR envelope or the $g_{||}$ of an axial EPR spectrum, we can be certain that within experimental error, one the principal values of the hyperfine tensor is coaxial with that direction with the **g** tensor. Going back to the work by Rist and Hyde (*14*), the narrow features of the ^{14}N ligand ENDOR are observed at $g_{||}$ of Cu(II) precisely because the directions of the **g** tensor are determined by the molecular orbitals involved in the Cu-N bond, which also determine both the hyperfine and quadrupolar axes of the N-donor atom. One of the reasons this concept breaks down with dipolar hyperfine tensors is because the tensors of these nuclei usually are not tied directly to the ligand-field orbitals that determine the g anisotropy, unlike donor-atom hyperfine tensors.

For dipolar-coupled nuclei, the traditional approach to understanding orientation selection that relies so heavily on the single-crystal-like spectra that are expected from g_1 and g_3, has literally broken down. In addition, because the hyperfine interaction is nearly traceless, it is impossible to measure the breadth of the ENDOR frequency envelope at most fields across the EPR spectrum because when the A values approach zero the ENDOR intensity is minimal. The question arises, how do we attack this type of ENDOR simulation?

The ironic answer lies in the mathematics surrounding orientation-selection in the g_2 region of the EPR envelope, which is the intensity maximum of the EPR spectrum and is usually considered to be the least selective position. Because of the higher intensity of the EPR spectrum, an ENDOR spectrum at g_2 is almost always the first spectrum taken in any new study, after which the experimenter tends to head out to the extrema, where the signal intensity is substantially lower, in a quest for the single-crystal like spectrum. Yet, the intensity functions of equation 12 predict that there can be an additional 'single-crystal-like' peak at g_2 that will occur if one of the principal axes of the **A** tensor is coaxial with the g_2 axis. Under such a condition, there are no off-diagonal xy or yz elements in the ^{g}**A** matrix that would give rise to the effects described by equations 13 and 14 as $y \rightarrow 1$. This retains the 'double-divergence' ($x = 0, z = 0$) that is no different in mathematical form from the 'double-divergence' that is produced at g_1 ($x \rightarrow 1$) or g_3 ($z \rightarrow 1$) when the **A** tensor is coaxial with those directions. The result is a set of peaks that arise from a single orientation that dominates the appearance of the spectrum.

In addition, the g_2 orientation-selective set of magnetic field vectors that solve the EPR resonance condition is a great circle containing points from (0, 1, 0) to a point on the xz plane (x_2, 0, z_2) that is easily calculated using equation 6. Any intrinsically axial hyperfine interaction such as a point-dipole has a perpendicular plane of turning points. This requires that for any axial hyperfine matrix, the ENDOR spectrum at g_2 will contain an $A_\perp$ turning point, regardless of the relative orientations of g and A. The term intrinsically axial refers to the fact that in the presence of an anisotropic g tensor, the measured dipolar interaction is rhombic and not traceless, but its rhombicity is tied to the principal values and directions of the g tensor. In addition, the maximum observed hyperfine interaction at g_2 indicates the smallest angle, θ_{min}, between the $A_{||}$ direction and the set of magnetic field orientations. A simulated spectrum is shown in **Figure 10**, with the g tensor (2.2, 2.0, 1.8) and the $A_{||}$ direction at (0.866, 0.353, 0.353) in the g tensor frame. If there is no isotropic component to hyperfine matrix, then the value of θ_{min} can be obtained by a ratio of $|A(\theta_{min})|/|A_\perp| = 3\cos^2\theta_{min} - 1$.

This approach was first used in studies of the orientation of the reactive guanidino nitrogen of a L-arginine substrate to the ferriheme protein neuronal nitric oxide synthase (nNOS) (26). This protein catalyzes the formation of NO from L-arginine, but it was known that the substrate was not a ligand to the ferriheme. The five coordinate substrate-bound form of nNOS comprises a high spin Fe(III) with an effective g tensor (g_1, g_2, g_3) = (7.56, 4.19, 1.81), where g_3 is taken to be the heme normal (27). The Mims pulsed ENDOR spectrum (28) of [$^{15}N_g$]-L-Arg taken at g_{obs} = 4.19 is shown in **Figure 11** and exhibits a doublet, centered at the ^{15}N Larmor frequency and split by $|A_\perp|$ = 0.22 MHz, with outer shoulders defining $A(\theta_{min})$ ~0.3 MHz. Following the previous discussion, the approximate value of θ_{min} is 27°. The $|A_\perp|$ value sets the dipolar distance between the Fe(III) ion and the guanadino ^{15}N to be approximately 4.05 Å. The heme structure itself excludes most of the possible positions that could solve both the distance and angle criteria. This leaves a cone of possible positions in which the guanadino nitrogen lies over the FeN_4 core of ferriheme as shown in Figure 11(lower). This original 'back of the envelope' structure calculation was verified by more extensive orientation-selected work and by x-ray diffraction studies (29-31).

Orientation-Selection in ESEEM

Though not a significant area of research in our group, there have been a number of noteworthy advances in applications of orientation-selection in the analysis of ESEEM spectra. In its original form, 3-pulse (stimulated-echo) ESEEM is ill-suited to recover the broad lineshapes of anisotropic $S = 1/2$, $I = 1/2$

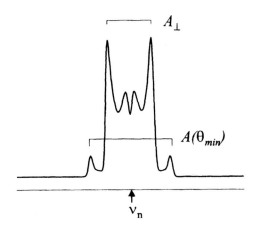

*Figure 10. Simulated 35 GHz ENDOR spectrum at $g_{obs} = g_2$ for a dipolar-coupled proton at a distance of 4.0 Å from a metal center with rhombic **g** tensor (2.20, 2.00, 1.80) and directional cosines (0.866, 0.353, 0.353) relative to the (x, y, z) directions, respectively. The $A_\perp$ peaks are the dominant spectral features and $A(\theta_{min})$ represent the smallest angle between the $g_{obs} = g_2$ orientation-selected field pathway and the $A_\parallel$ direction.*

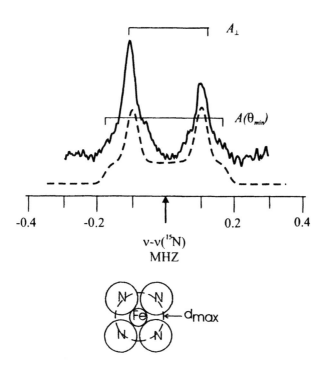

Figure 11. (Top) 35 GHz Mims pulsed ENDOR spectrum of [$^{15}N_g$]-L-Arg taken at g_{obs} = g_2 = 4.19 (solid line) and the simulation (dashed line) assuming no isotropic component using A(θ_{min}) to calculate the possible angles with respect to the heme normal. (Bottom) The maximum distance that the [$^{15}N_g$] nitrogen can be off the ferriheme normal based these ENDOR data. (Adapted with permission from reference 28. Copyright 1998 American Chemical Society).

interactions that are common in ENDOR spectroscopy (*32*). Some of the earliest work on understanding orientation-selective patterns in ESEEM focused on ^{14}N (*33*) rather than ^{1}H due to the fact that ESEEM can more easily obtain nuclear quadrupole and hyperfine tensor information than ENDOR in systems that exhibit hyperfine coupling constants in the range, 1 MHz < $|A(^{14}N)|$ < 10 MHz (*34*). The development of a two-dimensional 4-pulse variant of ESEEM called HYSCORE (*35,36*) (Hyperfine-selective Sublevel Correlation) allows for the recovery of the broad line information that is lost in the 1-dimensional ESEEM. The contour plots in HYSCORE contain much the same information (*37*) as an ENDOR lineshape function and have been used to extract **A** tensor values and orientations relative to the **g** tensor.

The Future of Orientation-Selection

In the past decade, a steadily increasing number of groups have been working on applying orientation selective analysis using both ENDOR and ESEEM (*38-40*) in a variety of different spin systems. Our most recent work in orientation-selection shows that there is a wealth of information that can be extracted near g_2, with no knowledge of the ENDOR spectra at either of the high-field or low-field edges of the EPR envelope. This information can be extracted quickly and without using any simulation program. It is nearly axiomatic that the first spectrum ENDOR spectrum taken on any system will be obtained at g_2, since the EPR intensity is greatest there. We are working on a more elaborate method based on the simple observations of $A_\perp$ and $A(\theta_{min})$ for multiple fields that will extract the Euler angles for any intrinsically axial tensor. A recently published paper by Shubin and Dikanov (*41*) sets up a procedure that analytically solves for the principal values and Euler angles of a general rhombic hyperfine tensor using only the three field positions g_1, g_2, and g_3, and is an excellent extension of the original process laid out by Hoffman and coworkers (*17*) and will likely be useful in analyzing ligand or metal-ion ENDOR patterns. What all groups are working towards is to be able to analyze structural data during the time of spectral acquisition, so that experimental conditions can be adjusted in real-time. The original method of doing orientation-selection saw the simulation process take more time than the data collection. We are now at a point where the simulations, using any approach, are now orders of magnitude faster than data collection, and we need to be able to take advantage of this speed to more effectively optimize valuable instrument time.

Conclusions

The original work in orientation-selection laid the foundation for ENDOR and ESEEM spectroscopists to obtain highly detailed structural information from randomly oriented samples (*14*). Early work that was heavily weighted towards analyzing the coordination environments of metal centers has given way to studies involving nuclei with no covalent pathway to these metal centers (*28*). Studies on resting states of enzymes have blossomed into studies on kinetically-trapped intermediates (*18*), radiolytically-reduced centers (*42,43*), and rapid-freeze quenched reactions (*44*). Though changes in instrumentation provide us with newer techniques that improve data collection and allow us to study this wider range of samples, it is still our ability to extract the relevant structural information out of these data that determines the success of the experiment. And, to a great extent, our understanding of this process is a tribute to Brian Hoffman's original work in this field.

Acknowledgements

The author gratefully acknowledges Prof. Brian M. Hoffman of Northwestern University for many years of animated discussions ranging from science to politics to the life of a Chicago White Sox fan on a side of town that loves the Cubs. The author also acknowledges the volume editor for his persistent encouragement to finish this chapter. This work has been supported through grants to BMH (NIH HL-13531 and NSF MCB-9904018).

References

1. Hoffman, B. M.; Martinsen, J.; Venters, R. A. *J. Magn. Reson.* **1984**, *59*, 110-123.
2. Hoffman, B. M.; Venters, R. A.; Martinsen, J. *J. Magn. Reson.* **1985**, *62*, 537-542.
3. Hoffman, B. M.; Venters, R. A.; Roberts, J. E.; Nelson, M.; Orme-Johnson, W. H. *J. Am. Chem. Soc.* **1982**, *104*, 4711-4712.
4 Venters, R. A.; Nelson, M. J.; McLean, P. A.; True, A. E.; Levy, M. A.; Hoffman, B. M.; Orme-Johnson, W. H. *J. Am. Chem. Soc.* **1986**, *108*, 3487-3498.
5. True, A. E.; McLean, P.; Nelson, M. J.; Orme-Johnson, W. H.; Hoffman, B. M. *J. Am. Chem. Soc.* **1990**, *112*, 651-657.
6. Hoffman, B. M.; Gurbiel, R. J. *J. Magn. Reson.* **1989**, *82*, 309-317.
7. Fan, C.; Kennedy, M. C.; Beinert, H.; Hoffman, B. M. *J. Am. Chem. Soc.* **1992**, *114*, 374-375.
8. Doan, P. E.; Fan, C.; Hoffman, B. M. *J. Am. Chem. Soc.* **1994**, *116*, 1033-1041.

9. Thomann, H.; Bernardo, M. *Methods Enzymol.* **1993**, *227*, 118-190.

10. Sturgeon, B. E.; Ball, J. A.; Randall, D. W.; Britt, R. D. *J. Phys. Chem.* **1994**, *98*, 12871-12883.

11. Schweiger, A.; Jeschke, G. *Principles of Pulse Electron Paramagnetic Resonance*; Oxford University Press: Oxford, UK, 2001.

12. Abragam, A.; Bleaney, B. *Electron Paramagnetic Resonance of Transition Metal Ions*; 2nd ed.; Clarendon Press: Oxford, 1970.

13. Kneubuhl, F. K. *J. Chem. Phys.* **1960**, *33*, 1074-1078.

14. Rist, G. H.; Hyde, J. S. *J. Chem. Phys.* **1970**, *52*, 4633-4643.

15. Sands, R. H. In *Multiple Electron Resonance Spectroscopy*; Dorio, M. M., Freed, J. H., Eds.; Plenum Press: New York & London, 1979, pp 331-374.

16. Thuomas, K.; Lund, A. *J. Magn. Res.* **1975**, *18*, 12-21.

17. Hoffman, B. M.; DeRose, V. J.; Doan, P. E.; Gurbiel, R. J.; Houseman, A. L. P.; Telser, J. *Biol. Magn. Reson.* **1993**, *13(EMR of Paramagnetic Molecules)*, 151-218.

18. Lee, H.-I.; Dexter, A. F.; Fann, Y.-C.; Lakner, F. J.; Hager, L. P.; Hoffman, B. M. *J. Am. Chem. Soc.* **1997**, *119*, 4059-4069.

19. Dexter, A. F.; Hager, L. P. *J. Am. Chem. Soc.* **1995**, *117*, 817-818.

20. Mims, W. B. *Proc. Roy. Soc. Lond.* **1965**, *283*, 452-457.

21. Huang, W.; Jia, J.; Cummings, J.; Nelson, M.; Schneider, G.; Lindquist, Y. *Structure* **1997**, *5*, 691-699.

22. Nelson, M. J.; Jin, H.; Turner, J., Ivan M.; Grive, G.; Scarrow, R. C.; Brennan, B. A.; Que, J., Lawrence. *J. Am. Chem. Soc.* **1991**, *113*, 7072-7073.

23. Jin, H.; Turner, I. M., Jr.; Nelson, M. J.; Gurbiel, R. J.; Doan, P. E.; Hoffman, B. M. *J. Am. Chem. Soc.* **1993**, *115*, 5290-5291.

24. Doan, P. E.; Nelson, M. J.; Jin, H.; Hoffman, B. M. *J. Am. Chem. Soc.* **1996**, *118*, 7014-7015.

25. Hutchison, C. A. Jr.; McKay, D. B. *J. Chem. Phys.* **1977**, *66*, 3311-3330.

26. Kerwin, J. F. J.; Lancaster, J. R. J.; Feldman, P. L. *Med. Res. Rev.* **1994**, *14*, 23-74.

27. Salerno, J. C.; Martásek, P.; Roman, L. J.; Masters, B. S. S. *Biochemistry* **1996**, *35*, 7626-7630.

28. Tierney, D. L.; Martásek, P.; Doan, P. E.; Masters, B. S.; Hoffman, B. M. *J. Am. Chem. Soc.* **1998**, *120*, 2983-2984.

29. Tierney, D. L.; Huang, H.; Martásek, P.; Roman, L. J.; Silverman, R. B.; Hoffman, B. M. *J. Am. Chem. Soc.* **2000**, *122*, 7869-7875.

30. Tierney, D. L.; Huang, H.; Martásek, P.; Masters, B. S. S.; Silverman, R. B.; Hoffman, B. M. *Biochemistry* **1999**, *38*, 3704-3710.

31. Li, H.; Raman, C. S.; Glaser, C. B.; Blasko, E.; Young, T. A.; Parkinson, J. F.; Whitlow, M.; Poulos, T. L. *J. Biol. Chem.* **1999**, *274*, 21276-21284.

32. de Groot, A.; Evelo, R.; Hoff, A. J. *J. Magn. Reson.* **1986**, *66*, 331-343.

33. Flanagan, H. L.; Gerfen, G. J.; Lai, A.; Singel, D. L. *J. Chem. Phys.* **1988**, *88*, 2161-2168.
34. Lee, H.-I.; Doan, P. E.; Hoffman, B. M. *J. Magn. Reson.* **1999**, *140*, 91-107.
35. Höfer, P.; Grupp, A.; Nebenfur, H.; Mehring, M. *Chem. Phys. Lett.* **1986**, *132*, 279-282.
36. Shane, J.J.; Höfer, P.; Reijerse, E.J.; de Boer, E. *J. Magn. Reson.* **1992**, *99*, 596-604.
37. Dikanov, S. A.; Bowman, M. K. *J. Magn. Reson.* **1995**, *116*, 125-128.
38. Dikanov, S. A.; Davydov, R. M.; Gräslund, A.; Bowman, M. K. *J. Am. Chem. Soc.* **1998**, *120*, 6797-6805.
39. Madi, Z. L.; Van Doorslaer, S.; Schweiger, A. *J. Magn. Reson.* **2002**, *154*, 181-191.
40. Matar, K.; Goldfarb, D. *J. Magn. Reson. A* **1994**, *111*, 50-61.
41. Shubin, A. A.; Dikanov, S. A. *J. Magn. Reson.* **2002** *155*,100-105.
42. Davydov, R.; Makris, T. M.; Kofman, V.; Werst, D. W.; Sligar, S. G.; Hoffman, B. M. *J. Am. Chem. Soc.* **2001**, *123*, 1403-1415.
43. Telser, J.; Davydov, R.; Horng, Y. C.; Ragsdale, S. W.; Hoffman, B. M. *J. Am. Chem. Soc.* **2001**, *123*, 5853-5860.
44. Sturgeon, B. E.; Burdi, D.; Chen, S.; Huynh, B. H.; Edmondson, D. E.; Stubbe, J.; Hoffman, B. M. *J. Am. Chem. Soc.* **1996**, *118*, 7551-7557.

Chapter 4

Effects of Electron Spin Delocalization and Non-Collinearity of Interaction Terms in EPR Triplet Powder Patterns

Steven O. Mansoorabadi and George H. Reed

Department of Biochemistry, University of Wisconsin, 1710 University Avenue, Madison, WI 53726–4087

The origin of rhombicity in the zero-field splitting (zfs) interaction in "intermolecular" triplet spin systems is reviewed. Relationships between the D and E terms of the zfs interaction and structure are developed. The consequences of non-collinearity of principal axes of zfs and g or hyperfine interactions are examined in the allowed transitions of triplet powder spectra.

Analyses of magnetic dipole-dipole interactions are important in many applications of magnetic resonance spectroscopy. In electron paramagnetic resonance (EPR) spectroscopy, zero-field splitting (zfs) arising from the dipole-dipole interaction in triplet spin systems has provided unique insight into the disposition of the unpaired electrons (1,2). Interesting examples of "hybrid"-spin systems involving transition metal ions and nitroxide radicals have been reported (3). More recently, strongly coupled, hybrid triplet spin systems have been discovered as intermediates in reactions of coenzyme B_{12} dependent enzymes (4,5,6). It is desirous to extract distance and geometry information from the triplet EPR powder patterns that arise in these cases. Rhombicity in the zfs as well as non-collinearity of the principal axes of the zfs interaction with, for example, the g or hyperfine terms, are two complications that often arise in such systems. In the present chapter, we review the origin of rhombicity in the dipole-dipole interaction in terms of the point charge approximation. We also examine the "benefits" of g and/or hyperfine anisotropies in extracting geometrical information from EPR triplet powder patterns.

We focus our attention on the $\Delta M_s = 1$ (allowed EPR) transitions. In favorable cases, distance information can also be extracted from the position and intensities of the $\Delta M_s = 2$ (half-field) transitions (3,7).

The Origin and Analysis of Rhombicity in the Dipole-Dipole Interaction

The energy, U, of a magnetic dipole $\vec{m}$ in an external magnetic field $\vec{B}$ is (8):

$$U = -\vec{m}^T \cdot \vec{B}$$

The magnetic field produced at a point $\vec{r}$ by a magnetic dipole $\vec{m}$ at $\vec{r}'$ is (9):

$$\vec{B} = \frac{\mu_o}{4\pi} \left\{ \frac{3[(\vec{r} - \vec{r}')^T \cdot \vec{m}](\vec{r} - \vec{r}') - |\vec{r} - \vec{r}'|^2 \vec{m}}{|\vec{r} - \vec{r}'|^5} \right\} + \frac{2\mu_o}{3} \left\{ \delta(\vec{r} - \vec{r}')\vec{m} \right\}$$

where μ_o is the permeability of free space and $\delta(\vec{r} - \vec{r}')$ is a delta function located at $\vec{r}'$. The magnetic dipole moment of an electron can be written as:

$$\vec{m} = \beta \vec{\vec{g}} \cdot \vec{S}$$

where β is the Bohr magneton, $\vec{\vec{g}}$ is the electron g-tensor, and $\vec{S}$ is the electron-spin angular momentum (in units of $\hbar$). Thus, the energy of the electron-spin–electron-spin dipolar interaction is (10):

$$U = \frac{\mu_o \beta^2}{4\pi} \left\{ \frac{|\vec{r}_2 - \vec{r}_1|^2 (\vec{\vec{g}}_2 \cdot \vec{S}_2)^T \cdot (\vec{\vec{g}}_1 \cdot \vec{S}_1) - 3(\vec{r}_2 - \vec{r}_1)^T \cdot (\vec{\vec{g}}_1 \cdot \vec{S}_1)(\vec{\vec{g}}_2 \cdot \vec{S}_2)^T \cdot (\vec{r}_2 - \vec{r}_1)}{|\vec{r}_2 - \vec{r}_1|^5} \right\}$$

$$- \frac{2\mu_o \beta^2}{3} \left\{ \delta(\vec{r}_2 - \vec{r}_1)(\vec{\vec{g}}_2 \cdot \vec{S}_2)^T \cdot (\vec{\vec{g}}_1 \cdot \vec{S}_1) \right\}$$

The second term corresponds to the isotropic magnetic dipole-dipole interaction and will not be considered further. The first term corresponds to the anisotropic magnetic dipole-dipole interaction, and contains information on distance and orientation (11). The equation can be rewritten as:

$$U = \vec{S}_2^T \cdot \vec{g}_2^T \cdot \vec{D}'' \cdot \vec{g}_1 \cdot \vec{S}_1$$

where

$$\vec{D}'' = \frac{\mu_o \beta^2}{4\pi} \begin{pmatrix} \dfrac{\left|\vec{r}_2 - \vec{r}_1\right|^2 - 3(x_2 - x_1)^2}{\left|\vec{r}_2 - \vec{r}_1\right|^5} & \dfrac{-3(x_2 - x_1)(y_2 - y_1)}{\left|\vec{r}_2 - \vec{r}_1\right|^5} & \dfrac{-3(x_2 - x_1)(z_2 - z_1)}{\left|\vec{r}_2 - \vec{r}_1\right|^5} \\ \dfrac{-3(x_2 - x_1)(y_2 - y_1)}{\left|\vec{r}_2 - \vec{r}_1\right|^5} & \dfrac{\left|\vec{r}_2 - \vec{r}_1\right|^2 - 3(y_2 - y_1)^2}{\left|\vec{r}_2 - \vec{r}_1\right|^5} & \dfrac{-3(y_2 - y_1)(z_2 - z_1)}{\left|\vec{r}_2 - \vec{r}_1\right|^5} \\ \dfrac{-3(x_2 - x_1)(z_2 - z_1)}{\left|\vec{r}_2 - \vec{r}_1\right|^5} & \dfrac{-3(y_2 - y_1)(z_2 - z_1)}{\left|\vec{r}_2 - \vec{r}_1\right|^5} & \dfrac{\left|\vec{r}_2 - \vec{r}_1\right|^2 - 3(z_2 - z_1)^2}{\left|\vec{r}_2 - \vec{r}_1\right|^5} \end{pmatrix}$$

The Hamiltonian corresponding to the energy of the electron-spin–electron-spin dipolar interaction is

$$\hat{H}' = \hat{S}_2^T \cdot \vec{g}_2^T \cdot \vec{D}'' \cdot \vec{g}_1 \cdot \hat{S}_1$$

where the angular momentum vector, $\vec{S}$, in the classical expression for the energy is replaced by the corresponding angular momentum operator, $\hat{S}$. The energy of a particular state of this system is obtained from the expectation value of the Hamiltonian using the wave functions of the two electrons, which contain information about the spatial extent of their orbitals as well as their spin states (12). If the spatial part of the wave function, $\psi(\vec{r}_1, \vec{r}_2)$, is known, we can integrate first over the spatial extent of the electrons, converting the Hamiltonian of the system into a spin-Hamiltonian:

$$\hat{H} = \langle \psi(\vec{r}_1, \vec{r}_2) | \hat{S}_2^T \cdot \vec{g}_2^T \cdot \vec{D}'' \cdot \vec{g}_1 \cdot \hat{S}_1 | \psi(\vec{r}_1, \vec{r}_2) \rangle = \hat{S}_2^T \cdot \vec{g}_2^T \cdot \langle \psi(\vec{r}_1, \vec{r}_2) | \vec{D}'' | \psi(\vec{r}_1, \vec{r}_2) \rangle \cdot \vec{g}_1 \cdot \hat{S}_1$$
$$= \hat{S}_2^T \cdot \vec{g}_2^T \cdot \vec{D}' \cdot \vec{g}_1 \cdot \hat{S}_1 = \hat{S}_2^T \cdot \vec{D} \cdot \hat{S}_1$$

where

$$\vec{D}' = \int \psi^*(\vec{r}_1, \vec{r}_2) \vec{D}'' \psi(\vec{r}_1, \vec{r}_2) d^3 r_1 d^3 r_2$$

and

$$\vec{D} = \vec{g}_2^T \cdot \vec{D}' \cdot \vec{g}_1$$

Determination of the D'-tensor thus requires a precise knowledge of the molecular orbitals of the electrons and evaluation of complicated integrals, which can be computationally challenging (13). Therefore, approximation methods are generally used. A commonly used approximation is the point-dipole (or point-spin) approximation. This simplification assumes the wave function of each electron is localized at a single point – the average position of the electron, which is frequently the nucleus of the atom containing the electron (10). With this approximation, the D'-tensor has the form:

$$\ddot{D}' = \frac{\mu_0\beta^2}{4\pi}\begin{pmatrix} \dfrac{\left|\vec{\bar{r}}_2 - \vec{\bar{r}}_1\right|^2 - 3(\bar{x}_2 - \bar{x}_1)^2}{\left|\vec{\bar{r}}_2 - \vec{\bar{r}}_1\right|^5} & \dfrac{-3(\bar{x}_2 - \bar{x}_1)(\bar{y}_2 - \bar{y}_1)}{\left|\vec{\bar{r}}_2 - \vec{\bar{r}}_1\right|^5} & \dfrac{-3(\bar{x}_2 - \bar{x}_1)(\bar{z}_2 - \bar{z}_1)}{\left|\vec{\bar{r}}_2 - \vec{\bar{r}}_1\right|^5} \\[3ex] \dfrac{-3(\bar{x}_2 - \bar{x}_1)(\bar{y}_2 - \bar{y}_1)}{\left|\vec{\bar{r}}_2 - \vec{\bar{r}}_1\right|^5} & \dfrac{\left|\vec{\bar{r}}_2 - \vec{\bar{r}}_1\right|^2 - 3(\bar{y}_2 - \bar{y}_1)^2}{\left|\vec{\bar{r}}_2 - \vec{\bar{r}}_1\right|^5} & \dfrac{-3(\bar{y}_2 - \bar{y}_1)(\bar{z}_2 - \bar{z}_1)}{\left|\vec{\bar{r}}_2 - \vec{\bar{r}}_1\right|^5} \\[3ex] \dfrac{-3(\bar{x}_2 - \bar{x}_1)(\bar{z}_2 - \bar{z}_1)}{\left|\vec{\bar{r}}_2 - \vec{\bar{r}}_1\right|^5} & \dfrac{-3(\bar{y}_2 - \bar{y}_1)(\bar{z}_2 - \bar{z}_1)}{\left|\vec{\bar{r}}_2 - \vec{\bar{r}}_1\right|^5} & \dfrac{\left|\vec{\bar{r}}_2 - \vec{\bar{r}}_1\right|^2 - 3(\bar{z}_2 - \bar{z}_1)^2}{\left|\vec{\bar{r}}_2 - \vec{\bar{r}}_1\right|^5} \end{pmatrix}$$

where the bars denote average quantities.

If we place the first electron at the origin of our coordinate system and drop the labels on the second, we have:

$$\ddot{D}' = \frac{\mu_0\beta^2}{4\pi}\begin{pmatrix} \dfrac{\bar{r}^2 - 3\bar{x}^2}{\bar{r}^5} & \dfrac{-3\bar{x}\bar{y}}{\bar{r}^5} & \dfrac{-3\bar{x}\bar{z}}{\bar{r}^5} \\[3ex] \dfrac{-3\bar{x}\bar{y}}{\bar{r}^5} & \dfrac{\bar{r}^2 - 3\bar{y}^2}{\bar{r}^5} & \dfrac{-3\bar{y}\bar{z}}{\bar{r}^5} \\[3ex] \dfrac{-3\bar{x}\bar{z}}{\bar{r}^5} & \dfrac{-3\bar{y}\bar{z}}{\bar{r}^5} & \dfrac{\bar{r}^2 - 3\bar{z}^2}{\bar{r}^5} \end{pmatrix}$$

$\ddot{D}'$ can then be expressed using spherical coordinates as:

$$\ddot{D}' = \ddot{R}\cdot\ddot{D}'_d\cdot\ddot{R}^T$$

where $\ddot{R}$ is a rotation matrix that contains the spherical angles ζ and η:

$$R = \begin{pmatrix} \cos\eta\cos\zeta & -\sin\eta & \cos\eta\sin\zeta \\ \sin\eta\cos\zeta & \cos\eta & \sin\eta\sin\zeta \\ -\sin\zeta & 0 & \cos\zeta \end{pmatrix}$$

and $\ddot{D}_d$ is an axially-symmetric diagonal matrix:

$$\ddot{D}'_d = \frac{\mu_o\beta^2}{4\pi\bar{r}^3} \begin{pmatrix} 1 & 0 & 0 \\ 0 & 1 & 0 \\ 0 & 0 & -2 \end{pmatrix}$$

Experimentally it is found that the D'_d-tensor often exhibits rhombic symmetry, especially in cases where one or both of the unpaired electrons are extensively delocalized or are in close proximity (11). A rhombic D'_d-tensor is incompatible with the point-dipole approximation in this form. It can be shown that the inclusion of g-anisotropy into the zfs interaction generates a slight rhombicity in the D'_d-tensor, but this effect is small whenever g-anisotropy is modest and, alone, is typically insufficient to account for experimentally observed rhombicity (14). Thus, in a typical case of a rhombic zfs tensor, the point-dipole approximation is inadequate. The point-charge approximation is a logical extension of the point-dipole approximation (15). In this elaboration, each point-dipole is broken up into two or more individual points to reflect the delocalization of the electron within the molecule or complex of interest. With this approximation, the D'-tensor has the form:

$$D'_{ij} = \frac{\mu_o\beta^2}{4\pi} \sum_m \sum_n \rho_1^m \rho_2^n \frac{\left[\left|\vec{r}_2^n - \vec{r}_1^m\right|^2 \delta_{ij} - 3\left(\vec{r}_2^n - \vec{r}_1^m\right)_i \left(\vec{r}_2^n - \vec{r}_1^m\right)_j \right]}{\left|\vec{r}_2^n - \vec{r}_1^m\right|^5}$$

where δ_{ij} is the Kronecker delta, ρ_1^m and ρ_2^n are the spin densities of the m-th point-charge of the first molecule and the n-th point-charge of the second (which are located at $\vec{r}_1^m$ and $\vec{r}_2^n$) respectively, and the sums extend over all point-charges that constitute the interacting molecules (15). Approximate values for the spin densities on each atom can be obtained from electronic structure calculations (16). The D'-tensors obtained from spectral simulations of experimental EPR spectra can then be fit to the above equation to determine the relative positions of the two radicals.

As a simple example, consider a hypothetical radical pair wherein one radical has spin delocalized over two atoms with spin-densities ρ_1 and ρ_2, respectively. This delocalized radical interacts with a second radical located at the origin of a molecule-fixed axis system. We calculate the D'-tensor in the frame of reference in which one of the spin-bearing atoms of the delocalized radical lies along the z-axis a distance r_1 away. The other spin-bearing atom is positioned along the x-axis a distance r_2 away. The position vector of the second atom makes an angle ζ_2 with the z-axis. The D'-tensor then becomes:

$$\tilde{D}' = \frac{\mu_\circ \beta^2 \rho_1}{4\pi r_1^3} \begin{pmatrix} 1 & 0 & 0 \\ 0 & 1 & 0 \\ 0 & 0 & -2 \end{pmatrix} + \frac{\mu_\circ \beta^2 \rho_2}{4\pi r_2^3} \begin{pmatrix} 1-3\sin^2\zeta_2 & 0 & -3\sin\zeta_2\cos\zeta_2 \\ 0 & 1 & 0 \\ -3\sin\zeta_2\cos\zeta_2 & 0 & 1-3\cos^2\zeta_2 \end{pmatrix} =$$

$$\frac{\mu_\circ \beta^2}{4\pi} \begin{pmatrix} \dfrac{\rho_1}{r_1^3} + \dfrac{\rho_2}{r_2^3} - \dfrac{3\rho_2\sin^2\zeta_2}{r_2^3} & 0 & -\dfrac{3\rho_2\sin\zeta_2\cos\zeta_2}{r_2^3} \\ 0 & \dfrac{\rho_1}{r_1^3} + \dfrac{\rho_2}{r_2^3} & 0 \\ -\dfrac{3\rho_2\sin\zeta_2\cos\zeta_2}{r_2^3} & 0 & -\dfrac{2\rho_1}{r_1^3} + \dfrac{\rho_2}{r_2^3} - \dfrac{3\rho_2\cos^2\zeta_2}{r_2^3} \end{pmatrix}$$

Diagonalization of the D'-tensor leads to the principal values

$$(D'_d)_x = -\frac{\mu_\circ \beta^2}{4\pi} \left(\frac{\rho_1}{2r_1^3} + \frac{\rho_2}{2r_2^3} - \frac{3}{2} \sqrt{\frac{\rho_1^2}{r_1^6} + \frac{\rho_2^2}{r_2^6} + \frac{2\rho_1\rho_2(1-2\sin^2\zeta_2)}{r_1^3 r_2^3}} \right)$$

$$(D'_d)_y = \frac{\mu_\circ \beta^2}{4\pi} \left(\frac{\rho_1}{r_1^3} + \frac{\rho_2}{r_2^3} \right)$$

$$(D'_d)_z = -\frac{\mu_\circ \beta^2}{4\pi} \left(\frac{\rho_1}{2r_1^3} + \frac{\rho_2}{2r_2^3} + \frac{3}{2} \sqrt{\frac{\rho_1^2}{r_1^6} + \frac{\rho_2^2}{r_2^6} + \frac{2\rho_1\rho_2(1-2\sin^2\zeta_2)}{r_1^3 r_2^3}} \right)$$

Since the D'-tensor is traceless, only two independent variables, D' and E', are needed to construct the D'_d-tensor. Using the definitions

$$(D'_d)_x = -\frac{D'}{3} + E', \quad (D'_d)_y = -\frac{D'}{3} - E', \quad (D'_d)_z = \frac{2D'}{3}$$

we obtain the following expressions for the zero-field splitting parameters, D' and E':

$$D' = -\frac{3\mu_o\beta^2}{8\pi}\left(\frac{\rho_1}{2r_1^3} + \frac{\rho_2}{2r_2^3} + \frac{3}{2}\sqrt{\frac{\rho_1^2}{r_1^6} + \frac{\rho_2^2}{r_2^6} + \frac{2\rho_1\rho_2(1 - 2\sin^2\zeta_2)}{r_1^3 r_2^3}}\right)$$

$$E' = -\frac{3\mu_o\beta^2}{8\pi}\left(\frac{\rho_1}{2r_1^3} + \frac{\rho_2}{2r_2^3} - \frac{1}{2}\sqrt{\frac{\rho_1^2}{r_1^6} + \frac{\rho_2^2}{r_2^6} + \frac{2\rho_1\rho_2(1 - 2\sin^2\zeta_2)}{r_1^3 r_2^3}}\right)$$

The values for the zfs parameters obtained from spectral simulations can then be fit directly to the expressions for D' and E', obtaining approximate values for r_1, r_2, and ζ_2. The distance, d, between the spin-bearing atoms of the delocalized radical

$$d = \sqrt{r_1^2 + r_2^2 - 2r_1r_2\cos\zeta_2}$$

can be used as an additional constraint. From the above expression it is evident that whenever $\vec{r}_1$ and $\vec{r}_2$ point in the same direction (i.e. $\zeta_2 = 0°$, and the principal axis of the two component D'-tensors are therefore collinear), $E' = 0$, and there is no rhombicity in the zfs interaction. The delocalized radical can then be viewed as a single point-charge at a distance

$$\bar{r} = \frac{r_1 r_2}{\left(\rho_1 r_2^3 + \rho_2 r_1^3\right)^{1/3}}$$

If instead, $r_1 = r_2 = r$ and $\rho_1 = \rho_2$, it is easy to show that the rhombicity is

$$\frac{E'}{D'} = \frac{1 - \cos\zeta_2}{1 + 3\cos\zeta_2}$$

This expression can be rewritten in terms of the ratio $R = r/d$ as

$$\frac{E'}{D'} = \frac{1}{8R^2 - 3}$$

As R increases (i.e. the distance between the spin-bearing atoms of the delocalized radical decreases relative to the distance between the radicals), the rhombicity decreases (i.e. E' decreases relative to D').

The principal axes corresponding to the principal values are:

$$\bar{x}' = \begin{pmatrix} \dfrac{1}{\sqrt{1 + \dfrac{1}{4\sin^2\zeta_2\cos^2\zeta_2}\left(1 - 2\sin^2\zeta_2 + \dfrac{\rho_1 r_2^3}{\rho_2 r_1^3} - \sqrt{1 + 2(1 - 2\sin^2\zeta_2)\dfrac{\rho_1 r_2^3}{\rho_2 r_1^3} + \dfrac{\rho_1^2 r_2^6}{\rho_2^2 r_1^6}}\right)^2}} \\[4ex] 0 \\[2ex] \dfrac{\dfrac{1}{2\sin\zeta_2\cos\zeta_2}\left(1 - 2\sin^2\zeta_2 + \dfrac{\rho_1 r_2^3}{\rho_2 r_1^3} - \sqrt{1 + 2(1 - 2\sin^2\zeta_2)\dfrac{\rho_1 r_2^3}{\rho_2 r_1^3} + \dfrac{\rho_1^2 r_2^6}{\rho_2^2 r_1^6}}\right)}{\sqrt{1 + \dfrac{1}{4\sin^2\zeta_2\cos^2\zeta_2}\left(1 - 2\sin^2\zeta_2 + \dfrac{\rho_1 r_2^3}{\rho_2 r_1^3} - \sqrt{1 + 2(1 - 2\sin^2\zeta_2)\dfrac{\rho_1 r_2^3}{\rho_2 r_1^3} + \dfrac{\rho_1^2 r_2^6}{\rho_2^2 r_1^6}}\right)^2}} \end{pmatrix}$$

$$\bar{y}' = \begin{pmatrix} 0 \\ 1 \\ 0 \end{pmatrix}$$

$$\bar{z}' = \begin{pmatrix} -\dfrac{\dfrac{1}{2\sin\zeta_2\cos\zeta_2}\left(1 - 2\sin^2\zeta_2 + \dfrac{\rho_1 r_2^3}{\rho_2 r_1^3} - \sqrt{1 + 2(1 - 2\sin^2\zeta_2)\dfrac{\rho_1 r_2^3}{\rho_2 r_1^3} + \dfrac{\rho_1^2 r_2^6}{\rho_2^2 r_1^6}}\right)}{\sqrt{1 + \dfrac{1}{4\sin^2\zeta_2\cos^2\zeta_2}\left(1 - 2\sin^2\zeta_2 + \dfrac{\rho_1 r_2^3}{\rho_2 r_1^3} - \sqrt{1 + 2(1 - 2\sin^2\zeta_2)\dfrac{\rho_1 r_2^3}{\rho_2 r_1^3} + \dfrac{\rho_1^2 r_2^6}{\rho_2^2 r_1^6}}\right)^2}} \\[4ex] 0 \\[2ex] \dfrac{1}{\sqrt{1 + \dfrac{1}{4\sin^2\zeta_2\cos^2\zeta_2}\left(1 - 2\sin^2\zeta_2 + \dfrac{\rho_1 r_2^3}{\rho_2 r_1^3} - \sqrt{1 + 2(1 - 2\sin^2\zeta_2)\dfrac{\rho_1 r_2^3}{\rho_2 r_1^3} + \dfrac{\rho_1^2 r_2^6}{\rho_2^2 r_1^6}}\right)^2}} \end{pmatrix}$$

Thus, the effect of delocalizing a radical in the xz-plane is a rotation of the principal axes of the D'-tensor about the y-axis by an angle:

$$\beta = \cos^{-1}\left(\dfrac{1}{\sqrt{1 + \dfrac{1}{4\sin^2\zeta_2\cos^2\zeta_2}\left(1 - 2\sin^2\zeta_2 + \dfrac{\rho_1 r_2^3}{\rho_2 r_1^3} - \sqrt{1 + 2(1 - 2\sin^2\zeta_2)\dfrac{\rho_1 r_2^3}{\rho_2 r_1^3} + \dfrac{\rho_1^2 r_2^6}{\rho_2^2 r_1^6}}\right)^2}}\right)$$

When $r_1 = r_2$ and $\rho_1 = \rho_2$, $\beta = \zeta_2/2$ and the principal axes are rotated such that the z'-axis of the new principal axis system points mid-way between the two atoms of the delocalized molecule (Figure 1.).

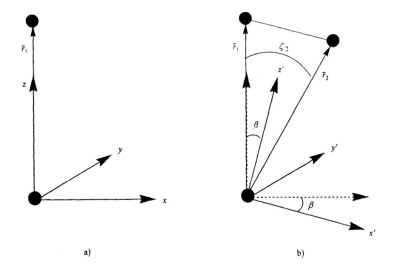

a)   b)

Figure 1. Effect on the principal axes of the D'-tensor when a single point-dipole is split into two point-charges in the xz-plane. The principal axes are rotated from their initial orientation a), by an angle $\beta = \zeta_2/2$ about the y-axis b).

Anisotropic Interactions and Non-Collinear Principal Axes

Anisotropies in the g and hyperfine interactions complicate the analysis of electron-spin–electron-spin dipolar interactions. However, such anisotropies provide a useful connection between the zfs interaction and a molecule fixed axis system.

In order to carry out energy calculations, each of the tensors in the spin Hamiltonian must be expressed in the same reference frame. In general, the principal axes of each tensor need not be collinear (17). To express each tensor in a common frame of reference we perform orthogonal transformations using rotation matrices. In any particular reference frame, say the laboratory frame defined by the external magnetic field of the EPR spectrometer, $\vec{g}_1$, $\vec{g}_2$, and $\bar{D}'$ will each, in general, have nine non-zero elements and the D-tensor will no longer be traceless and symmetric. The spin-Hamiltonian can therefore be expressed as:

$$\hat{H} = \hat{S}_2^T \cdot \left[\vec{g}_2^T \cdot \bar{D}' \cdot \vec{g}_1 \right] \cdot \hat{S}_1 = \hat{S}_2^T \cdot \bar{D} \cdot \hat{S}_1 = D_0 S_z^{(1)} S_z^{(2)} + D_{+1} S_+^{(1)} S_z^{(2)} + D_{-1} S_-^{(1)} S_z^{(2)} +$$
$$D_{+2} S_z^{(1)} S_+^{(2)} + D_{-2} S_z^{(1)} S_-^{(2)} + D_{+3} S_+^{(1)} S_-^{(2)} + D_{-3} S_-^{(1)} S_+^{(2)} + D_{+4} S_+^{(1)} S_+^{(2)} + D_{-4} S_-^{(1)} S_-^{(2)}$$

where

$$D_0 = D_{33} = \sum_m \sum_n g_{m3}^{(2)} D'_{mn} g_{n3}^{(1)}$$

$$D_{\pm 1} = \frac{1}{2}\left(D_{13} \mp iD_{23}\right) = \frac{1}{2}\sum_m \sum_n \left(g_{m1}^{(2)} D'_{mn} g_{n3}^{(1)} \mp i g_{m2}^{(2)} D'_{mn} g_{n3}^{(1)}\right)$$

$$D_{\pm 2} = \frac{1}{2}\left(D_{31} \mp iD_{32}\right) = \frac{1}{2}\sum_m \sum_n \left(g_{m3}^{(2)} D'_{mn} g_{n1}^{(1)} \mp i g_{m3}^{(2)} D'_{mn} g_{n2}^{(1)}\right)$$

$$D_{\pm 3} = \frac{1}{4}\left(D_{11} + D_{22} \mp i\left(D_{21} - D_{12}\right)\right) =$$

$$\frac{1}{4}\sum_m \sum_n \left(g_{m1}^{(2)} D'_{mn} g_{n1}^{(1)} + g_{m2}^{(2)} D'_{mn} g_{n2}^{(1)} \mp i\left(g_{m2}^{(2)} D'_{mn} g_{n1}^{(1)} - g_{m1}^{(2)} D'_{mn} g_{n2}^{(1)}\right)\right)$$

$$D_{\pm 4} = \frac{1}{4}\left(D_{11} - D_{22} \mp i\left(D_{21} + D_{12}\right)\right) =$$

$$\frac{1}{4}\sum_m \sum_n \left(g_{m1}^{(2)} D'_{mn} g_{n1}^{(1)} - g_{m2}^{(2)} D'_{mn} g_{n2}^{(1)} \mp i\left(g_{m2}^{(2)} D'_{mn} g_{n1}^{(1)} + g_{m1}^{(2)} D'_{mn} g_{n2}^{(1)}\right)\right)$$

These expressions allow for simultaneous g-anisotropy and non-collinearity of the principal axes of the g- and D'-tensors to be included in the electron-spin–electron-spin dipolar interaction. Although the expressions appear messy, the implied matrix manipulations are readily amenable to computer programming. If both g-tensors are isotropic with values of g_1 and g_2, respectively, the D-tensor becomes both symmetric and traceless, and the above spin-Hamiltonian reduces to the more familiar form (18):

$$\hat{H} = D_0\left(3S_z^{(1)}S_z^{(2)} - \vec{S}_1 \cdot \vec{S}_2\right) + D_{+1}\left(S_+^{(1)}S_z^{(2)} + S_z^{(1)}S_+^{(2)}\right) + D_{-1}\left(S_-^{(1)}S_z^{(2)} + S_z^{(1)}S_-^{(2)}\right) +$$
$$D_{+2}S_+^{(1)}S_+^{(2)} + D_{-2}S_-^{(1)}S_-^{(2)}$$

where now

$$D_0 = \frac{1}{2}D_{33} = \frac{1}{2}g_1 g_2 D'_{33}$$

$$D_{\pm 1} = \frac{1}{2}\left(D_{13} \mp iD_{23}\right) = \frac{1}{2}g_1 g_2\left(D'_{13} \mp iD'_{23}\right)$$

$$D_{\pm 2} = \frac{1}{2}\left(\frac{1}{2}(D_{11} - D_{22}) \mp iD_{12}\right) = \frac{1}{2}g_1 g_2\left(\frac{1}{2}(D'_{11} - D'_{22}) \mp iD'_{12}\right)$$

If we set $D = g_1 g_2 D'$ and $E = g_1 g_2 E'$ and use the rotation matrix $\vec{R}$, containing the spherical angles θ and ϕ, to align the principal axis of the D'-tensor with the laboratory frame, we obtain the usual expressions (18):

$$D_0 = \frac{1}{6}D(3\cos^2\theta - 1) + \frac{1}{2}E\sin^2\theta\cos 2\phi$$

$$D_{\pm 1} = \frac{1}{4}\sin 2\theta(E\cos 2\phi - D) \pm \frac{1}{2}iE\sin\theta\sin 2\phi$$

$$D_{\pm 2} = \frac{1}{4}\left(D\sin^2\theta + E\cos 2\phi(1 + \cos^2\theta)\right) \pm \frac{1}{2}iE\cos\theta\sin 2\phi$$

To illustrate the effects of non-collinearity of the zfs tensor and g and/or hyperfine tensors, we analyze a specific example of a strongly coupled triplet system using spectral simulations. (In a strongly coupled triplet system, the exchange interaction is large such that it no longer influences the number or positions of signals in the spectrum (19).) In the simulations that follow, the zfs parameters, D' and E', are 175 G and 17.5 G, respectively, and the microwave frequency, v, is 9.1 GHz. The g-tensor of the first electron has principal values $g^{(1)}_x = 2.15$, $g^{(1)}_y = 2.075$ and $g^{(1)}_z = 2$ and that of the second electron is isotropic with a value $g_2 = 2$. We first examine the case when the principal axes of the first g-tensor are collinear with that of the D'-tensor in the absence of any hyperfine interactions. Then we add a hyperfine tensor to the spin-Hamiltonian and rotate the D'-tensor using the rotation matrix, $\ddot{R}$, and examine the effects. Figure 2. shows the calculated spectrum when the principal axes of the g_1- and D'-tensors are collinear.

The splitting between the symmetry related peaks, moving from the center of the spectrum out, are proportional to the principal values of the D'-tensor corresponding to the x, y, and z directions, respectively. In addition, the peaks for each direction are centered around the average g-value for that direction. For example, the splitting between the two inner-most peaks are proportional to $(D'_d)_x$ and their average resonant field value occurs at the average g_x value of 2.075. Thus, increasing g_x, g_y, and g_z for either electron shifts the corresponding peaks to lower fields.

Figure 3. shows the effect as the spherical rotation angle, ζ, is varied from 0° to 90°, with η held fixed at 0°. Additionally, a hyperfine interaction between the first electron and a nucleus with a spin of 1 has been added to the simulated spectra. The principal axes of the hyperfine, A-tensor, are collinear with those of the g_1-tensor and the principal values are $A_x = A_y = 0$ G, and $A_z = 90$ G.

As ζ is varied, the principal axes of the zfs tensor in the x- and z- directions become admixed. Consequently, as ζ increases from 0° to 90°, the peaks corresponding to the zfs along the x-direction shift from being centered around the average g_x value to being centered at the average g_z value. Conversely, the peaks corresponding to the zfs along the z-direction shift from being centered at the average g_z value to being centered at the average g_x value. In addition, the hyperfine splitting initially associated with the z-direction becomes admixed with the hyperfine splitting in the x-direction, and vice versa, as ζ is varied. Thus, as ζ is increased from 0° to 90°, the hyperfine splitting on the peaks corresponding to

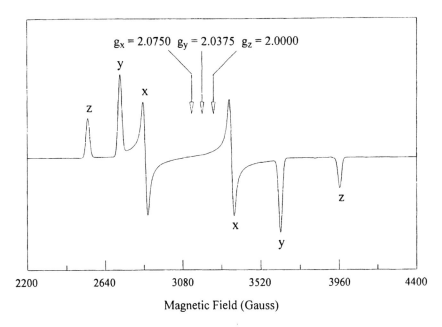

Figure 2: Calculated EPR powder spectrum illustrating the peaks that correspond to the zfs along the x, y, and z directions. The average g_x, g_y, and g_z values are indicated on the spectrum. The parameters used in the simulation were $D' = 175$ G, $E' = 17.5$ G, $g^{(1)}_x = 2.15$, $g^{(1)}_y = 2.075$ and $g^{(1)}_z = 2$, $g_2 = 2$, and $v = 9.1$ GHz.

the zfs along the z-direction is continuously transferred to the peaks corresponding to the zfs along the x-direction. Once $\zeta = 90°$, the hyperfine splitting is completely transferred to the peaks corresponding to the zfs along the x-direction.

Figure 4. shows the effect of varying η from 0° to 90°, with ζ fixed at 0°. The effects of varying η are analogous to those resulting from variations in ζ, only now the principal axes of the zfs tensor in the x- and y- directions are the ones that are mixed. The peaks corresponding to the zfs along the x-direction shift from being centered at the average g_x value to being centered at the average g_y value, and vice versa, as η is increased from 0° to 90°. Likewise, if there were any hyperfine splitting associated with the x-direction, it would be transferred to the y-direction, and vice versa, as η is varied.

Thus, the average g-values of the powder peaks corresponding to the x, y, and z principal axes of the zfs tensor, as well as the appearance of anisotropic hyperfine splitting in the signals, reflect the position of the second spin in the molecule fixed (g or A) axis system of the first spin.

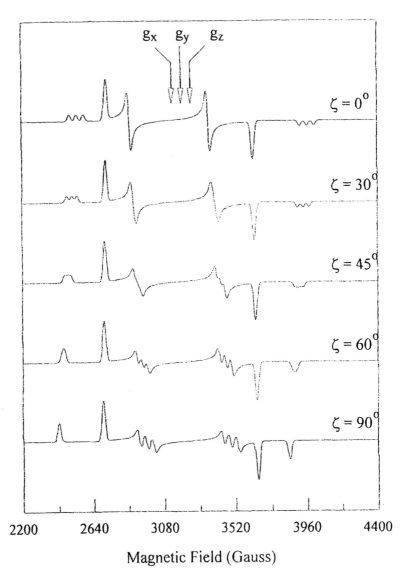

Figure 3: Calculated EPR powder spectra illustrating the effect of varying the spherical angle ζ. The average g_x, g_y, and g_z values are indicated on the spectrum. The parameters used in the simulation were $D' = 175$ G, $E' = 17.5$ G, $g^{(1)}_x = 2.15$, $g^{(1)}_y = 2.075$ and $g^{(1)}_z = 2$, $g_2 = 2$, $A_x = A_y = 0$ G, and $A_z = 90$ G, and $\nu = 9.1$ GHz.

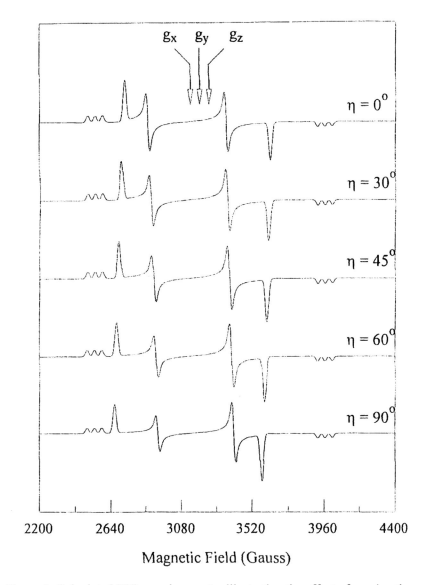

Figure 4: Calculated EPR powder spectra illustrating the effect of varying the spherical angle η. The average g_x, g_y, and g_z values are indicated on the spectrum. The parameters used in the simulation were $D' = 175$ G, $E' = 17.5$ G, $g^{(1)}_x = 2.15$, $g^{(1)}_y = 2.075$ and $g^{(1)}_z = 2$, $g_2 = 2$, $A_x = A_y = 0$ G, and $A_z = 90$ G, and $v = 9.1$ GHz.

Complicated powder spectra are a hallmark of cob(II)alamin-organic radical triplet states that occur as transient intermediates in the reactions of coenzyme B_{12} dependent enzymes (19). The g and A anisotropy of the low spin Co^{2+} may be exploited to determine the position of the spin of the organic radical within the molecular axis of the cob(II)alamin.

References

1. Luckhurst, G. R., in *Spin Labeling Theory and Practice,* Berliner, L. J., ed, Academic Press, New York, 1976, p 133.
2. Hutchison Jr., C. A.; Pearson, G. A.; J. Chem. Phys., 1967, 47, pp 520-533
3. Eaton, G. R.; Eaton, S. S.; *Biol. Magn. Reson.,* **1989**, pp 339-397.
4. Gerfen, G. J.; Licht, S.; Willems, J. P.; Hoffman, B. M.; Stubbe, J.; *J. Am. Chem. Soc.*, **1996**, *118*, pp 8192-8197.
5. Bothe, H.; Darley, D. J.; Albracht, S. P.; Gerfen, G. J.; Golding, B. T.; Buckel, W.; *Biochemistry*, **1998**, *37*, pp 4105-4113.
6. Padmakumar, R.; Banerjee, R.; *J. Biol. Chem.*, **1995**, *270*, pp 9295-9300.
7. Kottis, P.; Lefebvre, J.; J. Chem. Phys., **1963**, *39*, pp 393-403
8. *Introduction to Electrodynamics, Third Edition*; Griffiths, D. J.; Prentice-Hall, Inc., New Jersey, 1999.
9. *Classical Electrodynamics, Third Edition*; Jackson, J. D.; John Wiley & Sons, Inc., New York , 1999.
10. *EPR of Exchange Coupled Systems*; Bencini, A.; Gatteschi, D.; Springer-Verlag, Berlin, 1990.
11. *Electron Paramagnetic Resonance, Elementary Theory and Practical Applications;* Weil, J. A.; Bolton, J. R.; Wertz, J. E.; John Wiley & Sons, Inc., New York, 1994.
12. *Introduction to Quantum Mechanics*; Griffiths, D. J.; Prentice-Hall, Inc., New Jersey, 1995.
13. Lounsbury, J. B.; Barry, G. W.; *J. Chem. Phys.* **1966**, *44*, pp 4367-4372.
14. Bandarian, V.; Reed, G. H.; *Biochemistry*, **2002**, *41*, pp 8580-8588.
15. Mukai, K.; Sogabe, A.; *J. Chem. Phys.* **1980**, *72*, pp 598-601.
16. Perekhodtsev, G.D.; Tipikin, D. S.; Lebedev, Ya. S.; *Russ. J. Phys. Chem.*, **1994**, *68*, pp 1131-1134.
17. Bandarian, V.; Reed, G. H.; *Biochemistry*, **1999**, *38*, pp 12394-12402.
18. Reed, G. H.; Markham, G. D.; *Biol. Magn. Reson.*, **1984**, *6* , pp 73-142.
19. Gerfen, G. J., in *Chemistry and Biochemistry of B_{12}*, Banerjee, R., ed, Wiley-Interscience, New York, 1999, pp. 165-195.

Chapter 5

EPR Characterization of the Heme Oxygenase Reaction Intermediates and Its Implication for the Catalytic Mechanism

Masao Ikeda-Saito[1,3] and Hiroshi Fujii[2,3]

[1]Institute of Multidisciplinary Research for Advanced Materials, Tohoku University, Katahira, Aoba, Sendai 980–8577, Japan
[2]Institute for Molecular Science, Okazaki 444–8585, Japan
[3]Current address: Department of Physiology and Biophysics, Case Western Reserve University School of Medicine, Cleveland, OH 44106–4970

Heme oxygenase (HO) catalyzes the regiospecific degradation of heme to biliverdin by using three O_2 molecules and seven electrons. The enzyme binds one equivalent of heme to form the heme complex, and electron donation initiates the three stepwise oxygenase reactions through the two novel heme derivatives, α-hydroxyheme and verdoheme, during which CO and free Fe are also produced. EPR has been used to study electronic and coordination structures of the HO catalytic intermediates, including the ferric hydroperoxo active species generated by one-electron reduction of the ferrous oxy form. A combination of the novel characteristics of the reaction intermediates and the protein environment are responsible for the unique HO enzyme catalytic mechanism.

Biological heme catabolism is conducted by a family of enzymes termed as heme oxygenase (HO) which catalyzes oxidative degradation of iron protoporphyrin IX (heme hereafter) to biliverdin, iron and CO in the presence of electron donors (*1*). In mammalian systems where electrons are supplied by NADPH through NADPH-cytochrome P450 reductase (*2*), HO is the enzyme responsible for excess hemin excretion and iron recycling (*3*). The product CO has been implicated as a messenger molecule in vasodilation and neuronal transmission by activating soluble guanylyl cyclase (*4-6*) and in regulation of circadian rhythm by binding to a transcription factor, NAPS2 (*7*). In pathogenic bacteria which have to acquire iron from their hosts for their own survival, HO is the key component in the heme-based iron acquisition pathway so as to circumvent the low concentration of free extra-cellular iron (*8-9*).

In its catalytic cycle, HO first binds 1 equivalent of heme to form a heme-HO complex (**1** of Figure 1). The first electron donated from the reducing equivalent reduces the ferric iron to the ferrous state (**2**). Then O_2 binds to form a meta-stable oxy complex (**3**). Further electron donation to the oxy complex initiates the three-step conversion of oxyheme to Fe(III)-biliverdin (**7**) through ferric hydroperoxo heme (**4**), α-meso-hydroxyheme (**5**) and ferrous verdoheme (**6**) intermediates. The final step involves electron donation to convert Fe(III)-biliverdin (**7**) to the ferrous complex, and free iron and biliverdin (**8**) are liberated from the HO protein. HO is not a heme protein *per se* but utilizes heme as both a prosthetic group and a substrate. The salient aspect of HO catalysis is that the one electron reduction of the oxy form generates a ferric hydroperoxo species that leads to site-selective hydroxylation of the α-meso-carbon of the porphyrin ring (*10, 11*). This is different from the ferryl species, commonly found as an active catalytic intermediate in peroxidase and P450 enzymes (*12*). While major work has been conducted on the isoform-1 of mammalian HO termed as HO-1, recent studies have demonstrated that the bacterial counterparts convert heme to biliverdin by the same mechanism (*13-14*). In this article, we describe the active site structures of HO intermediates derived from our paramagnetic resonance studies and their implications for HO catalysis. The objective is to demonstrate the power of EPR spectroscopy, when combined with other techniques, in elucidating the HO reaction mechanism by providing detailed active site structures of reaction intermediates beyond the resolution currently attained by X-ray crystallography on the HO proteins.

Active Site Structure of the Heme-HO Complex

The first intermediate of HO catalysis, the ferric heme-enzyme complex (**1**), exhibits an EPR spectrum of high spin axial symmetry (Figure 2A). The $g = 2$ signal of the spectrum is broadened when measured in buffered $H_2^{17}O$ (Figure 2A, *inset*). The broadening is due to the unresolved hyperfine interaction with *I*

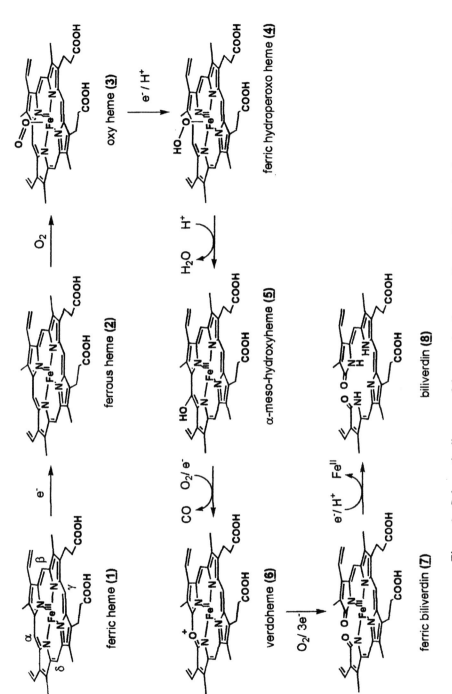

Figure 1. Schematic diagram of the reaction intermediates of HO catalysis.

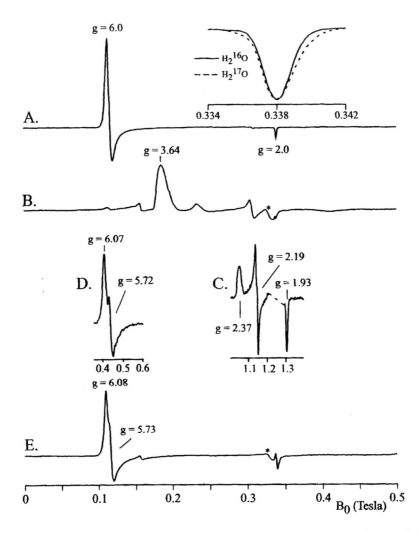

Figure 2. EPR spectra of the heme-HO-1 complex (A), the azide adduct of the heme-HO-1 complex (B), the ferric hydroperoxo spices (C), the annealed product of the ferric hydroperoxo species (D), and ferric α-meso-hydroxyheme-HO-1 (E). The inset is the $g = 2$ region of the heme-HO complex measured in buffered $H_2^{16}O$ (solid line) and $H_2^{17}O$ (broken line). Spectra (A), (B), and (E) were recorded at X-band (9.45 GHZ) at 6 K, while (C) and (D) were at Q-band (35 GHz) at 2.1 K. The asterisk (*) denotes signals due to cavity impurities.

= 5/2 of the water oxygen, indicating water coordination to the heme iron. EPR spectra of the ferrous NO complex of wild-type and histidine mutants show coordination of histidine as the proximal axial ligand (*15-17*). The axial coordination structure established by EPR has been confirmed by the crystal structures of both mammalian and bacterial HOs (*18-20*).

The ferric heme-enzyme complex binds azide to form a low spin species with a dissociation constant of ~10 μM for HO-1 at pH 7 and 20 °C. Different from low spin EPR spectra commonly seen in heme protein azide complexes (for example, $g \sim 2.78$, 2.22, and 1. 73 for MbN_3), the azide-bound HO-1 shows a so-called "large g_{max} EPR spectrum" (Figure 2B), $g=3.64$. The large g_{max} EPR spectrum is induced by the spin-orbit coupling of the t_{2g} ground state, mainly from degeneration of iron d_{xz} and d_{yz} orbitals. The d_{xz} and d_{yz} orbital energies are changed by the orientations of the two axial ligands, the proximal histidine imidazole and the bound-azide in this case. When the imidazole plane and azide axis eclipse the porphyrin meso carbon-Fe axes, the π-orbitals of the imidazole and azide are orthogonal to the d_{xz} and d_{yz} orbitals. This results in the degeneration of the d_{xz} and d_{yz} orbitals and exhibits the large g_{max} EPR spectrum. The large g_{max} spectra are seen in a heme model complex with plane axial ligands with relative perpendicular orientation eclipsing adjacent Fe-meso carbon axes (*21*). On the other hand, rotation of the imidazole and/or azide ligands allows the ligand π-orbitals to overlap with the iron d_{xz} and d_{yz} orbitals. This d-π interaction leads to the split of the degeneration and gives a rhombic EPR spectrum as seen for most ferric low spin heme proteins, where the proximal histidine imidazole orients to overlap with the d_{xz} and d_{yz} oribitals. Hence, the large g_{max} EPR spectrum of the azide-bound HO-1 indicates that the proximal imidazole plane lays over the porphyrin meso carbon-Fe axis and the bound azide has its $N=N=N$ vector perpendicular to the imidazole plane.

The HO crystal structures (*18-20*) show that the imidazole plane of the proximal histidine closely eclipses the heme meso β-δ axis, and that the kinked distal helix located close to the heme plane sterically controls direction of the bound ligand pointing towards the α-meso carbon of the porphyrin ring. These structural features predict that the projection of the bound N_3 onto the porphyrin plane is perpendicular to the plane of the imidazole ligand *trans* to the bound azide. The crystal structure of the azide-bound heme-HO-1 complex reported recently confirms the aforementioned axial ligand geometry of the HO azide complex (*22*). This coordination structure of HO-1 also explains the α-regioselectivity of the HO reaction as discussed in the later section on the hydroperoxo intermediate.

The first electron donated from the reductase reduces the ferric heme iron to the ferrous state. The deoxy (**2**) and oxy (**3**) forms are $S=2$ and 0, respectively, thus their EPR examination is either difficult or unfeasible. This unfavorable situation could be circumvented by replacing heme with EPR-visible cobalt porphyrins, as originally applied for Mb and Hb (*23*). Deoxy cobalt-HO-1 exhibits an EPR spectrum (Figure 3A) of axial symmetry ($g_\perp$ = 2.310, $g_{//}$ = 2.027) typical of a pentacoordinate Co(II) porphyrin complex with a nitrogenous base as the axial ligand (*24*). The $g_{//}$ signal has an octaplet hyperfine structure due to the hyperfine interaction with ^{59}Co (I = 7/2, $A_{//}(^{59}$Co) = 8.00 mT). Each of these hyperfine lines is further split into a triplet by hyperfine interaction with a ^{14}N (I = 1, $A_{//}(^{14}$N) = 1.80 mT) of the proximal imidazole axial ligand.

In the absence of crystal structures of the oxy form, EPR of oxy cobalt-HO provides information on the mode of the oxygen binding in HO. Upon oxygen binding, cobalt-HO-1 exhibits a free radical type spectrum (Figure 3B) centered around g = 2, indicating that O_2 binds to Co(II) in a manner similar to that in the oxy cobalt-Mb (*25*). The spectral parameters of oxy cobalt-HO is estimated by computer simulation as g_1= 2.104, g_2= 2.007, g_3= 1.990; A_1 = 1.76 mT, A_2 = 0.98 mT, A_3 = 0.83 mT. The $g_{//}$ component, g_1, of the cobalt-HO spectrum is more anisotropic than that of cobalt-Mb, while the $g_\perp$ component, g_2 and g_3, are slightly less anisotropic. This is an indication of the different Co-O-O geometry between HO and Mb and is consistent with the resonance Raman results of the iron counterparts (*26*). The spectrum of oxy cobalt-HO consists of a single paramagnetic species as opposed to the presence of at least two species in oxy cobalt-Mb spectrum (*27*). The bound dioxygen in oxy cobalt-HO-1 appears to assume a single geometry due to the strong distal pocket interactions in line with the conclusion based on the resonance Raman studies (*26*).

EPR could assess possible hydrogen bonding interactions involved with the bound O_2. In a Co^{2+}-O_2 adduct, the paramagnetic center is the bound O_2 (*23*). Protons located within hydrogen bonding distances to the O_2 cause line broadening of the EPR spectrum. Deuteration can reduce the broadened linewidth by a factor of 33 % because of the smaller nuclear magnetic moment of the deuteron. Thus, a reduction of the linewidth in the EPR spectrum of oxy cobalt-HO is expected upon deuteration, if an exchangeable proton(s) is located near the bound O_2 in oxy cobalt-HO. The EPR spectrum of oxy cobalt-HO recorded in buffered D_2O (Figure 3C) shows a sharpened hyperfine structure in comparison to that recorded in buffered H_2O. A similar spectral change has been observed for oxy cobalt-Mb and has been interpreted as evidence of the hydrogen bond formation between the bound O_2 and the distal His proton in Mb (*28-29*). Based on this, we infer that there exists a hydrogen bond interaction between the coordinated O_2 and a dissociable group in the HO distal pocket. In the absence of ionizable residues in the distal pocket, the hydrogen bonding partner of the O_2 is likely a non-coordinated distal pocket water molecule.

The hydrogen bonding interaction of the bound O_2 plays critical roles in HO catalysis. The hydrogen bonding stabilizes the oxy form, which not only prevents auto-oxidation but also increases the O_2 affinity by decreasing the O_2

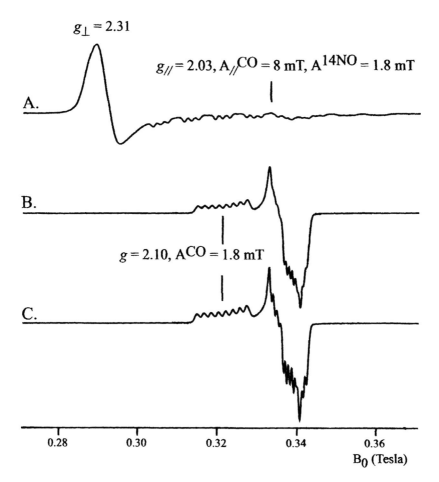

Figure 3. X-band EPR spectra of cobalt-HO-1 at 15 K. (A), deoxy form; (B), oxy form in buffered H_2O; (C), oxy form in buffered D_2O.

dissociation rate. The heme-HO complex has an extremely high affinity for O_2 (K_{O_2} = 30 μM^{-1}) which is roughly equal to that for CO. This effectively prevents product inhibition by CO that is generated during catalysis of heme degradation (*30*). Furthermore, the hydrogen bonding facilitates the efficient proton transfer in the formation of the hydroperoxo active intermediate, as described in the next section.

Ferric Hydroperoxo Intermediate

The one-electron reduction of the oxy form (**3**) at ambient temperatures rapidly generates the ferric α-meso-hydroxyheme species (**5**), and any intermediate between them cannot be detected by usual rapid-mixing methods. However, one-electron reduced oxy form could be generated stably by radiolytic irradiation of the oxy form at cryogenic temperatures. This method, cryo-reduction, has been used for studies on catalytic intermediates of heme enzymes including cytochrome P-450, NO synthase, and horseradish peroxidase (*31-33*). Upon irradiation of EPR-silent oxy HO at 77 K, a new EPR spectrum of ferric low spin appears as shown in Figure 2C, which has a spread in *g*-values (2.37, 2.19, and 1.93) that is substantially greater than that of all previously studied ferric peroxo-heme complexes (*11, 34*). ENDOR reveals an interaction of an exchangeable proton whose hyperfine coupling is comparable to that of the proton of the ferric-hydroperoxo-heme species (*11, 34*), leading to a conclusion that reduction of oxy HO at 77 K generates the ferric hydroperoxo species (**5**) rather than ferric peroxo species. As described above, the bound O_2 in the precursor oxy HO forms a hydrogen bonding with a distal pocket water molecule, and the proton in this hydrogen bond is highly likely to become the hydroperoxo-proton, the first proton delivered to reduced dioxygen during catalysis (*35*). Mutagenesis studies (*34-35*) have revealed a significant role of the carboxylate side chain of Asp140 in delivering the proton required for formation of the active hydroperoxo species. This has led to the model for the proton delivery as shown in Figure 4 (*35*). In this model, the side chain of Asp140 interacts with the bound O_2 through a bridging water molecule which is found in the crystal structure of HO-1 (*36*). Asp-140 is also involved in a hydrogen bonding network with other distal pocket water molecules. The bridging water forms a hydrogen-bonding interaction with the bound O_2 as described above. This hydrogen-bonding network not only stabilizes the bound O_2 against autooxidation but also functions as a channel for efficient proton delivery to the bound O_2. Thus, the oxy form is efficiently reduced to the hydroperoxo intermediate, which then oxidizes the α-meso carbon to form hydroxyheme. The presence of the distal pocket hydrogen bonding network has

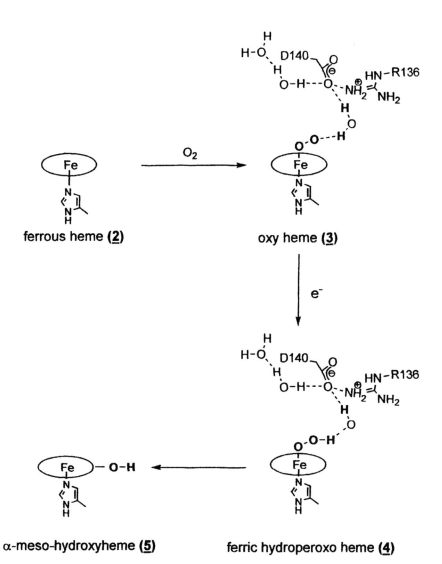

α-meso-hydroxyheme (**5**)   ferric hydroperoxo heme (**4**)

Figure 4. Proposed mechanism of the proton delivery and oxygen activation by HO.

been confirmed by recent crystallographic and NMR studies on HOs (20, 36-37), supporting the proton delivery mechanism depicted in Figure 4.

Interestingly, the EPR spectrum of the ferric hydroperoxo species of HO is more anisotropic than those of cytochrome P-450, NO synthase, hemoglobin, myoglobin, and horseradish peroxidase (31-33). The calculations of d-orbital splits based on these EPR g parameters indicate that the split of the d_{xz} and d_{yz} orbital energies for HO is much smaller than those for other heme proteins listed above. Thus, the π-orbitals of the proximal imidazole and the iron bound hydroperoxo species are orthogonal to the iron d_{xz} and d_{yz} orbitals, as found for the azide form of HO. This coordination structure decreases the push-effect of the proximal histidine imidazole and prevents breakage of the O-O bond of the hydroperoxo complex. Furthermore, the iron bound hydroperoxo species is directed toward the α-meso position due to the steric and hydrogen bonding effects of the distal helix. This orientation, together with the acute Fe-O-OH angle deduced from the resonance Raman studies on oxy HO, which is the immediate precursor of ferric hydroperoxo (26), places the distal oxygen atom of the iron bound hydroperoxo species sufficiently close to the α-meso carbon to regioselectively oxidize the α-meso position to form α-meso-hydroxyheme.

α-meso-Hydroxyheme Intermediate

Upon annealing the ferric hydroperoxo species (**4**) at 240 K, the ferric low spin EPR signal is replaced by a new rhombic high spin EPR signal, the g~6 region shown in Figure 2D. The high spin signal is identical to that of the HO complex with authentic ferric α-meso-hydroxyheme (Figure 2E) (**5**). Detailed analysis of this conversion reveals that the hydroperoxo species is the precursor of the ferric α-meso-hydroxyheme species (34), thereby supporting the early proposal that ferric hydroperoxo, but not ferryl oxo, is the active intermediate in the first oxygenation step of HO catalysis (10). The ferric α-meso-hydroxyheme HO species is five coordinate with the proximal histidine axial ligand (38), and the rhombic high spin EPR signal is consistent with this notion.

Model system studies have shown that the ferric α-meso-hydroxyheme is in a resonance structure including ferrous neutral radical species that is generated by intramolecular electron transfer between the iron and the macrocycle (39-40), as schematically illustrated in Figure 5. The ferrous character likely reduces ligand affinity of the ferric α-meso-hydroxyheme HO complex toward typical ferric heme ligands, such as cyanide, azide, and water. This explains why the ferric α-meso-hydroxyheme-HO complex is five coordinate, different from the water bound hexacoordinate heme complex. However, the pentacoordinate meso-hydroxyheme is only reported for the protein complexes, and suitable models for such a pentacoordinate complex of the meso-hydroxyheme are not

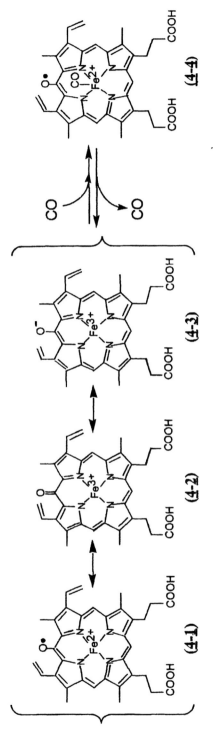

Figure 5. Schematic diagram for the ferric α-meso-hydroxyheme resonance structure. (**4-1**), ferrous neutral radical form; (**4-2**), keto-form of oxopholin; (**4-3**), enorate form of oxopholin; (**4-4**), the CO-bound ferrous neutral radical form.

available (*41*). The presence of the ferrous neutral form is further supported by the observation that CO, which is a well-established ligand to ferrous heme species, binds to the ferric α-meso-hydroheme-HO complex. The binding of CO shifts the electronic structure to the ferrous porphyrin neutral radical character from the ferric states and, thus, exhibits a free-radical type EPR signal at g = 2.004 (Figure 6F). The ferric α-meso-hydroxyheme is easily oxidized to ferric α-meso-oxophlorin radical even with oxygen (*42*). Because of the possible antiferromagnetic spin coupling between ferric iron and a pholrin radical, the α-meso-oxophlorin radical is EPR-silent. More importantly, the ferrous porphyrin neutral radical form seems to be involved in the verdoheme formation as an active species (*10, 38-40*). A porphyrin peroxy radical complex has been proposed as one of the reaction intermediates. However, the electronic structure of the peroxy radical species has yet to be established.

Verdoheme and Ferric Biliverdin intermediates

Reaction of ferric α-meso-hydroxyheme-HO with one electron and O_2 generates ferrous verdoheme (**6**) (*38*). The verdoheme-HO intermediate is ferrous low spin (*40, 43*), is diamagnetic, and thus does not exhibit an EPR signal (Figure 6G). It was reported that a low spin EPR signal with g = (2.57, 2.14, 1.86) was observed, when O_2 was added to ferric α-meso-hydroxyheme-HO (*44*). This signal was tentatively assigned to ferric verdoheme-HO. This reaction, however, does not yield ferric verdoheme-HO, when O_2 is added to HO containing authentic α-meso-hydroxyheme and verdoheme (*42*). In light of this, the assignment of the low spin EPR requires further confirmation.

The last oxygenation cycle in HO catalysis is the conversion of ferrous verdoheme to ferric biliverdin (**7**). The ferric biliverdin-HO complex exhibits a highly rhombic high spin EPR spectrum (g = 8.63, ~3, 1.47) (Figure 6H), indicating that cleavage of the porphyrin macrocycle at the α-meso position severely lowers in-plane symmetry, although the ferric biliverdin is still bound to HO. When the ferric biliverdin is further reduced by NADPH-cytochrome P450 reductase, the rhombic high spin EPR signal diminishes and an EPR signal appears at g = 4.3 due to the iron released from the biliverdin (**8**) (Figure 6I).

Conclusions and Future Prospects

The contribution of X-ray crystallography to recent advances in HO studies has been undisputable. However, as described in this article, detection of hydrogen bonding with bound O_2 and unequivocal demonstration of the active ferric hydroperoxo species are possible only by EPR in combination with cobalt substitution, cryo-reduction, and protein engineering. The reaction mechanism

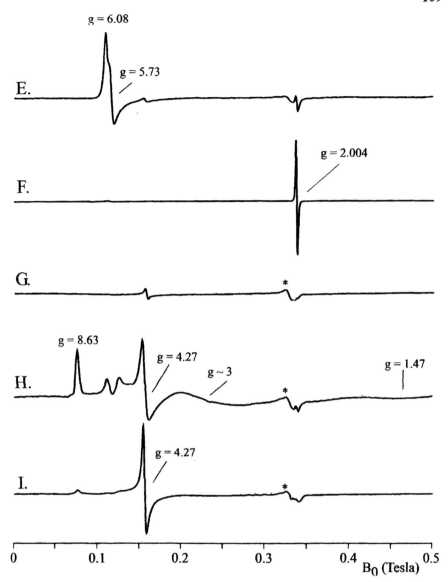

Figure 6. X-band EPR spectra of ferric α-meso-hydroxyheme-HO-1 (E), its CO complex (F), verdoheme-HO-1 (G), ferric bliveridin-HO-1 (H), and the final HO-1 reaction product. The ferric biliverdin-HO complex (H) was prepared by the aerobic reaction of the ferric heme-HO complex and ascorbic acid. The HO catalysis final product (I) was generated by single-turnover reaction of the heme-HO complex with NADPH and NADPH-cytochrome P450 reductase. The asterisk (*) denotes signals due to cavity impurities.

for the first step of oxygenation and formation of α–meso-hydroxyheme are now well-understood. There is a clear need for further studies on the second and third oxygenation steps. Future paramagnetic resonance studies on α-meso-hydroxyheme- and verdoheme-HO and their derivatives will delineate their electronic structures and more chemical details of the catalytic mechanism.

Acknowledgement

This work was supported by the National Institutes of Health (GM57272) and the Ministry of Education, Science, Technology, Culture, and Sport, Japan (1214702, 14380300, and 14340212). This article is based on our collaborative EPR studies on HO with Drs. G. C. Chu, R. M. Davydov, B. M. Hoffman, K. Mansfield Matera, C. T. Migita, and T. Yoshida, to whom many thanks are due.

References

1. Maines, M. D. *Annu. Rev. Pharmacol. Toxicol.* **1997**, *37*, 517-554.
2. Schacter, B. A.; Nelson, E. B.; Marver, H. S.; Masters, B. S. S. *J. Biol. Chem.* **1972**, *247*, 3601-3607.
3. Poss, K. D.; Tonegawa, S. *Proc. Natl. Acad. Sci. U.S.A.* **1997**, *94*, 10919-10924.
4. Zakhary, R.; Gaine, S. P.; Dinerman, J. L.; Ruat, M.; Flavahan, N. A.; Snyder, S. H. *Proc. Natl. Acad. Sci. USA* **1996**, *93*, 795-798.
5. Suematsu, M.; Ishimura, Y. *Hepatology* **2000**, *31*, 3-6.
6. Prabhakar, N. R. (1999) *Respir. Physiol.* **1999**, *115*, 161-168.
7. Dioum, E. M.; Rutter, J.; Tuckerman, J. R.; Gonzalez, G.; Gilles-Gonzalez, M.-A.; McKnight, S. L. *Science* **2002**, *298*, 2385-2387.
8. Wandersman, C.; Stojilijkovic, I. *Curr. Opin. Micobiol.* **2000**, *3*, 215-220.
9. Genco, C.; Dixon, D. W. *Mol. Microbiol.* **2000**, *39*, 1-11.
10. Ortiz de Montellano, P. R. *Curr. Opin. Chem. Biol.* **2000**, *4*, 221-227.
11. Davydov, R. M.; Yoshida, T.; Ikeda-Saito, M.; Hoffman, B. M. *J. Am. Chem. Soc.* **1999**, *121*, 10656-10657
12. Sono, M.; Roach, M. P.; Coulter, E. D.; Dawson, J. H. *Chem. Rev.* **1996**, *96*, 2841-2887.
13. Wilks, A.; Schmitt, M. P. *J. Biol. Chem.* **1998**, *273*, 837-841.
14. Chu, G. C.; Katakura, K.; Zhang, X.; Yoshida, T.; Ikeda-Saito, M. *J. Biol. Chem.* **1999**, *274*, 21319-21325.
15. Takahashi, S.; Wang, J.; Rousseau, D. L.; Ishikawa, K.; Yoshida, T.; Host, J. R.; Ikeda-Saito, M. *J. Biol. Chem.* **1994**, *269*, 1010-1014.

16. Ito-Maki, M.; Ishikawa, K.; Mansfield Matera, K.; Sato, M.; Ikeda-Saito, M.; Yoshida, T. *Arch. Biochem. Biophys.* **1995**, *317*, 253-258.
17. Chu, G. C.; Tomita, T.; Sonnichsen, F. D.; Yoshida, T.; Ikeda-Saito, M. *J. Biol. Chem.* **1999**, *274*, 24490-24496.
18. Schuller, D. J.; Wilks, A.; Ortiz de Montellano, P. R.; Poulos, T. L. *Nature Struct. Biol.* **1999**, *6*, 860-867.
19. Schuller, D. J.; Zhu, W.; Stojilijkovic, I.; Wilks, A.; Poulos, T. L. *Biochemistry* **1999**, *40*, 11552-11558.
20. Hirotsu, S.; Chu, G. C.; Lee, D.-S.; Unno, M.; Yoshida, T.; Park, S.-Y.; Shiro, Y.; Ikeda-Saito, M. manuscript in preparation.
21. Walker, F. A. *Coord. Chem. Rev.* **1999**, *185-186*, 471-543.
22. Sugishima, M; Sakamoto, H.; Higashimoto, Y.; Omata, Y.; Hayashi, S., Noguchi, M.; Fukuyama, K. *J. Biol. Chem.* **2002**, *277*, 45086-45090.
23. Hoffman, B. M.; Petering, D. H. *Proc. Natl. Acad. Sci. U.S.A.* **1970**, *67*, 637-643.
24. Fujii, H.; Dou, Y.; Zhou, H.; Yoshida, T.; Ikeda-Saito, M. *J. Am. Chem. Soc.* **1998**, *120*, 8251-8252.
25. Brucker E. A.; Olson, J. S.; Phillips, Jr. G. N.; Dou, Y.; Ikeda-Saito, M. *J. Biol. Chem.* **1996**, *271*, 25419-25422.
26. Takahashi, S.; Ishikawa, K.; Takeuchi, N.; Ikeda-Saito, M.; Yoshida, T.; Rousseau, D. L. *J. Am. Chem. Soc.* **1995**, *117*, 6002-6006.
27. Hori, H.; Ikeda-Saito, M.; Yonetani, T. *J. Biol. Chem.* **1982**, *257*, 3636-3642.
28. Ikeda-Saito, M.; Lutz, R. S.; Shelley, D. A.; McKelvey, E. J.; Mattera, R.; Hori, H. *J. Biol. Chem.* **1991**, *266*, 23641-23647.
29. Lee, H. C.; Peisach, J.; Dou, Y.; Ikeda-Saito, M. *Biochemistry* **1994**, *33*, 7609-77618.
30. Migita, C. T.; Mansfield Matera, M.; Ikeda-Saito, M.; Olson, J. S.; Fujii, H., Yoshimura, T.; Zhou, H.; Yoshida, T. *J. Biol. Chem.* **1998**, *273*, 945-949.
31. Davydov, R.; Macdonald, I. D. G.; Makris, T. M.; Sligar, S. G.; Hoffman, B. M. *J. Am. Chem. Soc.* **1999**, *121*, 10654-10655.
32. Davydov, R.; Ledbetter-Rogers, A.; Martasek, P.; Larukhin, M.; Sono, M.; Dawson, J. H., Masters, B. S. S.; Hoffman, B. M. *Biochemistry* **2002**, *41*, 10375-10381.
33. Denisov, I. G.; Makris, T. M.; Sligar, S. G. *J. Biol. Chem.* **2002**, *277*, 42706-427034.
34. Davydov, R. M.; Kofman, V.; Fujii, H.; Yoshida, T.; Ikeda-Saito, M.; Hoffman, B. M. *J. Am. Chem. Soc.* **2002**, *124*, 1798-1808.
35. Fujii, H.; Zhang, X.; Tomita, T.; Ikeda-Saito, M.; Yoshida, T. *J. Am. Chem. Soc.* **2001**, *123*, 6475-6484.
36. Lad, L.; Schuller, D. J.; Shimizu, H.; Friedman, J.; Li, H.; Ortiz de Montellano, P. R.; Poulos, T. L. *J. Biol. Chem.* **2003**, *278*, 7834-7843.

37. Sivitsky, R.; Li, Y.; Auclair, K.; Ortiz de Montellano, P. R.; La Mar, G. N. *J. Am. Chem. Soc.* **2002**, *124*, 14296-14297.
38. Mansfield Matera, K.; Takahashi, S.; Fujii, H.; Zhou, H.; Ishikawa, K.; Yoshimura, T.; Rousseau, D. L.; Yoshida, T.; Ikeda-Saito, M. *J. Biol. Chem.* **1996**, *271*, 6618-6624.
39. Sano, S.; Sano, T.; Morishima, I.; Shiro, Y.; Maeda, Y. *Proc. Natl. Acad. Sci. U. S. A.* **1986**, *83*, 531-535.
40. Morishima, I.; Fujii, H.; Shiro, Y.; Sano. S. *Inorg. Chem.* **1995**, *34*, 1528-1535.
41. Balch, A. L. *Coord. Chem. Rev.* **2000**, *200-202*, 349-377.
42. Migta, C. T.; Fujii, H.; Mansfield Matera, K.; Takahashi, S.; Zho, H.; Yoshida, T. *Biochim. Biophys. Acta* **1999**, 1432, 203-213.
43. Takahashi, S.; Mansfiled Matera, K.; Fujii, H.; Zhou, H.; Ishikawa, K.; Yoshida, T.; Ikeda-Saito, M.; Rousseau, D. L. *Biochemistry* **1997**, 36, 1402-1410.
44. Liu, Y.; Moënne-Locoz, P.; Loehr, T. M.; Ortiz de Montellano, P. R. *J. Biol. Chem.* **1997**, *272*, 6909-6917.

Chapter 6

Paramagnetic Resonance in Mechanistic Studies of Fe-S/Radical Enzymes

Joan B. Broderick[1], Charles Walsby[2], William E. Broderick[1],
Carsten Krebs[3], Wei Hong[1], Danilo Ortillo[1], Jennifer Cheek[1],
Boi Hanh Huynh[3], and Brian M. Hoffman[2]

[1]Department of Chemistry, Michigan State University,
East Lansing, MI 48864
[2]Department of Chemistry, Northwestern University, Evanston, IL 60203
[3]Department of Physics, Emory University, Atlanta, GA 30322

Paramagnetic resonance studies have provided critical insight into the structural and mechanistic properties of the "radical-SAM" enzymes, a superfamily of enzymes which utilize iron-sulfur clusters and S-adenosylmethionine (SAM or AdoMet) to initiate radical catalysis. Here we report recent studies of pyruvate formate-lyase activating enzyme (PFL-AE), which functions to generate the catalytically essential glycyl radical of pyruvate formate-lyase (PFL). Quantitative EPR studies show that the $[4Fe-4S]^+$ cluster on PFL-AE is the electron source required for generation of the glycyl radical on PFL. Electron-nuclear double resonance (ENDOR) studies show that the co-substrate AdoMet sits close to the $[4Fe-4S]^+$ cluster of PFL-AE in the ES complex, with the methyl carbon approximately 4-5 Å, and the methyl hydrogen approximately 3-4 Å, from the closest iron of the cluster. This close association of AdoMet and the Fe-S cluster has significant implications regarding the catalytic mechanism.

Radical-SAM Enzymes

Pyruvate formate-lyase activating enzyme is a member of the recently identified "radical-SAM" superfamily of enzymes (1,2). These enzymes are believed to have in common the requirement for an iron-sulfur cluster and S-adenosylmethionine to initiate radical catalysis. Among the reactions catalyzed by this family of enzymes are the insertion of sulfur into dethiobiotin to generate biotin, the insertion of sulfur into octanoic acid to make lipoic acid, the rearrangement of lysine, the repair of UV-induced DNA damage, and the generation of catalytically essential glycyl radicals (Figure 1) (1). Although the reactions are diverse, there is thought to be a common mechanistic step which involves the generation of an AdoMet-derived 5'-deoxyadenosyl radical intermediate, which serves to abstract a hydrogen atom from the substrate.

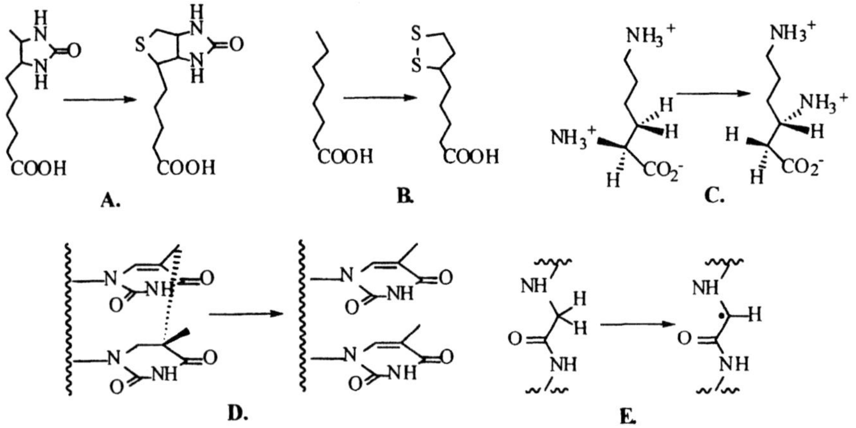

Figure 1. Representative reactions of the Radical-SAM enzymes, including A. biotin synthase, B. lipoate synthase, C. lysine 2,3-aminomutase, D. spore photoproduct lyase, and E. activating enzymes.

Pyruvate Formate-Lyase Activating Enzyme

Pyruvate formate-lyase activating enzyme (PFL-AE) generates a catalytically essential radical at G734 on pyruvate formate-lyase, thereby activating PFL for its central function in anaerobic glucose metabolism in *E. coli* (Figure 2) (3). The glycyl radical of PFL, once generated, is stable under anaerobic conditions. Under aerobic conditions the radical reacts rapidly with O_2, resulting in peptide cleavage and thus irreversible inactivation (3). *In vivo,*

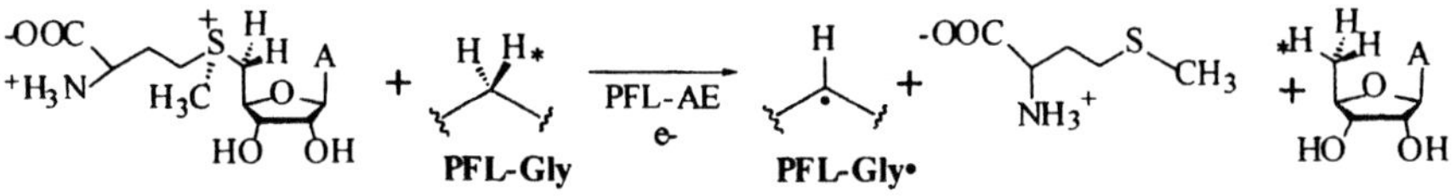

Figure 2. The reaction catalyzed by pyruvate formate-lyase activating enzyme.

however, the onset of oxic conditions causes the PFL deactivating enzyme to add a hydrogen atom to the glycyl radical to regenerate inactive PFL and prevent irreversible inactivation (4). The PFL/PFL-AE system is not the only glycyl radical enzyme system known; *E. coli* also utilize a glycyl radical-containing ribonucleotide reductase and its associated activase under anaerobic growth conditions (1). In addition, other glycyl radical enzymes have been identified in other organisms.

Spectroscopic Studies of PFL-AE

PFL-AE was identified early on as an iron-requiring enzyme. However it was not until relatively recently that spectroscopic studies of anaerobically isolated enzyme identified PFL-AE as an iron-sulfur cluster-containing enzyme (5). Initial resonance Raman and EPR studies suggested the presence of $[2Fe-2S]^{2+}$ and $[4Fe-4S]^{2+}$ clusters in isolated PFL-AE, with only [4Fe-4S] clusters remaining after reduction with dithionite (5). Spectroscopic studies on reconstituted PFL-AE showed the presence of [4Fe-4S] clusters (6). As shown in Figure 3, isolation of PFL-AE under more strictly anaerobic conditions results in enzyme containing $[3Fe-4S]^+$ as the primary cluster form (accounting for 66% of total iron), with minor contributions from $[2Fe-2S]^{2+}$ and $[4Fe-4S]^{2+}$ (12% and 8% of total iron, respectively), as determined from a combination of EPR and Mössbauer spectroscopy (7,8). As with the earlier preparations, reduction of this "as-isolated" protein resulted in the presence of [4Fe-4S] as the only cluster form. Dithionite reduction produces primarily $[4Fe-4S]^{2+}$ (66% of total iron) with small amounts of $[4Fe-4S]^{1+}$ (12% of total iron), while extended photoreduction in the presence of 5-deazariboflavin converts most of the clusters (up to 80-95% of total iron) to the $[4Fe-4S]^+$ state (8).

$[4Fe-4S]^+$ is the Catalytically Relevant Cluster

Quantitative EPR spectroscopy under single turnover conditions has been used to demonstrate that the $[4Fe-4S]^+$ cluster of PFL-AE is the catalytically active cluster, and that it provides the electron necessary for the reductive cleavage of AdoMet (9). PFL-AE samples were photoreduced in the presence

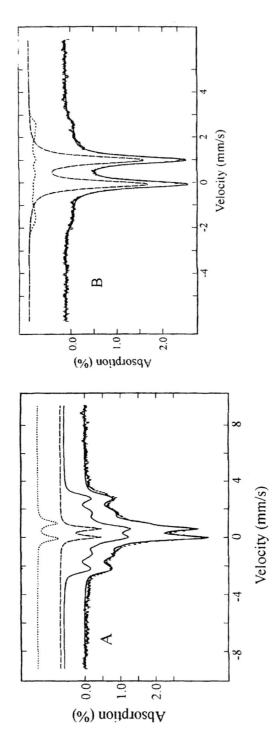

Figure 3. Mössbauer spectra of purified PFL-AE before (A) and after (B) reduction with dithionite. A. The data is simulated as a summation of contributions (solid line overlaying the data) from [3Fe-4S]⁺ (solid line, 66%), [2Fe-2S]²⁺ (dashed line, 12%), and [4Fe-4S]⁺ (dotted line, 8%). A small contribution from a linear [3Fe-4S]⁺ cluster is not shown. B. The solid line overlaying the data is a sum of simulations of [4Fe-4S]²⁺ (dashed line, 66%) and [4Fe-4S]⁺ (dotted line, 12%). (Reproduced from reference 10. Copyright 2002 American Chemical Society.)

of 5-deazariboflavin for varying amounts of time followed by addition of AdoMet. The resulting EPR spectra (Figure 4A) show that the amount of [4Fe-4S]$^+$ increases with illumination time, accounting for approximately 85% of the total iron after 30 minutes illumination. The PFL-AE/AdoMet samples taken at each time point were split in two, and to half of each sample was added an equimolar amount of PFL; the resulting spectra are shown in Figure 4B. As can be seen in Figure 4, the increase in glycyl radical generated on PFL bears a 1:1 correspondence with the increase in [4Fe-4S]$^+$ present on PFL-AE prior to reaction with PFL (9). Furthermore, the EPR signal due to the [4Fe-4S]$^+$ is absent after addition of PFL and generation of the glycyl radical on PFL. Together, these results demonstrate that the [4Fe-4S]$^+$-PFL-AE in the presence of Adomet is catalytically competent to generate the glycyl radical on PFL (9). Furthermore, the data suggests that the [4Fe-4S]$^+$ is oxidized by one electron to the EPR-silent [4Fe-4S]$^{2+}$ state upon generation of the PFL glycyl radical. The [4Fe-4S]$^+$, therefore, is the source of the electron required for reductive cleavage of Adomet, which generates the intermediate adenosyl radical responsible for H atom abstraction from PFL G734 (9).

Does AdoMet Interact Directly with the [4Fe-4S]$^+$?

The results of the single-turnover experiments demonstrated that the [4Fe-4S]$^+$ of PFL-AE provides the electron necessary for reductive cleavage of AdoMet. This reductive cleavage generates the putative intermediate adenosyl radical

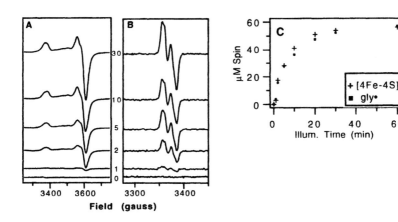

Figure 4. Quantitative X-band EPR measurements of single-turnover experiments demonstrate that the PFL-AE [4Fe-4S]$^-$ in the presence of AdoMet generates the glycyl radical of PFL. A. EPR spectra of PFL-AE/AdoMet taken after the indicated minutes of photoreduction. B. The samples in A after addition of PFL. C. Spin quantitation of the spectra in A and B show that one [4Fe-4S]$^-$ generates one glycyl radical.
(Reproduced from reference 13. Copyright 2002 American Chemical Society.)

which abstracts a hydrogen atom from PFL G734 to generate the active, radical form of PFL. However, these experiments did not address the question of whether the [4Fe-4S]$^+$ cluster of PFL-AE serves simply as an electron-transfer center, to transfer an electron to a remote AdoMet, or whether the [4Fe-4S]$^+$ serves a more intimate role in generating the 5'-deoxyadenosyl radical intermediate. X-band EPR studies were suggestive of a close association between AdoMet and the [4Fe-4S]$^+$ cluster, as addition of AdoMet to [4Fe-4S]$^+$/PFL-AE dramatically perturbs the signal of the S=1/2 [4Fe-4S]$^+$ cluster (Figure 5) (10). This perturbation of the EPR signal could result from a direct interaction between AdoMet and the [4Fe-4S]$^+$; however, it could also arise via an allosteric effect of AdoMet binding at a location remote from the cluster. In order to definitively resolve this question, we undertook electron-nuclear double resonance (ENDOR) studies of PFL-AE/[4Fe-4S]$^+$ in complex with specifically isotopically labeled AdoMet.

^{1}H and ^{13}C ENDOR Studies of the PFL-AE/[4Fe-4S]-AdoMet Complex

Proton ENDOR studies of unlabeled AdoMet complexed to the [4Fe-4S]$^+$/PFL-AE show numerous overlapping and unresolved proton resonances.

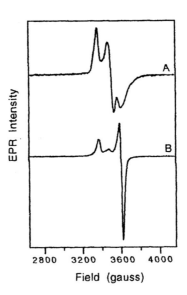

Figure 5. X-band EPR of [4Fe-4S]$^+$-PFL-AE in the absence (A) and presence (B) of AdoMet.
(Reproduced from reference 14. Copyright 2002 American Chemical Society.)

Proton ENDOR of methyl-D_3-AdoMet with $[4Fe-4S]^+$/PFL-AE shows some slight perturbations relative to the spectrum with unlabeled AdoMet, but still the proton peaks are unresolved. Deuteron ENDOR at $g_\parallel$ of methyl-D_3-AdoMet in complex with $[4Fe-4S]^+$/PFL-AE, however, shows two well-resolved peaks corresponding to the ν_+ and ν_- resonances of the methyl deuterons of AdoMet (Figure 6) (10). The spectra collected at $g_\parallel$ (g=2.01) and $g_\perp$ (g=1.88) are shown in Figure 6. The breadth of the signal at $g_\perp$ corresponds to a proton coupling of ~6-7 MHz; this substantial coupling, approximately half that of water bound to low-spin heme (11), could not be explained by a remote binding of AdoMet allosterically perturbing the EPR spectrum of the $[4Fe-4S]^+$. At a minimum, this coupling requires that AdoMet lie close to the $[4Fe-4S]^+$ cluster. The field-dependence of the deuteron ENDOR spectra, which show a monotonic decrease in the breadth of the signal from $g_\perp$ to $g_\parallel$, is most consistent with the presence of a through-space dipolar interaction between the methyl deuterons and a single iron of the $[4Fe-4S]^+$ cluster. Modeling the outer features of the deuteron pattern as the interaction of a single deuteron with one iron site of the $[4Fe-4S]$ cluster leads to an estimated distance of ~3.0-3.8 Å (10).

^{13}C-ENDOR spectra obtained for $[4Fe-4S]^+$/PFL-AE/methyl-^{13}C-AdoMet are shown in Figure 7 and provide information consistent with that obtained for the deuterated samples described above (10). In both the $g_\parallel$ and $g_\perp$ spectra, doublets centered at the ^{13}C Larmor frequency are seen. Modeling the field-dependence of the ^{13}C spectra requires both isotropic and dipolar terms, the former providing evidence for a "local" interaction with spin density on ^{13}C. The local interaction requires that there be orbital overlap between AdoMet and an iron of the $[4Fe-4S]^+$ cluster in order to provide a pathway for delocalization of spin density onto AdoMet, and we have proposed that this overlap occurs via a close contact between the sulfonium sulfur of AdoMet and a μ_3-bridging sulfide of the $[4Fe-4S]$ cluster (Figure 8). Using the ^{13}C ENDOR data, we have estimated the distance between the methyl carbon and the closest iron of the $[4Fe-4S]^+$ cluster to be ~4-5 Å, consistent with the deuteron distance discussed above (10).

The experiments just described probed the catalytically competent and paramagnetic $[4Fe-4S]^+$ state of PFL-AE. However, after turnover and before re-reduction, the $[4Fe-4S]$ cluster presumably exists in the 2+ state. In order to probe the interaction of AdoMet with the oxidized $[4Fe-4S]^{2+}$ state of PFL-AE, we have used radiolytic cryoreduction. Labeled AdoMet was added to the $[4Fe-4S]^{2+}$ state of PFL-AE, the sample was frozen, and then at 77 K the sample was γ-irradiated to reduce the iron-sulfur clusters. The resulting ^2H and ^{13}C ENDOR spectra are shown in Figures 6 and 7, respectively (10). The results clearly demonstrate that AdoMet interacts with the oxidized $[4Fe-4S]^{2+}$ in substantially the same manner as with the $[4Fe-4S]^+$ state.

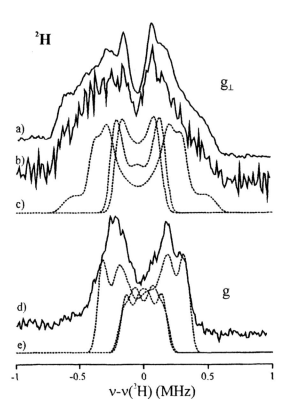

^{2}H

$g_\perp$

a)

b)

c)

d)

g

e)

-1   -0.5   0   0.5   1

$\nu - \nu(^2H)$ (MHz)

Figure 6. 35 GHz Mims pulsed-ENDOR spectra of PFL-AE with methyl-D₃-AdoMet: (a) and (d) are photoreduced samples, (b) is a cryoreduced sample, (c) and (e) are simulations.
(Reproduced from reference 14. Copyright 2002 American Chemical Society.)

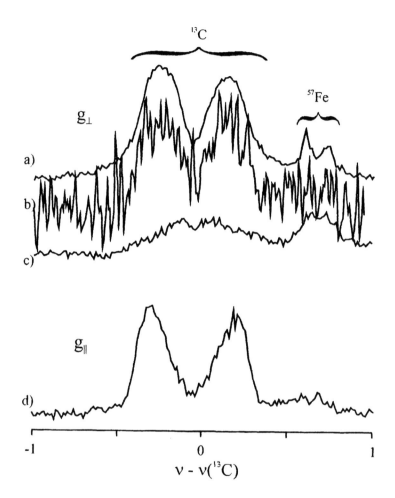

*Figure 7. 35 GHz Mims pulsed-ENDOR spectra of PFL-AE: (a) and (d) are
photoreduced samples in the presence of methyl-^{13}C-AdoMet, (b) is a
cryoreduced sample in the presence of methyl-^{13}C-AdoMet, and (c) is a
photoreduced sample in the presence of unlabeled AdoMet.
(Reproduced from reference 8. Copyright 2000 American Chemical Society.)*

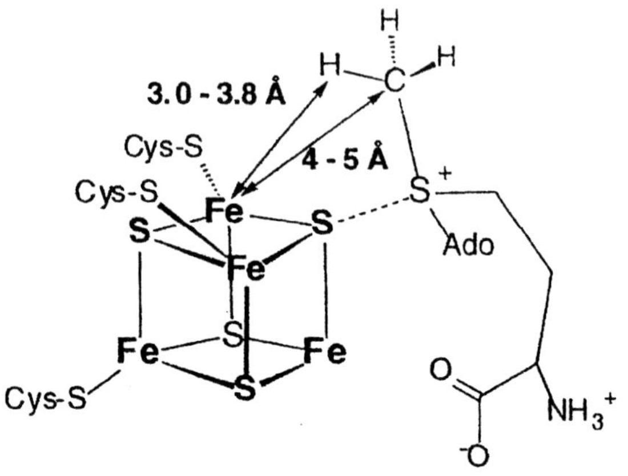

Figure 8. Interaction of AdoMet with the [4Fe-4S] cluster as determined by ^2H and ^{13}C ENDOR.

The Unique Iron Site in the Radical-SAM Enzymes

The radical-SAM enzymes have in common a three-cysteine motif which has been shown in several cases to provide thiol ligands to the iron-sulfur clusters of these enzymes. That only three cysteines are found in this motif, but a [4Fe-4S] cluster has been demonstrated to be the catalytically relevant cluster, suggests the involvement of a site-differentiated [4Fe-4S] cluster in catalysis. The unique iron of the site-differentiated cluster was expected to be of functional significance, since the three-cysteine motif is conserved throughout the superfamily. A number of possible roles for a unique site in a [4Fe-4S] cluster could be envisaged, including interaction of the unique site with the sulfonium sulfur or the 5'-carbon of AdoMet, or coordination of the unique iron by the amino acid moiety or the ribose hydroxyls of AdoMet. In fact, previous X-ray absorption studies on lysine 2,3-aminomutase, a member of the radical-SAM superfamily, provided evidence for a short distance between the product methionine sulfur and an iron of the cluster (12). Based on these results it was proposed that the sulfonium sulfur of AdoMet is in close proximity to an iron of the [4Fe-4S] of lysine aminomutase, and that this close interaction provides a pathway for electron transfer from the cluster to Adomet. However our model in which the methyl group of AdoMet is in close proximity to one of the irons of the cluster, and the sulfonium sulfur is close to, and has orbital overlap with, one of the μ_3-bridging sulfides of the cluster, appears to preclude direct interaction of the sulfonium with the unique iron site in PFL-AE (10).

Investigation of the Unique Site using Mössbauer Spectroscopy

In order to investigate the presence of a unique iron site in PFL-AE and to probe its interaction with substrate AdoMet, we undertook a dual iron isotope Mössbauer study (13). Protein containing natural abundance iron was oxidized to generate the $[3Fe-4S]^+$ form of the cluster, excess iron was removed by gel filtration, and then $^{57}Fe^{2+}$ and reductant were added to reconstitute only the unique site with the Mössbauer-active isotope. Alternatively, ^{57}Fe-enriched PFL-AE could be substituted at the unique site with natural-abundance iron using an analogous procedure. The results are shown in Figure 9. The quadrupole splitting and isomer shift of the ^{57}Fe substituted into the unique site are typical for iron in $[4Fe-4S]^{2+}$ clusters. Addition of AdoMet dramatically perturbs these parameters, with the increase in isomer shift suggesting an increase in coordination number and/or binding of a more ionic ligand to the iron (13). The change in the isomer shift is *not* consistent with coordination of a sulfonium sulfur to the unique site. Therefore, these results point to the presence of a unique site in the [4Fe-4S] of PFL-AE, and a role for the unique site in interacting with AdoMet, but point to an interaction different than that previously proposed for lysine 2,3-aminomutase.

Identification of the Role of the Unique Site Using ENDOR Spectroscopy

A possible interpretation of the Mössbauer results just described is that the amino acid moiety of AdoMet coordinates to the unique iron site in the PFL-AE/AdoMet complex. In order to examine this possibility, we undertook ENDOR studies of the PFL-AE/AdoMet complex with AdoMet labeled with ^{17}O or ^{13}C at the carboxylate or ^{15}N at the amino group (14). The results of these investigations are shown in Figure 10. The coupling to the ^{17}O- AdoMet is 12.2 MHz, which is essentially the same (after correction for electron-spin coupling) to the values obtained for coupling of ^{17}O-citrate to the [4Fe-4S] cluster of aconitase (15,16), and of labeled amino-cyclopropane carboxylate (ACC) coordinated to the non-heme iron of its oxidase ACCO (17). The ^{13}C hyperfine coupling in the presence of carboxylato-^{13}C-AdoMet (0.71 MHz) is also similar to that observed between the carboxylato-^{13}C of citrate and the [4Fe-4S] of aconitase (14-16). The PFL-AE-$[4Fe-4S]^+$ complex made with unlabeled AdoMet shows a broad feature extending to 8 MHz which disappears upon substition with^{15}N at the amino of AdoMet. The ^{15}N-AdoMet sample exhibits a new feature assigned to the v_+ peak of ^{15}N coupled to the $[4Fe-4S]^+$ cluster. This ^{15}N coupling is similar to those observed between ^{15}N-histidines and the Rieske [2Fe-2S] cluster (18) and to that of ACC bound to ACCO (17).

Together, these results support a model in which the AdoMet binds to the [4Fe-4S] cluster of PFL-AE via coordination of the amino and carboxylato groups of AdoMet to a unique iron of the cluster to form a classical five-member chelate ring (Figure 11). The unique iron site in PFL-AE therefore serves as an

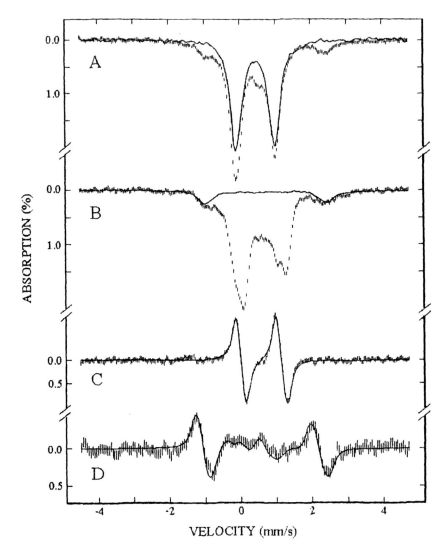

Figure 9. Mössbauer spectra of ^{56}Fe PFL-AE reconstituted with 57Fe and DTT in the absence (A) and presence (B) of AdoMet, both at 50 mT. The difference spectrum is shown in (C) and the difference spectrum at 8 T is shown in (D). The solid line in (A) is the experimental spectrum of $[4Fe-4S]^{2+}$ clusters in PFl-AE, and the solid line in (B) is the spectrum of a control sample containing only the reconstitution ingredients and AdoMet but without PFL-AE. (Reproduced from reference 8. Copyright 2000 American Chemical Society.)

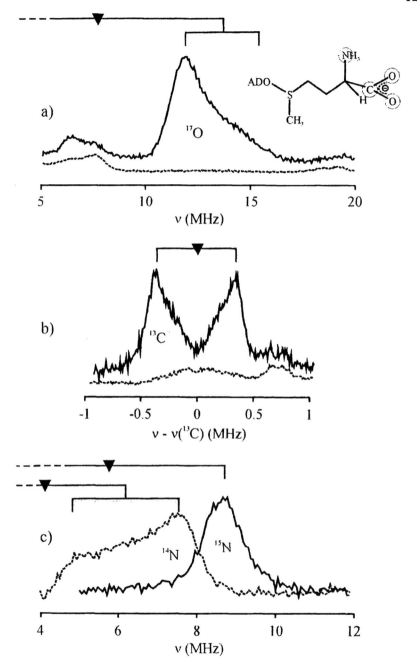

Figure 10. 35 GHz pulsed ENDOR spectra of PFL-AE with (a) ^{17}O-carboxylato, (b) ^{13}C-carboxylato, and (c) ^{15}N-amino-labeled AdoMet. (a) and (c) are Davies ENDOR, and (b) is Mims ENDOR. For further details see reference 14. (Reproduced from reference 9. Copyright 2000 American Chemical Society.)

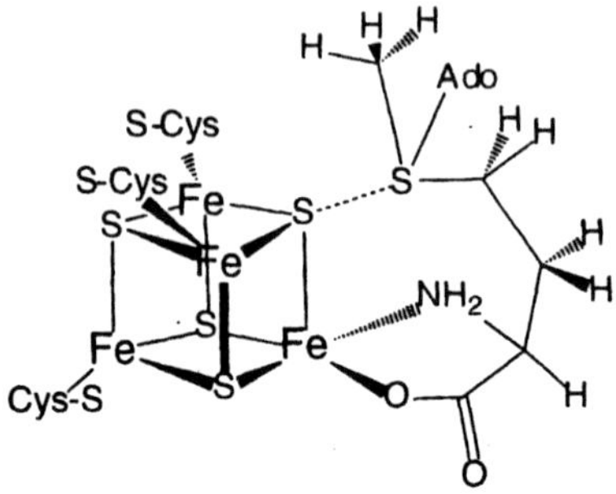

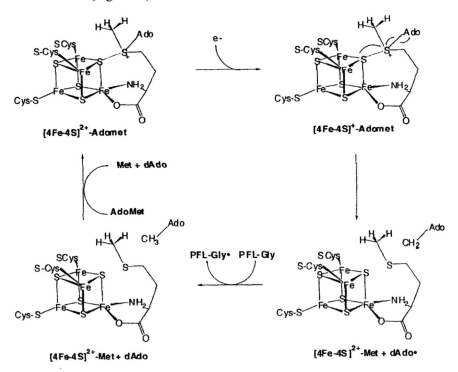

Figure 11. AdoMet interacts directly with the [4Fe-4S] cluster of PFL-AE, as determined by a combination of ENDOR and Mössbauer spectroscopy. (Reproduced from reference 9. Copyright 2000 American Chemical Society.)

anchor, holding AdoMet in the active configuration and thus allowing for the subsequent radical chemistry. Based on our evidence for orbital overlap between the sulfonium of AdoMet and a μ-3 bridging sulfide of the cluster, we propose that the radical chemistry is initiated via an inner-sphere electron transfer from the reduced cluster to AdoMet upon binding of the second substrate PFL (Figure 12).

Figure 12. Proposed mechanism for the activation of PFL by PFL-AE. The reduced [4Fe-4S]⁻ of PFL-AE reduces a coordinated AdoMet via inner-sphere electron transfer to generate a bound methionine and an intermediate adenosyl radical. The latter abstracts H• from PFL to generate the activated enzyme.

Acknowledgements

The work described was supported by the NIH (GM54608 to J.B.B., HL13531 to B.M.H., and GM47295 to B.H.H.).

Literature Cited

1. Cheek, J.; Broderick, J. B. *J. Biol. Inorg. Chem.* **2001**, *6*, 209-226.
2. Sofia, H. J.; Chen, G.; Hetzler, B. G.; Reyes-Spindola, J. F.; Miller, N. E. *Nucleic Acids Res.* **2001**, *29*, 1097-1106.
3. Wagner, A. F. V.; Frey, M.; Neugebauer, F. A.; Schäfer, W.; Knappe, J. *Proc. Natl. Acad. Sci. USA* **1992**, *89*, 996-1000.
4. Kessler, D.; Herth, W.; Knappe, J. *J Biol Chem* **1992**, *267*, 18073-9.
5. Broderick, J. B.; Duderstadt, R. E.; Fernandez, D. C.; Wojtuszewski, K.; Henshaw, T. F.; Johnson, M. K. *J. Am. Chem. Soc.* **1997**, *119*, 7396-7397.
6. Külzer, R.; Pils, T.; Kappl, R.; Hüttermann, J.; Knappe, J. *J. Biol. Chem.* **1998**, *273*, 4897-4903.
7. Broderick, J. B.; Henshaw, T. F.; Cheek, J.; Wojtuszewski, K.; Trojan, M. R.; McGhan, R.; Smith, S. R.; Kopf, A.; Kibbey, M.; Broderick, W. E. *Biochem. Biophys. Res. Commun.* **2000**, *269*, 451-456.
8. Krebs, C.; Henshaw, T. F.; Cheek, J.; Huynh, B.-H.; Broderick, J. B. *J. Am. Chem. Soc.* **2000**, *122*, 12497-12506.
9. Henshaw, T. F.; Cheek, J.; Broderick, J. B. *J. Am. Chem. Soc.* **2000**, *122*, 8331-8332.
10. Walsby, C. J.; Hong, W.; Broderick, W. E.; Cheek, J.; Ortillo, D.; Broderick, J. B.; Hoffman, B. M. *J. Am. Chem. Soc.* **2002**, *124*, 3143-3151.
11. Fann, Y.-C.; Gerber, N. C.; Osmulski, P. A.; Hager, L. P.; Sligar, S. G.; Hoffman, B. M. *J. Am. Chem. Soc.* **1994**, *116*, 5989-5990.
12. Cosper, N. J.; Booker, S. J.; Ruzicka, F.; Frey, P. A.; Scott, R. A. *Biochemistry* **2000**, *39*, 15668-15673.
13. Krebs, C.; Broderick, W. E.; Henshaw, T. F.; Broderick, J. B.; Huynh, B. H. *J. Am. Chem. Soc.* **2002**, *124*, 912-913.
14. Walsby, C. J.; Ortillo, D.; Broderick, W. E.; Broderick, J. B.; Hoffman, B. M. *J. Am. Chem. Soc.* **2002**, *124*, 11270-11271.
15. Kennedy, M. C.; Werst, M.; Telser, J.; Emptage, M. H.; Beinert, H.; Hoffman, B. M. *Proc. Natl. Acad. Sci. U.S.A.* **1987**, *84*, 8854-8858.
16. Werst, M. M.; Kennedy, M. C.; Beinert, H.; Hoffman, B. M. *Biochemistry* **1990**, *29*, 10526-10532.
17.. Rocklin, A. M.; Tierney, D. L.; Kofman, V.; Brunhuber, N. M. W.; Hoffman, B. M.; Christoffersen, R. E.; Reich, N. O.; Lipscomb, J. D.; Que, L., Jr. *Proc. Natl. Acad. Sci. U.S.A.* **1999**, *96*, 7905-7909.
18. Gurbiel, R. J.; Doan, P. E.; Gassner, G. T.; Macke, T. J.; Case, D. A.; Ohnishi, T.; Fee, J. A.; Ballou, D. P.; Hoffman, B. M. *Biochemistry* **1996**, *35*, 7834-3845.

Chapter 7

EPR and ENDOR Studies of [NiFe] Hydrogenase: Contributions to Understanding the Mechanism of Biological Hydrogen Conversion

Wolfgang Lubitz[1], Marc Brecht[2], Stefanie Foerster[2], Maurice van Gastel[1], and Matthias Stein[2]

[1]Max-Planck-Institut für Strahlenchemie, Stiftstraße 34–36, D–45470 Mülheim an der Ruhr, Germany
[2]Max-Volmer-Laboratorium für Biophysikalische Chemie, Fakultät für Mathematik und Naturwissenschaften, Technische Universität Berlin, Straße des 17, Juni 135, D–10623 Berlin, Germany

EPR experiments are performed on the paramagnetic intermediates of the enzymatic cycle of a [NiFe] hydrogenase in (frozen) solution and single crystals. For states Ni-A, Ni-B, Ni-C and Ni-L the g-tensor magnitudes and orientations were obtained. Pulse EPR and ENDOR techniques give information on the electron-nuclear hyperfine and nuclear quadrupole couplings. The data are compared with those derived from density functional calculations performed on a geometry-optimized model cluster for the active NiFe-center. Thereby detailed information of the electronic and geometrical structure of the intermediates is obtained. Based on these data a reaction mechanism is proposed for this enzyme.

Introduction

Hydrogenases catalyze the reversible heterolytic cleavage of molecular hydrogen $H_2 \rightleftharpoons H^+ + H^-$. The most commonly found enzyme (*1*) contains two subunits. A heterodinuclear NiFe center (active site) is present in the large subunit (Figure 1) and three iron-sulfur centers (electron transfer components) are found in the small subunit. The [NiFe] hydrogenase from *Desulfovibrio (D.) vulgaris* Miyazaki F studied in this work has been crystallized and high resolution structures are available for the oxidized and the reduced enzyme (*2,3*). Details of the related electronic structures are, however, still not fully understood (*4*).

The structure of the active site and its unusual ligation sphere is shown schematically in Figure 1. In the as-isolated (oxidized) state the Fe is six- and the Ni five-coordinated. The nickel is ligated by four cysteines, two of them act as bridging ligands between Ni and Fe. The iron, in addition, carries three small inorganic diatomic ligands. They were identified by FTIR spectroscopy as two CN^- and one CO ligand (*4-6*). For *D. vulgaris* Miyazaki F the presence of one SO has been proposed (*2,3*). In the oxidized enzyme a third bridging ligand between the nickel and the iron is present, whose chemical nature is still a matter of debate; it may be either an oxygen (*7*) or a sulfur species (*2,8*), although for *D. gigas* hydrogenase an oxygen has been fairly well proven (*9*). In the reduced enzyme this ligand is absent in the X ray structure (*3*). It can, however, not be excluded that a hydrogenic species is bound to the metal center.

In the catalytic cycle of hydrogenase various redox states occur and several of them are paramagnetic. These are termed Ni-A, Ni-B, Ni-C and can be studied by EPR spectroscopy (*10*). In particular, the Ni-C state is believed to be directly involved in the catalytic turnover and to carry the substrate hydrogen (*11,12*). This state is light-sensitive and can be converted reversibly at low temperatures into the paramagnetic Ni-L state, which is spectroscopically distinct from Ni-C (*4,10*). The enzyme is inhibited by CO which is believed to react directly with the NiFe center (*10*).

Although much work has been devoted to the investigation of the [NiFe] hydrogenases many open questions remain. These concern the formal oxidation states of the Ni and Fe in all steps of the reaction cycle, the electron spin and charge density distribution of the NiFe center, the effect of the protein environment, the exact binding site of the substrate hydrogen and, finally, the detailed mechanism of hydrogen conversion. In this paper we demonstrate how these problems can be approached by a combination of cw/pulse EPR/ENDOR techniques and density functional theory (DFT) calculations.

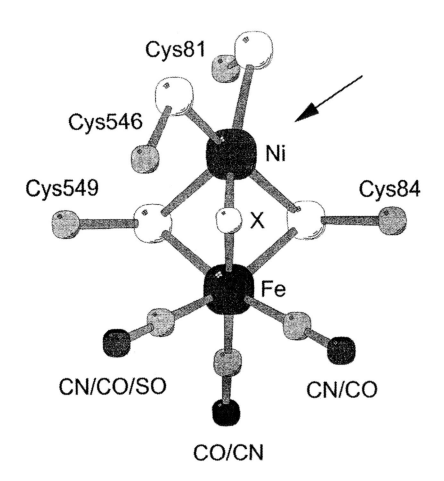

Figure 1. Active Site of the [NiFe] hydrogenase from D. vulgaris Miyazaki F (DvM). Two CN⁻ and one CO ligand at the Fe were determined by FTIR (4-6), one possible SO ligand was also proposed (2,3). In the oxidized enzyme X is O or S (7,2,8), in the reduced enzyme no bridging ligand is detected in the X-ray structure (3). The arrow indicates the open coordination site at the nickel. For the DFT calculations a model cluster was chosen with 40-42 atoms depending on the type of bridge X (see text). The cysteines were represented by ⁻S-CH₂CH₃, and two CN⁻ and one CO were chosen as Fe ligands.

Materials and Methods

Experiments have been performed predominantly on frozen solutions and single crystals of the "standard" [NiFe] hydrogenase of *D. vulgaris* Miyazaki F (DvM). The isolation and purification of the hydrogenase has been described earlier (*13*). In single crystals of DvM (*2*) the Ni-A/B as well as the Ni-C and Ni-L states were generated and angular dependent cw EPR experiments at X-band were performed as described (*14-17*). X-band pulse EPR and ESEEM measurements were done on a Bruker ESP 380 E spectrometer equipped with a Bruker dielectric ring cavity (ESP 380-1052 DLQ-H) and an Oxford CF 935 liquid helium cryostat. Pulse ENDOR experiments were carried out on the same instrument by use of a Bruker ESP 360-D-P ENDOR system. For the data analysis of the frozen solution and single crystal spectra, simulation and fit programs were used that were described previously (*17*). DFT calculations were performed by using the ADF program (SCM, Vrije Universiteit, Amsterdam), for details of the used methods see (*18-21*).

Results and Discussion

Determination of g-tensors

The EPR spectra of Ni-A/Ni-B, Ni-C and Ni-L obtained from frozen solutions of DvM are depicted in Figure 2. The three principal values of the rhombic g-tensors obtained from simulations of these powder-type EPR spectra (Table 1) are in good agreement with data obtained earlier on standard [NiFe] hydrogenases (*10*). They are typical for species with an $S = \frac{1}{2}$ ground state. Labeling of the enzyme with the magnetic nickel isotope ^{61}Ni ($I = \frac{3}{2}$) shows (*22,23*) that the major part of the unpaired spin is located at the nickel. Labeling with ^{57}Fe ($I = \frac{1}{2}$) does not lead to significant changes of the spectra (*24*). Apparently, the iron is always kept in a low-spin Fe^{II} ($S = 0$) state. Recent ^{57}Fe ENDOR experiments by Hoffman et al. (*9*) fully support this view.

For Ni-A, Ni-B and Ni-C a formal d_{z^2} ground state is indicated by the fact that the smallest g tensor component is close to $g_e \cong 2.00$ (*25*); in case of Ni-L g_z slightly deviates from g_e. The magnitude of the three g-tensor principal values of all states can be understood on the basis of a simple model in which only the d-orbitals of Ni are taken into account. The part of the wavefunction at the Ni is expected to make the dominant contribution to the g values since the spin orbit coupling parameter of Ni ($\varsigma_{Ni} = 649$ cm^{-1}) is much larger than those of the ligand

132

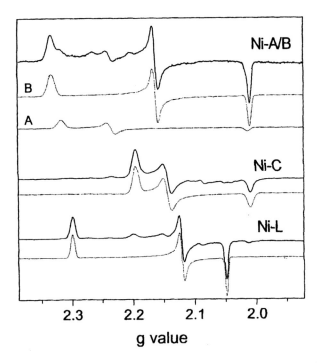

*Figure 2. Comparison of X-band EPR spectra in frozen solution of the DvM
[NiFe] hydrogenase in the oxidized state (mixture of Ni-A and Ni-B), the
reduced Ni-C and the light-induced Ni-L state. Simulations are shown as dotted
lines. The g-tensor principal values are collected in Table 1.*

atoms ($\varsigma_S = 191$ cm^{-1}, $\varsigma_O = 151$ cm^{-1}). Note that effectively no spin density is present at the iron (9,19,24). In our model, we describe the Ni character of the singly occupied molecular orbital by $\Psi_{Ni} = c_1\, d_{x^2-y^2} + c_2\, d_{z^2}$ and we assume that the energy of the other molecular orbitals with Ni(3d) character is lower by a quantity ΔE_{dd}. We can then derive an expression for the g-tensor using perturbation theory (perturbation parameter $k = \varsigma_{Ni}/\Delta E_{dd}$) and a formula adopted from Stone (26)

$$g^2 = \begin{pmatrix} 4g_ek\left(1+2c_2^2+\sqrt{3}c_1c_2\right) & 0 & 0 \\ 0 & 4g_ek\left(1+2c_2^2-\sqrt{3}c_1c_2\right) & 0 \\ 0 & 0 & 4g_ek\left(4-4c_2^2\right) \end{pmatrix} + g_e^2\,\mathbf{1} \qquad (1)$$

The respective graph shown in Figure 3 indicates that the g values for all paramagnetic redox states can, to first order, be described by a wavefunction that is dominated by the d_{z^2} orbital. The deviation from the pure d_{z^2} case (axial symmetry) leads to increasing rhombicity as observed here for the four paramagnetic nickel states. The strongest deviation is clearly observed for the Ni-L state, indicating the largest admixture of $d_{x^2-y^2}$, while d_{z^2} remains dominant.

In order to obtain more structural information on the active center the g-tensor principal axes have been obtained. Single crystal EPR experiments have been performed on all states of the DvM hydrogenase (14-16). The crystals are of orthorhombic symmetry (2,3) with space group $P2_12_12_1$ and four magnetically inequivalent proteins (sites) per unit cell (cf. Figure 4a). Rotation of a single crystal mounted in an EPR tube (Figure 4b) gave angular dependent EPR spectra as shown in Figure 4c for the as-isolated oxidized enzyme (mixture of Ni-A and Ni-B) (15). Reduction of the single crystal with H_2 in the presence of methyl viologen led to the Ni-C state and subsequent illumination at low temperature (77 K) to Ni-L (16). The reduction did not significantly change the crystallographic parameters (3). For Ni-C and Ni-L angular dependent EPR spectra were also obtained (not shown) (16).

In all cases a simultaneous fit of the effective g^2 values of all four sites yielded the g-tensor magnitudes as well as the orientations of the g-tensor principal axes in the crystal axis system a, b, c (Figure 4a). The correct assignment of the crystal axes for the investigated single crystal was determined beforehand by an X-ray diffraction experiment (14,15,17). In the next step, an assignment of the g-tensor axes to a particular site (one of the four possible sites in the unit cell) was made. This could be achieved based on molecular orbital and crystal field theory considerations and was supported by additional structural information derived from ENDOR and ESEEM experiments (see below). In Figure 5 the orientation of the g-tensor axes in the active site of all

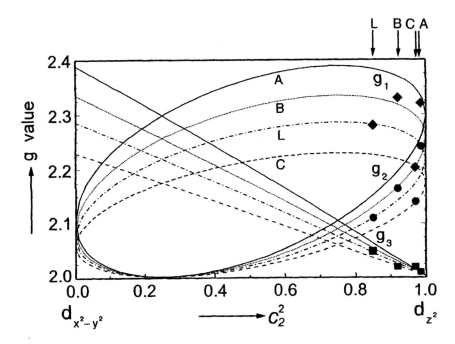

Figure 3. Dependence of the values of the g-tensor components on d orbital mixing (d_{z^2} and $d_{x^2-y^2}$) for four cases with different perturbation parameters $k = \varsigma_{Ni}/\Delta E_{dd}$ which were chosen to be: k(Ni-A) = 0.053 (solid line), k(Ni-B) = 0.045 (dotted line), k(Ni-C) = 0.030 (dashed line) and k(Ni-L) = 0.038 (dashed/dotted line). The experimental values of the four states are indicated by ♦ (g_1), • (g_2), and ■ (g_3).

paramagnetic states of the enzyme is depicted. Since the largest part of the spin density of the [NiFe] cluster is at the nickel (19) and the Ni ion determines the g anisotropy via its dominant spin orbit coupling (25), the displayed g-tensors are centered at this atom. It is interesting to note that the spatial structure of the complex and the g-tensor orientation are closely related since the interaction of the unpaired electron spin at the metal with the surrounding ligands largely determines the orientation of the axes (27). In the oxidized forms, the Ni coordination is close to a slightly distorted octahedron where one ligand position is empty.

In Ni-A and Ni-B the orientation of g-tensor axes is very similar. The smallest value g_3 is close to g_e and points to the *empty* coordination site of the Ni (see arrow in Figure 1), i.e., along the d_{z^2} orbital ($g_3 = g_z$). This agrees with a formal oxidation state of Ni(III) and a $3d_{z^2}$ ground state. In the reduced Ni-C state the magnitude and orientation of g_3 are similar to those of the oxidized states but the other two g values are smaller and the axes are exchanged. Most probably, the ground state has *not* been changed here, however, a difference in the in-plane ligand structure is more likely. The axes for Ni-L are not very much changed as compared to those of Ni-C, but the g-tensor values are different, indicating admixture of another state (see Figure 3), most probably $d_{x^2-y^2}$. This interpretation is in agreement with EXAFS experiments that also do not show changes of the metal oxidation states (12,28).

The elucidation of the correct formal oxidation states, the electronic ground states of the Ni, and the type of the bridging ligand X in all the intermediate states can be approached by density functional calculations.

DFT calculations on model structures for Ni-A, Ni-B, Ni-C, Ni-L and Ni-CO

Relativistic density functional calculations within the ZORA approximation were performed on a large (40-42 atoms) model cluster of the active site of [NiFe] hydrogenase (19). Based on the X-ray coordinates a full geometry optimization has been performed with the BP86 functional. Various oxidation states of the Ni and the Fe and different bridging ligands X were investigated. For all models the charge and spin density distributions were calculated (19,21) as well as the g-tensor magnitudes and orientations (16,20,21,29), which were compared with the experimental results (Table 1). For the oxidized state the models with a formal Ni(III) state and a μ-oxo and a μ-hydroxo bridge (X) between the Ni and Fe for Ni-A and Ni-B, respectively, give best agreement with the experimental data. The Ni-C state could also be reproduced best with a formal Ni(III) state. The calculations indicate that the bridge is occupied by a hydride (X = H⁻) that most probably originates from the heterolytic splitting of

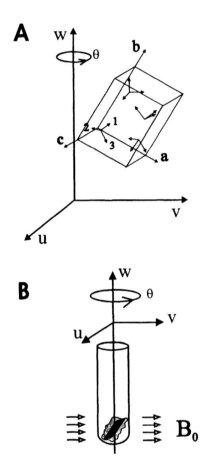

Figure 4. **A**: Orthorhombic unit cell with crystallographic axes a, b, c in the laboratory frame showing the four magnetically inequivalent sites with the g-tensor axes (1,2,3) indicated. **B**: Rotation by angle θ of single crystal mounted in an EPR tube in the magnetic field B₀ (laboratory frame u, v, w). **C**: Angular dependent cw-EPR spectra of the Ni-A/Ni-B state of the [NiFe] hydrogenase of DvM, taken in θ = 5° steps (for details see ref. 15). A maximum of four lines is obtained for each state (~ 30% Ni-A, ~ 70% Ni-B). The additional small lines are due to small crystallites with varying orientations in the tube.

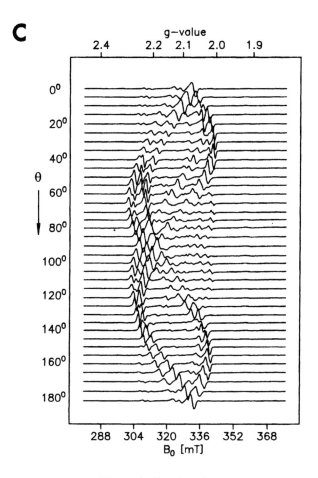

Figure 4. *Continued.*

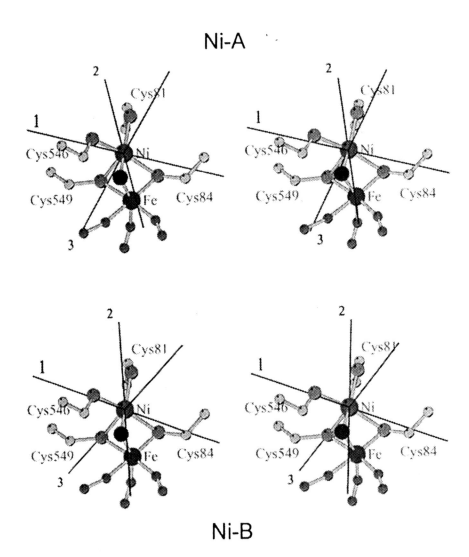

Figure 5. Stereo views of the orientations of the g-tensor axes of Ni-A, Ni-B, Ni-C and Ni-L as determined from EPR studies of single crystals of the DvM hydrogenase (14-16). The g-tensors are depicted in the respective X-ray crystallographic structures of the oxidized and the reduced form of the active center, taken from references (2) and (3), respectively. Note that the bridging position is not occupied in the reduced state (Figure 1). Adapted from (16).

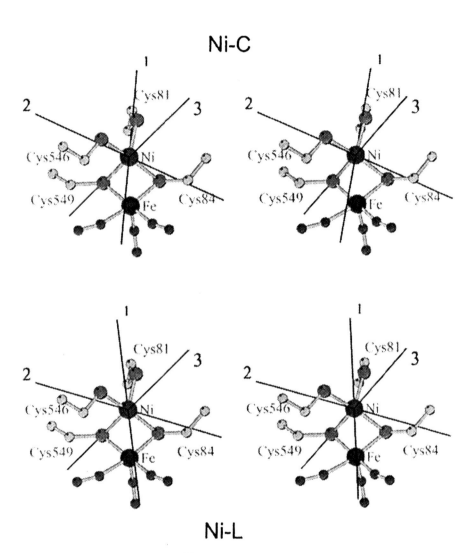

Figure 5. *Continued*

**Table I. Comparison of Experimental and Calculated g-Tensor Values
for Various States of [NiFe] Hydrogenase**

state	experiment			calculation		
	$g1$	$g2$	$g3$	$g1$	$g2$	$g3$
Ni-CO[a]	2.12	2.07	2.02	2.11	2.06	2.00
Ni-L	2.30	2.12	2.05	2.26	2.10	2.05
Ni-C	2.20	2.14	2.01	2.20	2.10	2.00
Ni-B	2.33	2.16	2.01	2.21	2.17	1.98
Ni-A[b]	2.32	2.24	2.01	(2.36)	(1.95)	(1.84)

[a] Exp. data from (39)

[b]For Ni-A with a µ-oxo bridge orbital degeneration occurs leading to large effects from spin-polarization on the calculated g-values (29), which are therefore not reliable in this approximation

hydrogen by the enzyme. This hydrogen is not detectable in X-ray experiments. Best agreement for Ni-L was obtained with a formal Ni(I) state and an empty bridging position. This indicates that upon illumination Ni-C looses a proton from the bridge, formally leading to a conversion Ni(III)H⁻Fe(II) $\xrightarrow{h\nu}$ Ni(I)Fe(II) + H⁺. Such a photodissociation has been proposed earlier based on ¹H ENDOR experiments by Hoffman et al. (12).

The DFT calculations suggest that the CO-inhibited state of the enzyme is formally also a Ni(I) state derived from Ni-L with the CO bound to the nickel (29). This agrees well with a recent X-ray crystallographic structure of the CO complex of the DvM hydrogenase (30). In all states the iron remains in a low spin Fe(II) state which is caused by the 3 strong inorganic ligands. The iron is thus not redox active and seems to be only indirectly involved in the catalytic process by providing a coordination site for the bridging ligand.

The calculated g-tensor principal values are given in Table 1. The models for Ni-B, Ni-C, Ni-L and Ni-CO give rise to g-values close to the experimental g-values. For Ni-A, the agreement is less good (29). The calculated g-tensor orientations agree well with the experimental ones, e.g. for Ni-C within a few degrees; only for Ni-L somewhat larger deviations are found (16). A population analysis of the DFT calculations supports a predominant d_{z^2} ground state for Ni-A, Ni-B and Ni-C. For Ni-L a larger contribution of the $d_{x^2-y^2}$ orbital is obtained, but the ground state is still predominantly $3d_{z^2}$ in this analysis (16). This explains why the g-tensor orientation of Ni-L and Ni-C are similar.

The calculated spin density distribution in the model clusters show that in Ni-A, Ni-B and Ni-C about 50%, and in Ni-L 75% of the spin is localized at the nickel, whereas only a vanishing amount is at the iron (16,21). The remaining spin density is distributed over the sulfurs of the cysteine ligands with the largest part on one of the bridging sulfurs. Results of calculations performed with the proposed SO ligand at the iron instead of CO/CN (2,3) show poor agreement with the experiment (19). Also, replacement of X = O^{2-} by X = S^{2-} in our calculation of the oxidized state yields poor results (19). An oxygen (O^{2-} or OH⁻) ligand in the Ni-A state has recently been directly observed by ¹⁷O-ENDOR experiments (9).

ENDOR and ESEEM experiments

Further corroboration of the proposed structural models for the paramagnetic states of the [NiFe] hydrogenase comes from application of advanced EPR techniques that are able to resolve the electron-nuclear hyperfine (hf) and nuclear quadrupole (nq) interactions of the magnetic nuclei. Such experiments have been performed in our laboratory, e.g., on the Ni-A and Ni-B states in single crystals (*21,32*) and the other states in frozen solutions (*31,33*). An example is given in Figure 6 that shows the field-swept echo-detected X-band EPR of Ni-A/B in single crystals, which allows species-selective and site-selective experiments to be performed. This is illustrated for a Davies Pulse ENDOR and a 3-Pulse ESEEM experiment in Figure 6 (right).

From the ENDOR spectra three ^{1}H hf tensors could be deduced for Ni-B and assigned to the CH_2 group of the bridging cysteine carrying the large spin density and to the exchangeable OH proton in the bridge (*32*). The ESEEM experiments (*31*) yielded one ^{14}N nq tensor that could be assigned by experiments on a genetically modified related system and concomitant DFT calculations to a histidine residue in the surrounding of the [NiFe] cluster that forms a hydrogen bond to the sulfur of one of the bridging cysteines (*34*). These experiments support the calculated spin density distribution described above.

Pulse EPR and ENDOR measurements on the Ni-C state of DvM are of particular importance since this species is supposed to carry the substrate hydrogen, which can be identified by H/D exchange and should be photolabile (*12*). Such experiments turned out to be difficult since in the reduced state spin coupling of Ni-C with the proximal reduced [4Fe-4S] cluster (*35*) leads to very fast relaxation and echo decays. The experiments have therefore been performed on a related (regulatory) hydrogenase (RH) from *Ralstonia (R.) eutropha*, which gives typical Ni-C EPR signals but does not couple to an Fe-S cluster (*36*). Since single crystals of the RH are not available the measurements were performed on frozen solutions using orientation-selection ENDOR (*37,33*). The crucial H/D exchange experiment led us to identify one exchangeable proton that, based on magnitude and orientation of the anisotropic hf tensor of (+20, −6, −14 MHz, a_{iso} = −4 MHz) could be assigned to the hydride in the bridge between Ni and Fe (*31*). This assignment has become possible due to the previously assigned g-tensor axes in Ni-C (*16*). A DFT calculation of this hf tensor in Ni-C yields (+18.6, −7.5, −11.1 MHz, a_{iso} = −8.7 MHz) (*20*) in good agreement with the ENDOR analysis.

In the exchanged Ni-C sample the deuterium in the bridge should be detectable in the ^{2}H ENDOR spectrum. Such experiments are, however, difficult to perform at X-band frequencies due to the small ^{2}H Larmor frequency and the reduced hfc. An alternative approach is to perform an ESEEM experiment. Figure 7a shows a 4-Pulse ESEEM (HYSCORE) spectrum of Ni-C in the RH activated with D_2 in D_2O. The simulation of this powder-type spectrum yields a ^{2}H hf tensor that agrees well with that obtained from the ^{1}H ENDOR analysis (*31*).

Conversion of this Ni-C state to Ni-L by low temperature illumination removes this coupling from the HYSCORE spectrum indicating that the respective deuteron is indeed photodissociated from the complex (Figure 7b). It is interesting to note that in Ni-L a ^{2}H hfc can still be detected but it is about one order of magnitude smaller than in Ni-C. Annealing of the sample at 200 K in the dark fully restores the Ni-C signal showing the reversibility of the process. It is suspected that the proton is transferred to a nearby amino acid.

Reaction mechanism

In summary, the paramagnetic states of the active center of the [NiFe] hydrogenase can be formulated as follows. Ni-A is best described as a μ-oxo bridged NiFe-cluster, Ni-B is μ-hydroxo bridged. In both oxidized states the nickel is in its formal Ni(III) oxidation state. Ni-A and Ni-B are viewed as states prior to catalytic activity that differ in their activation rates. The bridging ligand OH$^-$ in Ni-B is more easily lost (e.g., via protonation and water release) and replaced by a hydride bridge in Ni-C. The Ni-C form is also in a formal Ni(III) state. The heterolytic dissociation of H$_2$ must already have occurred before Ni-C formation. Ni-L is derived from Ni-C by photodissociation of the bridging atom, here H$^+$ release leads to a formal Ni(I) oxidation state. The inhibited state Ni-CO is obtained from the Ni-L precursor upon axial CO binding at the nickel.

Based on the previously determined redox steps of the enzyme and their pH-dependence (*38*) and the evolving knowledge about the structure of the paramagnetic intermediates a reaction mechanism can be proposed (Figure 8). In a first step, the bridging oxygen of the oxidized enzyme is protonated, leading to a water molecule bound to the complex. Subsequently, an electron enters the site and leads to the diamagnetic Ni-Si state. Then the heterolytic H$_2$ splitting occurs, leaving the hydride in the bridging position between Ni and Fe (Ni-C) and releasing the proton as H$_3$O$^+$ via the liberated bridging ligand or via a proton channel. The fully reduced state Ni-R is again diamagnetic and is thought to carry an extra proton either at a terminal cysteine or at the nickel. In this model the catalytic cycle involves the states Ni-SIa, Ni-C and Ni-R as indicated in Figure 8. The development of a more detailed mechanism based on various spectroscopic data and DFT calculations, which also include the immediate protein surrounding of the catalytic center, is currently under way in our laboratory.

Acknowledgement

The authors are grateful to Prof. Y. Higuchi and H. Ogata (Himeji University, Japan) for providing the hydrogenase single crystals. The work on the regulatory hydrogenase was done in collaboration with the group of Prof. B.

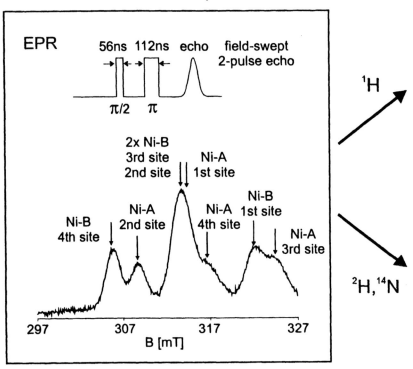

Figure 6. Left: Field-swept two-pulse echo-detected EPR spectrum at X-band of an arbitrarily oriented oxidized single crystal of the [NiFe] hydrogenase of DvM; six of the eight possible lines for Ni-A and Ni-B are resolved.

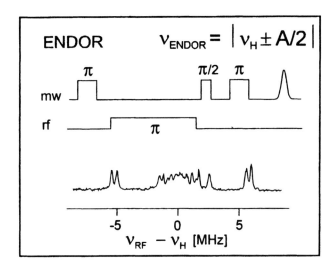

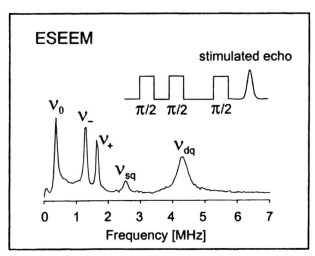

Right: Davies Pulse ^{1}H-ENDOR spectrum of Ni-B for one specific site and rotation angle θ (32); 3-Pulse electron spin echo envelope modulation (ESEEM) spectrum of Ni-B for a specific site and angle (31). Spectra obtained for different rotation angles yield magnitude and orientation of the hyperfine and quadrupole coupling constants (see text).

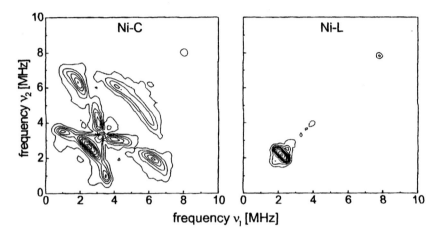

Figure 7. Hyperfine sublevel correlation (HYSCORE) spectra in the deuterium range for Ni-C and Ni-L recorded at g_2 for H/D exchanged frozen samples of the RH of R. eutropha. Note that the pattern yielding the ^2H hfc detected for Ni-C disappears from the spectrum upon conversion to the Ni-L state using light. The same effect is seen for the other g orientations (31).

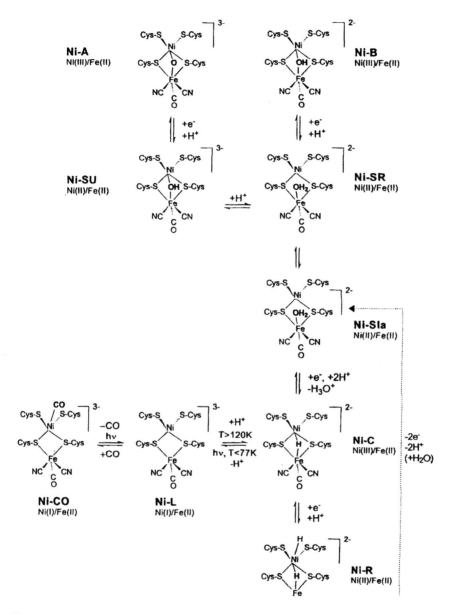

Figure 8. Reaction mechanism proposed for the activation and catalytic cycle of [NiFe] hydrogenase based on the known redox states of the enzyme (38) and the structures of intermediates determined from spectroscopic data and DFT calculations (see text). Note that the Ni-SR to Ni-SIa transition involves a structural change triggend by approach of hydrogen.

148

Friedrich (Humboldt University, Berlin). We are also grateful to our former collaborators Dr. Ch. Geßner and Dr. O. Schröder who were involved in the early stages of the project, and to Prof. E. J. Baerends and Dr. E. van Lenthe (Free University, Amsterdam) for their help with the ADF program. This work was supported by Deutsche Forschungsgemeinschaft (SfB 498, TP C2) and European Union (COST Action 841).

References

1. Vignais, P. M.; Billoud, B.; Meyer, J. *FEMS Microbiol. Rev.* **2001**, *25*, 455.
2. Higuchi, Y.; Yagi, T.; Yasuoka, N. *Structure* **1997**, *5*, 1671.
3. Higuchi, Y.; Ogata, H.; Miki, K.; Yasuoka, N.; Yagi, T. *Structure* **1999**, *7*, 549.
4. Albracht, S. P. J.; in: *Hydrogen as a Fuel;* Cammack, R.; Frey, M.; Robson, R. Eds.; Taylor & Francis: London and New York, **2001**, *Chapt. 7*, p 110-158.
5. Happe, R. P.; Roseboom, W.; Pierik, A. J.; Albracht, S. P. J.; Bagley, K. A. *Nature* **1997**, *385*, 126.
6. de Lacey, A. L.; Hatchikian, E. C.; Volbeda, A.; Frey, M.; Fontecilla-Camps, J. C.; Fernandez, V. M. *J. Am. Chem. Soc.* **1997**, *119*, 7181.
7. Volbeda, A.; Garcin, E.; Piras, C.; de Lacey, A. L.; Fernandez, V. M.; Hatchikian, E. C.; Frey, M.; Fontecilla-Campfs, J.-C. *J. Am. Chem. Soc.* **1996**, *118*, 12989.
8. Matias, P. M.; Soares, C. M.; Saraiva, L. M.; Coelho, R.; Morais, J.; Le Gall, J.; Carrondo, M. A. *J. Biol. Inorg. Chem.* **2001**, *6*, 63.
9. Carepo, M.; Tierney, D. L.; Brondino, C. D.; Yang, T. C.; Pamplona, A.; Telser, J.; Moura, I.; Moura, J. J. G.; Hoffman, B. M. *J. Am. Chem. Soc.* **2002**, *124*, 281.
10. Albracht, S. P. J. *Biochim. Biophys. Acta* **1994**, *1188*, 167.
11. Fan, C.; Teixeira, M.; Moura, J.; Moura, I.; Huynh, B. H.; LeGall, J.; Peck, H. D. Jr; Hoffman, B. M. *J. Am. Chem. Soc.* **1991**, *113*, 20.
12. Whitehead, J. P.; Gurbiel, R. J.; Bagyinka, C.; Hoffman, B. M.; Maroney, M. J. *J. Am. Chem. Soc.* **1993**, *115*, 5629.
13. Yagi, T.; Kimura, K.; Daidoji, H.; Sakai, F.; Tamura, S.; Inokuchi, H. *J. Biochem. (Tokyo)* **1976**, *79*, 661.
14. Geßner, C.; Trofanchuk, O.; Kawagoe, K.; Higuchi, Y.; Yasuoka, N.; Lubitz, W. *Chem. Phys. Lett.* **1996**, *256*, 518.
15. Trofanchuk, O.; Stein, M.; Geßner, C.; Lendzian, F.; Higuchi, Y.; Lubitz, W. *J. Biol. Inorg. Chem.* **2000**, *5*, 36.
16. Foerster, S.; Stein, M.; Brecht, M.; Ogata, H.; Higuchi, Y.; Lubitz, W. *J. Am. Chem. Soc.* **2002**, *125*, 83.
17. Geßner, C., doctoral thesis, Technische Universität Berlin, 1996.

18. Stein, M.; van Lenthe, E.; Baerends, E. J.; Lubitz, W. *J. Phys. Chem. A* **2001**, *105*, 416.
19. Stein, M.; Lubitz, W. *Phys. Chem. Chem. Phys.* **2001**, *3*, 2668.
20. Stein, M.; Lubitz, W. *Phys. Chem. Chem. Phys.* **2001**, *3*, 5115.
21. Stein, M., doctoral thesis, Technische Universität Berlin, 2002.
22. Albracht, S. P. J.; Graf, E. G.; Thauer, R. K. *FEBS Lett.* **1982**, *140*, 311.
23. Moura, J. J. G.; Moura, I.; Huynh, B. H.; Krüger, H. J.; Teixeira, M.; DuVarney, R. G.; DerVartanian, D. V.; Xavier, A. V.; Peck, H. D. Jr; LeGall, J. *Biochem. Biophys. Res. Commun,* **1982**, *108*, 1388.
24. Huyett, J. E.; Carepo, M.; Pamplona, A.; Franco, R.; Moura, I.; Moura, J. J. G.; Hoffman, B. M. *J. Am. Chem. Soc.* **1997**, *119*, 9291.
25. Salerno, J. C. in: *Bioinorganic Chemistry of Nickel;* Lancaster, J. R. Jr., Ed.; VCH New York: VCH Verlagsgesellschaft mbH, Weinheim, **1988**; *Chapt. 3*, p 53-71.
26. van Gastel, M.; Canters, G. W.; Krupka, H.; Messerschmidt, A.; de Waal, E.; Warmerdam, G. C. M.; Groenen, E. J. J. *J. Am. Chem. Soc.* **2000**, *121*, 2322.
27. Moura, J. J. G.; Teixeira, M.; Moura, I.; LeGall, J. in: *The Bioinorganic Chemistry of Nickel.* Lancaster, J. R. Jr. Ed.; VCH, New York: VCH Verlagsgesellschaft mbH, Weinheim, **1988**, *Chapt. 9*, p 191-226.
28. Davidson, G.; Choudhury, S. B.; Gu, Z.; Bose, K.; Roseboom, W.; Albracht, S. P. J.; Maroney, M. J. *Biochemistry* **2000**, *39*, 7468.
29. Stein, M.; van Lenthe, E.; Baerends, E. J.; Lubitz, W. *J. Am. Chem. Soc.* **2001**, *123*, 5839.
30. Ogata, H.; Mizoguchi, Y.; Mizuno, N.; Miki, K.; Adachi, S.; Yasuoka, N.; Yagi, T.; Yamauchi, O., Hirota, S. Higuchi, Y. *J. Am. Chem. Soc.* **2002**, *124*, 11628.
31. Brecht, M., doctoral thesis, Technische Universität Berlin, 2001.
32. Stein, M.; Trofanchuk, O.; Brecht, M.; Lendzian, F.; Bittl. R.; Higuchi, Y.; Lubitz, W. 2003, in preparation.
33. Geßner, Ch.; Stein, M.; Albracht, S. P. J.; Lubitz, W. *J. Biol. Inorg. Chem.* **1999**, *4*, 379.
34. Buhrke, T.; Brecht, M.; Lubitz, W.; Friedrich, B. *J. Biol. Inorg. Chem.* **2002**, *7*, 897.
35. Dole, F. Medina, M.; More, C. Cammack, R.; Bertrand, P.; Guigliarelli, B. *Biochemistry* **1996**, *35*, 16399.
36. Bernhard, M.; Buhrke, T. Bleijlevens, B.; DeLacey, A. L.; Fernandez, V. M.; Albracht, S. P. J.; Friedrich, B. *J. Biol. Chem.* **2001**, *276*, 15592.
37. Hoffman, B. M.; DeRose, V. J.; Doan, P. E.; Gurbiel, R. J.; Houseman, A. L. P.; Telser, J. in: *Biological Magnetic Resonance;* Berliner, L. J.; Reuben, J., Eds.; Plenum Press, New York and London, **1993**, p 151-218.
38. Cammack, R.; Patil, D. S.; Hatchikian, E. C.; Fernandez, V. M. *Biochim. Biophys. Acta,* **1987**, *912*, 98.
39. Happe, R. P.; Roseboom, W.; Albracht, S. P. J. *Eur. J. Biochem.,* **1999**, *259*, 602.

Chapter 8

Q-Band ENDOR Studies of the Nitrogenase MoFe Protein under Turnover Conditions

Substrate–Inhibitor-Binding to and Metal–Ion Valencies of the FeMo-Cofactor

Hong-In Lee[1,*], Linda M. Cameron[2], Jason Christiansen[3], Patricia D. Christie[4], Robert C. Pollock[5], Rutian Song[5], Morten Sørlie[2], W. H. Orme-Johnson[4], Dennis R. Dean[3,*], Brian J. Hales[2,*], and Brian M. Hoffman[5,*]

[1]Department of Chemistry Education, Kyungpook National University, Daegu 702–701, Korea
[2]Department of Chemistry, Louisiana State University, Baton Rouge, LA 70803
[3]Department of Biochemistry, Virginia Polytechnic Institute and State University, Blacksburg, VA 24061
[4]Department of Chemistry, Massachusetts Institute of Technology, Cambridge, MA 02139
[5]Department of Chemistry, Northwestern University, Evanston, IL 60208

The resting state of nitrogenase shows an $S = 3/2$ electron paramagnetic resonance (EPR) signal originating from the FeMo-cofactor, the active site of the enzyme. This signal disappears under turnover conditions. It has long been suggested that substrates bind to the FeMo-cofactor under these conditions and there has been much discussion of possible binding modes of substrates. However, these are put forth without any direct evidence as to how substrates interact with the protein-bound cofactor. In the present study, we first generate EPR spectroscopic signatures of inhibitor- or substrate-bound states of the cofactor by turning the enzyme over under CO or C_2H_2 atmosphere. Secondly, the electron spins of these EPR-active states of the cofactor are used as reporter groups to identify and characterize the CO- and C_2H_2-bound states. This is accomplished by carrying out electron-nuclear double resonance (ENDOR) measurements which detect hyperfine interactions between the electron spins and the surrounding nuclear spins. We demonstrate how to analyze orientation-selective ^{13}C and ^{57}Fe ENDOR spectra obtained from the inhibitor- and substrate-bound states of the MoFe protein to derive detailed information of the binding modes of these molecules to and the charge of the cofactor.

Nitrogenase plays the key role in the first step of the nitrogen cycle in Nature, in which dinitrogen is reduced to ammonia (*1-3*). The reduction process requires electrons and the hydrolysis of ATP is coupled to the process (*4*). Nitrogenase accommodates this chemical process by providing two proteins: the Fe protein, which functions as an electron source and ATP binding site and the MoFe protein, which functions as a substrate binding and reduction site (*5*). The Fe protein contains a [4Fe-4S] cluster, the starting point of electron supply for the nitrogen fixation process in nitrogenase (*5*). The MoFe protein contains a [8Fe-7S] cluster (P-cluster), the mediating site of electron flow, and a [Mo-7Fe-9S-homocitrate] cluster (FeMo-cofactor), the substrate binding and reduction site (*5*). X-ray crystallographic studies of nitrogenases extracted from *Azotobacter vinelandii* (*Av*), *Klebsiella pneumoniae* (*Kp*), and *Clostridium pasteurianum* (*Cp*) revealed the structures of the proteins and the metal-sulfur clusters in great detail (*6-11*), greatly increasing our understanding of biological nitrogen fixation. The FeMo-cofactor can be viewed as two metal cubanes (Mo3Fe3S and 4Fe3S) linked by three μ-2 sulfides (**Figure 1**). In spite of the wealth of structural data, the mechanism of substrate reduction remains largely unclear. There are four major outstanding questions yet to be completely solved (*4*): the role of ATP; the electron pathways in the Fe protein, in the MoFe protein, and between the proteins; the binding modes of the substrates to the FeMo-cofactor; and the redox levels of the metalloclusters. All these questions are of course intimately related.

Among these questions, the most interesting question is the third and the fourth questions combined: how do the substrates or reaction intermediates bind to the FeMo-cofactor, and at which redox level? Resting-state MoFe protein exhibits an $S = 3/2$ electron paramagnetic resonance (EPR) signal with $g = [4.33$

152

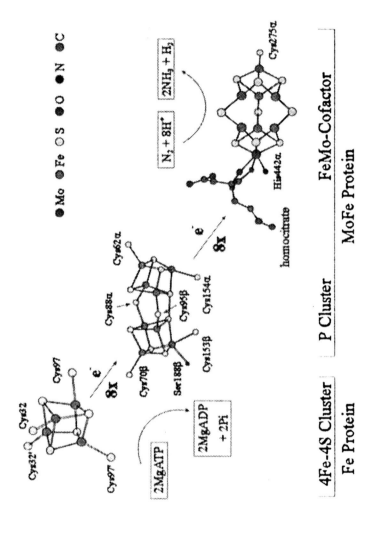

4Fe-4S Cluster	P Cluster	FeMo-Cofactor
Fe Protein	MoFe Protein	

Figure 1. Metal-sulfur clusters in nitrogenase Fe and MoFe proteins (6-11). Reduction of one dinitrogen molecule requires eight electrons and hydrolysis of ATP.

3.77 2.01] originating from the FeMo-cofactor (*12-18*). Many methods of EPR spectroscopy have been intensively employed to address the above question and to investigate the electronic structure and the characteristics of the constituent and surrounding nuclei of the cofactor. However, the resting state is known not to be a substrate-binding state (*1-5*), therefore, the study of this state provides no information as to the mode of substrate binding. As we will show, EPR-active states can be generated that are directly relevant to substrate binding during catalysis.

Determination of the redox states of the cofactor during catalysis requires assignment of the metal-ion valencies in the cofactor. ^{57}Fe and ^{95}Mo electron-nuclear double resonance (ENDOR) studies on the $S = 3/2$ EPR signal of the resting-state MoFe protein revealed five magnetically distinct iron sites and one diamagnetic molybdenum site (*19-21*). However, the observed ^{57}Fe hyperfine coupling constants could not at that time, nor even now, be linked to the charge of each Fe ion.

The resting state $S = 3/2$ EPR signal disappears under turnover conditions, indicating electron flow to the FeMo-cofactor. It has long been suggested that substrates bind to the FeMo-cofactor under these conditions and there has been much discussion of possible modes of substrates binding the cofactor (*9,22-26*). However, these are put forth without any direct evidence as to how substrates interact with either protein-bound cluster. The main reason for this difficulty is that, until recently, there have been no significant spectroscopic signatures associated with a substrate-bound form of the enzyme.

When the MoFe protein turns over in the presence of certain inhibitors or substrates, new EPR signals are observed while the resting-state $S = 3/2$ signal disappears (*27-33*). These new EPR-active states allow EPR and related spectroscopies to scrutinize substrate- or inhibitor-bound nitrogenase. The present paper describes how ENDOR methodologies are applied to explore the binding modes of inhibitors or substrates to the cofactor, and to determine the cofactor charge in the turnover states of the MoFe protein under CO or C_2H_2 atmosphere (*31,34-37*).

Sample Preparation

Turnover state of MoFe protein under CO: MoFe protein was prepared by standard procedures described elsewhere (*38*). The CO-bound forms of the turnover state MoFe protein were prepared by adding Fe protein (in 25 mM Tris, 0.35M NaCl, pH = 7.4) to a solution of MoFe protein that had been equilibrated with CO at a partial pressure of 0.08 atm (denoted lo-CO) or 0.5 atm (denoted hi-CO) in a serum-capped Wheaton vial (*31*). The initial concentrations of the proteins and reagents in the turnover mixture were as follows: 0.28 mM MoFe protein, 0.14 mM Fe protein, 50 mM MgCl$_2$, 100 mM Na$_2$ATP, 300 mM Na(phosphocreatine), 100 mM HEPES, 2 mg/ml creatine kinase, and 100 mM

$Na_2S_2O_4$. The reaction was allowed to proceed at 25°C for 5 minutes, then rapidly frozen by immersing the sample tube into liquid nitrogen.

   Turnover state of MoFe protein with ^{57}Fe-labeled metal clusters under CO: MoFe protein, in which the metal clusters had either natural isotopic abundance or uniform labelling with ^{57}Fe, were prepared from *Azotobacter vinelandii (Av)* by standard procedures (*38*). The selectively labelled proteins were prepared through the additional use of protein isolated from the *Av* UW45 strain, which is completely unable to make the FeMo-cluster. To make the MoFe protein selectively labelled with ^{57}Fe in the FeMo-cofactor, *Av* UW45 was grown on natural-abundance Fe and then reconstituted with ^{57}Fe-enriched M-center extracted from protein isolated from *Av* grown on ^{57}Fe. The MoFe protein selectively labelled in the P-cluster was prepared by inverting the procedure. This protocol eliminates all possibilities of label-scrambling (*39*). The CO-bound forms of turnover-state MoFe proteins were prepared as in the above section (*31,34-36*).

   Turnover state of α-Gln^{195} MoFe protein under C_2H_2: The α-Gln^{195} MoFe protein was purified from *Azotobacter vinelandii* strain DJ997. Cells were grown at 30°C with pressurized sparging (80 L/min at 5 psi) and 125 rpm agitation in a 150-L custom-built fermenter (W. B. Moore, Inc. Easton, PA) in modified Burk medium containing 10 mM urea as a the sole nitrogen source (*40*). After reaching a density of 220 Klett units (red filter), the cells were derepressed for *nif* gene expression by concentration (6-fold) using a custom-built AG Technologies tangential-flow concentrator and resuspended in Burk medium with no added nitrogen. All protein manipulations were performed under anaerobic conditions maintained using either a Schlenk apparatus (*38*) or an anaerobic glovebox. The α-Gln^{195} MoFe protein was purified using a combination of immobilized metal-affinity chromatography (IMAC) and DEAE-Sepharose anion exchange chromatography as previously described (*41*). Protein was quantified using a modified biuret assay with bovine serum albumin as the standard (*42*) and purity was monitored by SDS-PAGE electrophoresis (*43*). For 360 g of wet-weight cells, purification yielded approximately 1.1 g of purified α-Gln^{195} MoFe protein. Nitrogenase assays were performed as previously described (*44,45*) and activities for the α-Gln^{195} MoFe protein used in the current work were similar to those previously reported (*44*). Turnover samples consisted of 20 μM Fe protein, 100 μM α-Gln^{195} MoFe protein, 0.1 atm C_2H_2, 10 mM ATP, 25 mM $MgCl_2$, 20 mM $Na_2S_2O_4$, and 50 mM TES-KOH pH 7.4. Prior to turnover, the above mixture (without the Fe protein) was preincubated for 20 min at 30°C with 0.1 atm of the appropriate experimental gas (i.e. C_2H_2, C_2D_2, etc.) under a 1.0 atm of Ar. After initiation of turnover by the addition of Fe protein, a 100 μL sample was transferred to a Q-band ENDOR tube where it was rapidly frozen in liquid N_2. The interval between turnover initiation and final freezing was approximately 2 min. For experiments performed in D_2O, protein was first exchanged into buffered D_2O (initially 99.8 % in 38 mM TES-KOD at pD 7.4) to yield a final solution of about 95% D_2O. The ATP regenerating solution for these experiments was prepared in 99.8% D_2O for a final concentration (including protein) in the turnover mixture of about 98% D_2O.

ENDOR Spectroscopy

Continuous wave (CW) Q-band (35 GHz) ENDOR spectra were recorded on a modified Varian E-110 spectrometer equipped with a helium immersion dewar. The spectra were obtained in dispersion mode using 100 kHz field modulation under 'rapid passage' conditions (46-48). Spectra shown represent the absorption-shape spectrum, not the derivative. The bandwidth of radiofrequency (RF) was sometimes broadened to 100 kHz to improve the signal to noise ratio (49). Q-band pulsed-ENDOR data were collected on a spectrometer described in detail elsewhere (50). A stimulated-echo microwave pulse sequence, $\pi/2$-τ-$\pi/2$-T-$\pi/2$, and a modified stimulated-echo microwave pulse sequence, $\pi/2$-τ-$\pi/2$-T-$\pi/2$-t-π, were employed for Mims ENDOR (51-53) and Refocused-Mims (ReMims) ENDOR (54) experiments, respectively. The $\pi/2$ and π pulses were 28 and 56 ns. For a single orientation of a paramagnetic center, the first-order ENDOR spectrum of a nucleus with $I = 1/2$ in a single paramagnetic center consists of a doublet with frequencies given by (55):

$$\nu_{\pm} = |\nu_N \pm A/2| \tag{1}$$

Here, ν_N is the nuclear Larmor frequency and A is the orientation-dependent hyperfine coupling constant of the coupled nucleus. The doublet is centered at the Larmor frequency and separated by A when $\nu_N > |A/2|$, or centered at $|A/2|$ and separated by $2\nu_N$ when $|A/2| > \nu_N$. The full hyperfine tensor of a coupled nucleus can be obtained by analyzing a '2-D' dataset of ENDOR spectra collected across the EPR envelope, as described elsewhere (51,56-59).

Results and Discussion

CO Binding to the FeMo-Cofactor in CO-inhibited Nitrogenase: When nitrogenase turns over under a CO atmosphere, the EPR signal of the $S = 3/2$ resting state cofactor disappears and two new $S = 1/2$ EPR signals appear: one under low pressure of CO (denoted lo-CO; 0.08 atm) with $g = [2.09\ 1.97\ 1.93]$ and the other under high pressure of CO (denoted hi-CO; 0.5 atm) with $g = [2.06\ 2.06\ 2.17]$ (31,34).

Figure 2A shows a full '2-D' set of Q-band ^{13}C CW-ENDOR spectra taken at fields across the EPR envelope of lo-CO. **Figure 3A** displays Q-band ^{13}C CW and

ReMims ENDOR (*54*) spectra collected across the EPR envelope of hi-CO prepared with [13]CO. These signals are absent when the MoFe protein is incubated under natural abundance CO, indicating that the ^{13}C ENDOR signals are arising from [13]CO molecules associated with the $S = 1/2$ EPR signals. Because ^{13}C ENDOR studies of 'pulse-chase' experiments and [57]Fe ENDOR of the MoFe protein have disclosed that both the lo- and hi-CO EPR signals seen under turnover are arising from the FeMo-cofactor (*34*), the observed ^{13}C ENDOR implies that CO molecules are bound to the FeMo cofactor in both lo-CO and hi-CO states. This is the first ever direct observation of a small molecule bound to the active site, the FeMo-cofactor, of nitrogenase (*31,34*).

[13]C ENDOR obtained at a 'single-crystal like' field position (g_1) of lo-CO shows one doublet centered at the [13]C Larmor frequency with A(^{13}C, g_1) ~ 3 MHz, indicating a single CO bound to the cofactor. (**Figure 2A**) The spectrum taken at $g_{||}$ of hi-CO shows a doublet centered at the [13]C Larmor frequency with A([13]C1, $g_{||}$) = 5.8 MHz, (**Figure 3A**) which represents a single CO [CO(1)] interacting with the cofactor. A second feature at $g_{||}$ that is best visualized by Mims ENDOR employing $\tau = 400$ ns, (**Figure 3B**) comes from an additional CO [CO(2)] bound to the cofactor. Previous ^{13}C ENDOR studies of 'pulse-chase' experiments discovered that the bound CO of lo-CO form becomes the CO(1) of hi-CO (*31*).

The ^{13}C hyperfine couplings of lo-CO are highly field dependent, with a maximum value of A ~ 3.5 MHz near g_1. (**Figure 2A**) This pattern is well simulated (**Figure 2B**) through calculations that use a hyperfine tensor with comparable isotropic and dipolar components: A = [-2.0, 3.5, 2.0] = 1.2 + [-3.2, 2.3, 0.8] MHz = $A_{iso} + A_{aniso}$. The field dependance of the CW and ReMims ENDOR data (dotted lines) for [13]C1 (**Figure 3A**) is typical for a hyperfine interaction that is dominated by the isotropic component, with a dipolar contribution whose principal axis is perpendicular to $g_{||}$: A(C1) = [5.80, 5.80, 4.50] = 5.37 + [0.43, 0.43, -0.87] MHz (**Table 1**). At all fields, the weakly coupled [13]C ([13]C2) shows a sharp doublet with $0.6 \leq A \leq 0.9$ MHz that rides on a broad background doublet with A ~ 1.5 MHz. (**Figure 3B**) The sharp [13]C2 ENDOR pattern (dotted lines) in **Figure 3B** is also dominated by an isotropic contribution, with a small axial hyperfine term parallel to $g_{||}$: A = [0.6, 0.6, 0.9] = 0.7 + [-0.1, -0.1, 0.2] MHz (**Table 1**). The broad feature in **Figure 3B** could be yet a third CO or a distribution of alternate, poorly defined binding geometries of CO(2).

The characteristics of the ^{13}C hyperfine tensor as determined by the ENDOR measurements provide information about the mode of binding of the bound diatomic inhibitor (CO) in lo-CO and hi-CO. The ^{13}C hyperfine interaction of metal-bound [13]CO is the sum of isotropic and anisotropic interactions: $A = A_{iso} + A_{aniso}$. The isotropic interaction originates from direct spin delocalization into the carbon 2s orbital through the M-C σ-bond between a metal d-orbital and a carbon sp-hybrid. There can also be a contribution from spin polarization of the C 2s

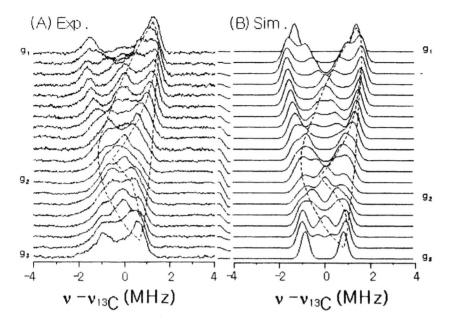

Figure 2. *Q-band CW ^{13}C ENDOR spectra taken at fields across the EPR envelope of (A) lo-^{13}CO and (B) their simulations. The spectra are centered at the ^{13}C Larmor frequencies. Dotted lines represent the ^{13}C ENDOR pattern of one electron spin manifold. Simulation parameters are in **Table 1**. Experimental conditions: microwave frequency, 34.992 GHz; modulation amplitude, 0.67 G; RF power, 30W; RF sweep speed, 0.5 MHz/s; temperature, 2K. The bandwidth of the RF excitation was broadened to 100 kHz. (Adapted with permission from reference 35. Copyright 1997 ACS)*

orbital, either by spin density on the metal ion acting on the σ-bond, or by spin density that is delocalized into carbon p-orbitals. The anisotropic interaction is the sum of contributions from *(i)* spin density in the C p_σ-orbital that is directed toward the metal because of the M-C σ-bond or polarization by the metal-ion spin, *(ii)* spin density in the C p_π-orbital that is generated by the back-donation of the metal d-electrons, and *(iii)* by the direct dipole-dipole interaction between the carbon nuclear spin and the electron spin on the metal (*60*). All three of these anisotropic contributions have axial symmetry, but the first and third would be coaxial, with their symmetry axis lying along the M-C bond, while in the second case, the symmetry axis would be perpendicular to the bond.

Previous ^{13}C ENDOR studies of ^{13}CO bound to Fe-S clusters of hydrogenase I and II showed $A_{iso} = 20 \sim 35$ MHz (**Table 1**) (*61,62*). Hyperfine tensors were essentially isotropic, with a relatively small and axial anisotropic contribution ($A_{aniso} = [-T/2, -T/2, T]$, where $|T/A_{iso}| = 0.1 \sim 0.2$). This suggests that the CO binds

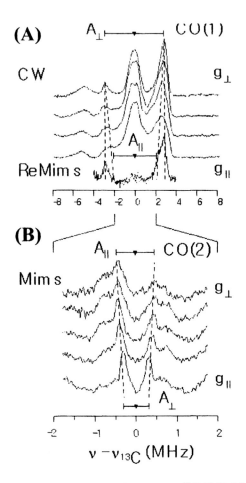

Figure 3. (A) Q-band CW, ReMims, and (B) Mims ^{13}C ENDOR spectra taken at several fields across the EPR envelope of hi-CO state of the MoFe protein under ^{13}CO for (A) CO(1) and (B) CO(2), respectively. The spectra are centered at the ^{13}C Larmor frequencies. ^{13}C ENDOR patterns are depicted by dotted lines. Experimental conditions: microwave frequency, (A-CW) 35.073, (A-ReMims) 34.636, and (B) 34.646 GHz; modulation amplitude, (A-CW) 0.67 G; RF power, (A-CW) 20W; RF sweep speed, (A-CW) 0.5 MHz/s; τ, (A-ReMims) 100 and (B) 400 ns; repetition rate, (A-ReMims, B) 33 Hz; RF pulse width, (A-ReMims, B) 30 μs; number of transients, (A-ReMims) 480 and (B) 200~500; temperature, 2K. The bandwidth of the RF excitation was broadened to 100 kHz for the CW experiments. (Adapted with permission from reference 35. Copyright 1997 ACS)

Table 1. ^{13}C hyperfine tensors of ^{13}C labeled intermediates bound to metal clusters in proteins.

Protein (Substrate/ Inhibitor)	EPR (g-tensor)	^{13}C Hyperfine coupling tensors $[A_{iso}]$ (MHz)	Reference
Nitrogenase MoFe protein[a] (CO)	lo-CO (2.09, 1.97, 1.93)	-2.0, 3.5, 2.0[b] [1.2]	(35)
	hi-CO (2.17, 2.06, 2.06)	(C1) 5.8, 5.8, 4.2 [5.4]	(35)
		(C2) 0.6, 0.6, 0.9 [0.7]	(35)
Nitrogenase MoFe protein[a] (CS$_2$)	a (2.035, 1.982, 1.973)	[4.9][c]	(33)
	b (2.111, 2.022, 1.956)	[1.8][c]	(33)
	c (2.211, 1.996, 1.978)	[2.7][c]	(33)
Nitrogenase α-Gln^{195} MoFe protein[a] (C$_2$H$_2$)	S$_{EPR1}$[d] (2.12, 1.98, 1.95)	(C1) 3.8, 1.8, 1.8 [2.5]	(37)
		(C2) 3.2, 2.0, 1.8 [2.3]	(37)
		(C3) [<0.5][c]	(37)
Hydrogenase I, Cp W5[f] (CO)	(2.074, 2.074, 2.009)	[~21][e]	(61)
Hydrogenase II, Cp W5[f] (CO)	(2.032, 2.017, 1.997)	[35.3][e]	(62)
Pyrococcus furiosus[g] (CN$^-$)	(2.09, 1.95, 1.92)	-4.5, -4.5, +0.1 [-3.0]	(97)

[a] Under turnover conditions. [b] Euler angle of the hyperfine tensor with respect to g-tensor frame: (α, β, γ) = (67.5°, 17.5°, 0°). The isotropic hyperfine couplings are absolute values unless indicated. [c] Mostly isotropic ^{13}C hyperfine tensor. [d] S$_{EPR2}$ and S$_{EPR3}$ EPR signals are simultaneously observed, but no ^{13}C ENDOR signals are found from the EPR signals (37). [e] Tensor components not determined. [f] Cp W5 - Anaerobic N$_2$-fixing bacterium *Clostridium pasteurianum* (Cp) W5. [g] Hyperthermophilic archeon *Pyrococcus furiosus* reduced 4Fe ferredoxin.

terminally to a single Fe, and that the ^{13}C anisotropic term is a combination of the coaxial contributions from spin density in the C σ-bonding p-orbital and the through-space dipole coupling to spin on Fe.

The ^{13}C hyperfine interaction tensors of both ^{13}CO molecules bound to hi-CO are similar to those of the ^{13}CO-bound hydrogenases, in that they are dominated by the isotropic component, with a relatively small, axially symmetric anisotropic contributions ($|T/A_{iso}| \sim 0.2$). The magnitude of the couplings to ^{13}CO in hi-CO are much less than seen with other systems (**Table 1**). The form of the ^{13}C hyperfine tensor suggests to us that the CO molecules in hi-CO likewise are terminally bound. The small magnitude could arise in part because the CO's are bound to metal ions with small spin-projection coefficients, K (63-65).

The ^{13}C hyperfine tensor of lo-CO shows quite different characteristics. Here, the hyperfine tensor, $A = A_{iso} + A_{aniso} = 1.2 + [-3.2, 2.3, 0.8]$ MHz, is dominated by the anisotropic interaction ($|T/A_{iso}| = 2.7$), not the isotropic interaction as in hi-CO and hydrogenase-CO. Furthermore, the anisotropic component of the ^{13}CO hyperfine tensor of lo-CO does not have the simple axial form, $[-T/2, -T/2, T]$, as seen for the hi-CO and hydrogenase-CO, but is rhombic. Because each of the observed tensor components is multiplied by the same K (36,64), the unusual character of the ^{13}C hyperfine tensor of lo-CO implies to us that the site hyperfine tensor, and thus the binding of CO in lo-CO, is quite different from that in hydrogenase and hi-CO. The rhombicity of A_{aniso} in lo-CO implies that it is the resultant of a summation of at least two of the three types of noncoaxial contributions listed above (66,67). One might suggest that the observed tensor is associated with a terminal CO where the M-C σ-bond gives a minimal contribution to both isotropic and anisotropic terms, while the anisotropic term is a combination of the noncoaxial M-C point-dipole and the local p_π terms, but this does not seem plausible for CO terminally bound to a high-spin ferrous or ferric ion, where there should be a significant spin contribution from direct Fe-C σ-bonding. It might arise from a terminal CO bound to a low-spin Fe, but such an iron would have a strongly anisotropic ^{57}Fe hyperfine interaction and nothing of this sort has been seen (36). We suggest instead an alternative model where the CO in lo-CO forms some type of bridge between two metal ions in the cluster. In this case, the point-dipole interactions and local terms from σ-spin density due to the bond from a single ion would be coaxial, but the interactions with one ion would not be coaxial with those from the other. The result plausibly gives rise to a tensor such as is seen.

In the FeMo-cofactor, there are *a priori* eight possible CO-binding metal sites, including Mo. However, $^{1,2}H$ ENDOR results suggest that CO binds near the waist of the cofactor (35), and thus away from Mo. (Note also, one of the ligands in the saturated ligand-coordination environment of Mo would have to be displaced for the CO-binding during the turnover.) Moreover, a Mo-CO bond needs d-electron back donation from Mo to π^*-orbitals of CO, which requires a low oxidation state of Mo. It is proposed that Mo in CO-inhibited FeMo-cofactor is in the same Mo(IV) state as in the resting state (36), in accordance with Mo EXAFS measurements which indicate that Mo does not change oxidation state nor bind

CO in CO-inhibited MoFe protein (*68*). We thus propose that the Mo site is not involved either in terminal or bridging CO-binding.

These considerations lead to a model in which, for lo-CO, a single CO binds as a bridge between two of the coordinatively unsaturated waist Fe ions, while in hi-CO, two CO's bind each as a terminal ligand to different Fe ions, as in **Figure 4**. The crystal structure and EXAFS studies of the cofactor in the resting state of *Av1* revealed the average distance between two such iron sites is ~ 2.5 Å (*6-9,69*), the same as the average distance between irons in binuclear iron carbonyl compounds which have bridging-carbonyls (*70*). The previous ^{13}C ENDOR study of lo-CO and hi-CO showed that lo-CO and hi-CO are interconvertible and CO(1) of hi-CO originates from the CO of lo-CO (*31*). This implies that the cofactor's CO binding site in lo-CO is retained when it is transformed into hi-CO, and that this process involves conversion of bridging into terminally bound CO.

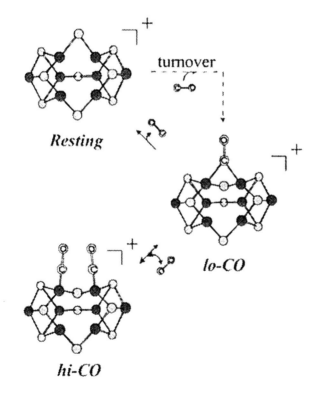

Figure 4. *Proposed CO-binding modes to the FeMo-cofactor in the lo-CO and hi-CO states and the cofactor cluster charges of the resting, lo-CO, and hi-CO states (35,36).*

Metal-Ion Valencies of the FeMo-Cofactor in CO-inhibited and Resting State Nitrogenase: As stated in the previous section, when the protein turns over under CO, two EPR signals (S = 1/2) associated with the cofactor appear; one is formed under low pressure of CO (lo-CO) by binding of a single CO, and has g = [2.09 1.97 1.93]; the other forms under high pressure (hi-CO) by binding of a second CO, and has g = [2.06, 2.06, 2.17]. This section illustrates how the valencies and d-electron count of the metal ions in the FeMo cofactor in the lo-CO form are obtained by complete orientation-selective [57]Fe ENDOR measurements (*56-59*). The lo-CO, hi-CO, and resting states all are interconvertible without redox or catalytic processes, which lead us to infer that the cofactor in all three states is at the same redox level, hence the conclusions apply to all three protein forms (*31,71*).

Full '2-D' sets of [57]Fe ENDOR spectra were taken at fields across the EPR spectrum of lo-CO; **Figure 5** is such a set. The spectra display a rich array of features from ~ 6 to ~ 22 MHz, corresponding to $15 \leq |A| \leq 41$ MHz. Although the features are severely overlapped, essentially all of them can be assigned. As indicated, there are clearly three $\nu_{\pm}$ doublets, separated by $\sim 2\nu_{Fe}$, that appear across the EPR envelope. These correspond to three iron sites, $Fe_{\beta 1}$, $Fe_{\alpha 1}$, and $Fe_{\alpha 2}$. The hyperfine tensors obtained for these three sites by a simulation of the 2-D ENDOR pattern are largely isotropic, as is common for Fe ions in Fe-S clusters; the isotropic components are listed in **Table 2**. A single ν_- feature of a fourth site, $Fe_{\beta 2}$, also can be followed through the envelope. Its presence among more intense peaks makes it impossible to get complete hyperfine information for this site, but the tensor must also be largely isotropic, and an estimate for the isotropic interaction is contained in **Table 2**. There are no additional [57]Fe signals with higher values of $|A_{iso}|$; if present, these would have been detected easily. Careful examination of the low-frequency region around ν_{Fe} also failed to disclose any signals from sites with small values of $|A_{iso}|$.

Spin coupling in multimetallic Fe clusters can be analyzed through examination of the isotropic [57]Fe coupling constants. **Table 2** contains the [57]Fe isotropic couplings for the S = 1/2 lo-CO form of the FeMo-cofactor as well as for selected 4Fe-4S clusters with the total spin of 1/2. The seven iron ions of the lo-CO cofactor give rise to four distinct types of [57]Fe signals, two with an isotropic coupling of $|A_{iso}| \sim 31$ MHz ($Fe_{\alpha 1}$, $Fe_{\alpha 2}$) and two with $|A_{iso}| \sim 16$ MHz ($Fe_{\beta 1}$, $Fe_{\beta 2}$).

The observed hyperfine interaction, $A_{exp}(Fe_x)$, for a given iron ion in a spin-coupled cluster is proportional to the projection of the ion's local spin onto the total spin, $K(Fe_x)$,(*63-65*),

$$A_{exp}(Fe_x) = K(Fe_x)a_{site}(Fe_x). \tag{2}$$

where $a_{site}(Fe_x)$ is the isotropic hyperfine constant for the x-th site when not spin-coupled and $K(Fe_x)$, the spin-projection coefficient of the site, is given by (*63,64*)

$$K(Fe_x) = \langle S(Fe_x) \cdot S_t \rangle / \langle S_t \cdot S_t \rangle. \tag{3}$$

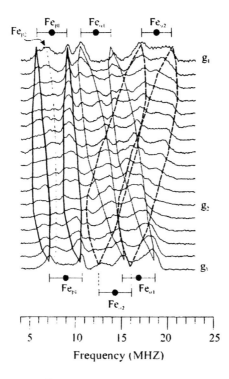

Figure 5. *Q-band CW ^{57}Fe ENDOR spectra taken at fields across the EPR envelope of the lo-CO states of the globally ^{57}Fe-enriched MoFe protein. The doublet patterns of the Fe$_{\beta_1}$, Fe$_{\alpha_1}$, and Fe$_{\alpha_2}$ sites in lo-CO are indicated by "goal-post" marks and their experimental variations with magnetic field are indicated by dot, solid, and dash lines, respectively. Fe$_{\beta_2}$ is indicated by dash-dot line. Experimental conditions: microwave frequency, 35.160 GHz; modulation amplitudes, 0.67 G; RF power, 30W; RF sweep speed, 1 MHz/s; temperature, 2K. The band width of the RF excitation was broadened to 100 kHz. (Adapted with permission from reference 36. Copyright 1997 ACS)*

Here, $S(Fe_x)$ is the electron spin of the site, S_t is the total cluster spin of the spin-coupled cluster, and $\langle S_t \cdot S_t \rangle = S_t(S_t + 1)$ when total spin is conserved.

We now show how the valencies of the Fe ions of the lo-CO FeMo-cofactor, and hence the d-electron count of the cofactor, can be deduced from the [57]Fe ENDOR data. Note that CW ENDOR signal intensities, particularly for [57]Fe in a complex spin-coupled system such as the FeMo cofactor, are not proportional to the number of nuclei giving rise to the signal. We begin with the assumption that the Mo ion is in the Mo(IV; $S = 0$) state, as we inferred for the resting state cofactor (21). If Mo had been reduced under turnover to a paramagnetic state, Mo(III; $S = 1/2$), then the coupling to [95,97]Mo would likely be large, say 10 MHz or greater, and Mo ENDOR signals should be detectable in both natural-abundance ([95,97]Mo, total abundance 25.5 %) and [95]Mo enriched CO-inhibited samples. However, no such signals were detected in the natural-abundance samples (34,36), and the same is true for preliminary measurements on [95]Mo enriched CO-inhibited samples. This assumed valence state of Mo is in accordance with Mo EXAFS measurements which indicate that Mo does not change valence nor bind CO in CO-inhibited MoFe protein (68). Given that a ferrous ion has an even-electron count (d[6]; $S = 2$) and a ferric ion has an odd (d[5]; $S = 5/2$), the first consequence of Mo being in an even-spin state is that the seven iron ions must include an odd number of Fe^{3+} ions in order to generate a total cluster spin of $S = 1/2$ for lo-CO MoFe protein. In the highly reducing nitrogenase environment it is inconceivable that the cofactor can be in a 'HiPIP-like' state, with more than half the Fe ions in the oxidized, ferric form. Thus, we conclude that an even-spin Mo in the FeMo-cofactor can be accompanied only by 1 or 3 ferric ions. In such reduced clusters, the ferric ion generally occurs as a mixed-valence pair, $2Fe^{2.5+}$, which corresponds to one Fe^{3+} and one Fe^{2+} ion that have one odd-electron (hole) delocalized between them (72-75). As a result, we infer that the FeMo-cofactor of lo-CO has either 1 or 3 such delocalized (mixed-valence) pairs, so that the metal ions in the cofactor are arranged as $[(2Fe^{2.5-})_{1 \text{ or } 3}, Fe^{2+}_{5 \text{ or } 1}, Mo^{4+}]$.

To proceed, we consider the ENDOR-determined hyperfine coupling constants (**Table 2**), which involve the unknown spin-projection coefficients (**Eq. 3**). At the most primitive level, one can make direct comparisons between the [57]Fe hyperfine couplings for different Fe-S centers that have the same total spin. The experimental isotropic coupling, $|A_{iso}| \sim 31$ MHz, seen for two sites in lo-CO is common for the mixed-valence pairs of $S = 1/2$ [4Fe-4S] clusters, which contain only a single $Fe^{2.5+}$ pair together with two ferrous ions (**Table 2**) (63). The occurrence of similar values in different clusters is not unexpected because such pairs have a high spin (ranging from 5/2 to 9/2) (64) and in general would have a relatively large spin-coupling coefficient, K, in any reduced cluster. In contrast, ferrous ([57]Fe[2+]) ions typically show smaller K and $|A_{iso}|$ values, comparable to those seen for the other two distinct sites of lo-CO (**Table 2**) (63). This suggests that $Fe_{\alpha 1}$ and $Fe_{\alpha 2}$ of lo-CO represent one or more such $Fe^{2.5+}$ pairs, while $Fe_{\beta 1}$ and $Fe_{\beta 2}$ of lo-CO are Fe^{2+} ions. In this case, the presence of 3 pairs is ruled out because only one of the seven Fe ions would exist as an $^{57}Fe^{2+}$ ion, while we observe resonances from two ($Fe_{\beta 1}$ and $Fe_{\beta 2}$). Thus, the two sites of lo-CO, $Fe_{\alpha 1}$ and $Fe_{\alpha 2}$, most likely correspond to the two members of a single $^{57}Fe^{+2.5}$ pair; by difference, $Fe_{\beta 1}$ and $Fe_{\beta 2}$ would represent five $^{57}Fe^{2+}$ ions, conceivably existing as two $^{57}Fe^{2+}$ pairs and an isolated $^{57}Fe^{2+}$.

To test this suggestion, we consider the parameter a_{test} introduced by Mouesca et al. (63):

Table 2. ^{57}Fe **Isotropic hyperfine coupling constants of** ^{57}Fe **enriched Fe-S clusters with** $S = 1/2$ **spin state.**

Enzyme	^{57}Fe Site	Isotropic Hyperfine Coupling Constant (MHz)[a]	Reference
FeMo-CO $AvI^{(b)}$ (lo-CO)	$Fe_{\alpha 1}$	-30	(36)
	$Fe_{\alpha 2}$	-31	(36)
	$Fe_{\beta 1}$	16[c]	(36)
	$Fe_{\beta 2}$	~17[c,d]	(36)
Aconitase (E)[e]	$Fe^{+2.5}$	-39, -37	(62,63)
	$Fe^{+2.0}$	~+16, +33	(62,63)
Aconitase (ES)[f]	$Fe^{+2.5}$	-36, -40	(62,63)
	$Fe^{+2.0}$	~+16, +29	(62,63)
Pf-Fd-CN[g]	$Fe^{+2.5}$	-32, -24	(62,98)
	$Fe^{+2.0}$	+17, +15	(62,98)
Pf-Fd[g]	$Fe^{+2.5}$	-31, -35	(98)
	$Fe^{+2.0}$	+21(x2)	(98)

[a] Clusters refer to an $[Fe_4S_4]^+$ core except for the $[MoFe_7S_9]$ core of FeMo-CO. Signs are indicated where available. (x2) indicates a spin delocalized iron pair. [b] AvI - MoFe protein of nitrogenase from *Azotobacter vinelandii* (Av). [c] The two Fe sites of lo-CO with A_{iso} ~ 16 MHz are taken to represent 5 Fe ions; see text. [d] Estimated magnitude of isotropic hyperfine coupling constant. [e] Substrate-free reduced aconitase. [f] Substrate-bound aconitase. [g] Pf-Fd-CN - Hyperthermophilic archeon *Pyrococcus furiosus* reduced 4Fe ferredoxin with bound CN^-. Pf-Fd -reduced 4Fe ferredoxin.

$$a_{test} = \Sigma A_{exp}(Fe_x) = \Sigma K(Fe_x)a_{site}(Fe_x), \tag{4}$$

where the sum runs over all seven Fe ions and a_{site} is the intrinsic isotropic coupling constant. If the Mo has $S = 0$, and thus does not participate in the spin coupling scheme, then $\Sigma K(Fe_x) = 1$ and a_{test} is the weighted average of the intrinsic isotropic constants for the Fe sites. Values of $a_{test} = -(16 \sim 25)$ MHz have been calculated by Mouesca et al. for a number of smaller clusters (63). If the $S = 1/2$ state of lo-CO arises from spin coupling among one $Fe^{2.5+}$ pair with $|A_{iso}| \sim 31$ MHz for each Fe, plus five Fe^{2+} ions, each with $|A_{iso}| \sim 16$ MHz, then the only way to achieve such a value of a_{test} is to have a large positive value of K for the pair, a positive K for one of the Fe^{2+} ions, and negative values of K for four of the Fe^{2+} ions. This yields, $A_{iso}(Fe^{2.5+}) \sim -31$ MHz (x2), $A_{iso}(Fe^{2+}) \sim 16$ MHz (x4), and $A_{iso}(Fe^{2+}) \sim -16$ MHz (x1), yielding $a_{test} \sim -14$ MHz for lo-CO. While this value is slightly outside the range of the reported values, the discussions of Mouesca et al. (63) show that a decrease in the magnitude of a_{test} as seen for lo-CO is consistent with a highly reduced cofactor cluster having a preponderance of Fe^{2+} ions, which are expected to have a lower value of intrinsic isotropic coupling constant and hence of a_{test}. Thus, the value of a_{test} for lo-CO supports the suggestion that the FeMo-cofactor contains only mixed-valence pair. Taking into account the proposed (IV) oxidation state for molybdenum, this leads to a formulation of the inorganic portion of the cluster as $[Mo, Fe_7, S_9]^+ = [Mo^{4+}, Fe^{3+}_1, Fe^{2+}_6, S^{2-}_9]^+ = [Mo^{4+}, (2Fe^{2.5+})_1, Fe^{2+}_5, S^{2-}_9]^+$, with a formal d-electron count of 43.

A number of independent observations support the conclusion that the FeMo-cofactor is at the same redox level in resting-state, lo-CO, and hi-CO proteins, and that the proposed valency assignments thus apply to all three states as depicted in **Figure 4**. The most direct evidence is that the hi-CO state quenched with ethylene glycol (EG at 40%) can be converted into lo-CO in the absence of turnover by simply pumping off atmospheric CO. Subsequent back addition of CO to the atmosphere regenerates hi-CO while extensive pumping on quenched lo-CO converts it to the resting state, all in the absence of electron transfer, meaning that lo-CO, hi-CO and resting state all correspond to the same oxidation state of the cofactor (71). Our oxidation state assignments of Fe are the only acceptable ones that yield a cluster with a half-integer spin state and a net negative charge. Thus, when homocitrate is included in the cofactor structure and assuming a -3 charge on homocitrate, the overall charge on the cluster $[MoFe_7S_9 : homocitrate]$ is -2. In contrast, a paramagnetic state with three mixed-valence pairs would likely be neutral. Previous justifications (76,77) for the negative charge of the extracted cofactor (as revealed by DEAE-binding studies (78), counter-ion chromatography (77), and electrophoretic (79) measurements) have required the association of anionic solvent and/or exogenous ligands to a positively charged cluster. Our valency assignments automatically make the cluster negatively charged and thus eliminate the necessity of including such ligands in the structure.

Mössbauer experiments of the resting state MoFe protein ($Av1$) suggested that the resting state cofactor has the electronic state of $[Mo^{4+},Fe^{3+}_3,Fe^{2+}_4,S^{2-}_9]^+$, based on the average isomer shift of the cofactor (80). ^{57}Fe ENDOR- and Mössbauer-derived results of the cofactor charge (36,80) have been tested by Noodleman and coworkers and their recent density-functional theory (DFT) study favored the former, $[Mo^{4+},Fe^{3+}_1,Fe^{2+}_6,S^{2-}_9]^+$, as the oxidation state of the resting-state cofactor. (65). The DFT calculation with a single ferric ion achieved hyperfine couplings closer to the reported experimental hyperfine couplings from both ^{57}Fe ENDOR and Mössbauer measurements of resting state nitrogenase, a lower energy, and more appropriate cofactor geometries, than the calculation with three ferric ions (65).

Acetylene-Binding to the FeMo-cofactor in the Nitrogenase α-Gln^{195} MoFe Protein: Among all nitrogenase substrates, C_2H_2 is the best-studied and its reduction to ethylene (C_2H_4) is routinely used for *in vitro* assays of nitrogenase activity (81). Despite the many models that have been suggested for C_2H_2 binding to the FeMo-cofactor (82), until recently there has been no direct spectroscopic observation of C_2H_2 interaction with the nitrogenase FeMo-cofactor. When wild-type nitrogenase is incubated under turnover conditions there is a dramatic reduction in the intensity of the $S = 3/2$ EPR signal associated with the resting state of FeMo-cofactor. However, this situation does not result in the appearance of other paramagnetic species that can be correlated with the binding of substrates such as N_2 or C_2H_2. By using an altered MoFe protein for which the α-subunit His^{195} residue was substituted by Gln (designated α-Gln^{195} MoFe protein), we detected the first EPR signals that are elicited by the binding of C_2H_2 to the FeMo-cofactor under turnover conditions (32). The α-Gln^{195} MoFe protein does not significantly reduce N_2 but remains capable of reducing C_2H_2 and does so with kinetic parameters very similar to the wild-type enzyme (83). When incubated under turnover conditions in the presence of C_2H_2, the α-Gln^{195} MoFe protein exhibits three simultaneously generated EPR signals: a rhombic $g = [2.12\ 1.98\ 1.95]$ signal (designated S_{EPR1}); a mostly isotropic, $g = 2.00$ signal (designated S_{EPR2}); and a minority component with an inflection at $g \sim 1.97$ (designated S_{EPR3}). The spectrum obtained by use of isotopically labeled $^{13}C_2H_2$ indicated that S_{EPR1} originates from C_2H_2 intermediates bound to the FeMo-cofactor during enzymatic turnover (32). In this section, we present direct evidence regarding the mode of C_2H_2 binding to the FeMo-cofactor, as obtained through Q-band ^{13}C ENDOR spectroscopy of the nitrogenase S_{EPR1} turnover state.

Figure 6 shows Q-band ^{13}C ENDOR collected at fields across the EPR envelope of S_{EPR1} generated during turnover of a α-Gln^{195} MoFe protein sample in the presence of $^{13}C_2H_2$. The presence of one or more fragments arising from $^{13}C_2H_2$ bound to the EPR-active FeMo-cofactor is indicated by the ^{13}C signals associated

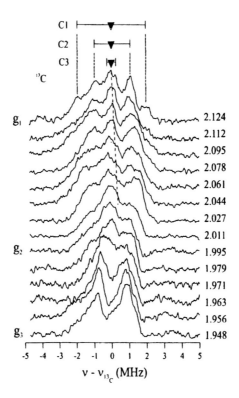

Figure 6. Q-band ^{13}C ENDOR spectra for the S_{EPR1} signal arising from α-Gln^{195} MoFe protein during turnover in the presence of $^{13}C_2H_2$ (0.1 atm). Spectra were taken at g-values across the EPR envelope of S_{EPR1}, as indicated. The spectra are centered at the ^{13}C Larmor frequency (ca. 13 MHz). The three ^{13}C doublets detected in the 'single-crystal-like' spectrum at g_1 are indicated. The development of these spectra with field, insofar as they can be traced clearly, are indicated by the dashed lines overlaid to guide the eye on the ν_+ branch of the spectra. Experimental Conditions; microwave frequency, 35.158 GHz; modulation amplitude, 1.3 G; radio frequency (RF) power, 20 W. (Adapted with permission from reference 37. Copyright 2000 ACS)

with S_{EPR1} that are absent in the ENDOR spectrum of the sample prepared with natural abundance C_2H_2 (data not shown) (37). The uppermost ^{13}C ENDOR spectrum in **Figure 6** was obtained at the low-field edge of the S_{EPR1} (g_1 = 2.12) spectrum. This spectrum represents a 'single-crystal-like' pattern associated with a single molecular orientation, with the magnetic field along g_1. In such a spectrum, each ^{13}C ENDOR doublet is associated with a single class of nuclei. Signals from three such classes can be recognized. Two of these yield well-resolved doublets, with coupling constants, $A(C1)$ = 3.9 MHz, $A(C2)$ = 2.0 MHz. The ENDOR intensity near the [13]C Larmor frequency, although not resolved into a doublet, nonetheless must originate from a third type of weakly-coupled ^{13}C nucleus, with $A(C3) \leq 0.5$ MHz. This spectrum therefore requires the presence of no fewer than two molecules of C_2H_2, or its reaction intermediates/products, bound to the S_{EPR1} FeMo-cofactor.

The strong overlap of the signals from the three types of ^{13}C nucleus prevents an accurate determination of their hyperfine tensors. However, the 2-D, field-frequency pattern of **Figure 6** can nonetheless be analyzed approximately. As indicated in the figure, the field-dependence of the v_+ branch of the pattern for C2 can be reasonably well followed. The modest change in hyperfine coupling with field is characteristic of a nucleus with a largely isotropic coupling, and the nature of the change indicates the anisotropic term is of roughly axial symmetry (56-59). The [13]C2 hyperfine tensor appears to be roughly coaxial with the g-tensor, and has principal values of ca. $A(C2) \sim [3.2, 2, 1.8]$ MHz, corresponding to an isotropic contribution, $A_{iso}(C2) \sim 2.3$ MHz, and anisotropic term, $2T(C2) \sim 0.9$ MHz.

The coupling to C1 at g_1 is $A(C1)$ ~3.6 MHz, significantly greater than that of C2, but the C1 coupling is more anisotropic and decreases with increasing field until the C1 signal becomes undistinguishable as g approaches g_2 (**Figure 6**). If, for heuristic purposes, we assume that C1 contributes, together with C2, to the intense doublet seen at g_3 ($A \sim 2$ MHz), then the 2-D pattern can be roughly described by a hyperfine tensor with principal values of, $A(C1) \sim [3.8, 1.8, 1.8]$ MHz, corresponding to an isotropic contribution, $A_{iso}(C1) \sim 2.5$ MHz, and anisotropic term, $2T(C1) \sim 1.4$ MHz. Thus, the hyperfine tensors of both C1 and C2 appear to be dominated by isotropic hyperfine couplings of comparable magnitude, but with a larger anisotropic term for C1.

As indicated in **Figure 6**, the weak coupling to C3 increases somewhat as the field is increased from g_1 to $g \sim 2.027$. It is impossible to determine whether the coupling continues to increase beyond this point, perhaps even reaching $A \sim 2$ MHz at g_3, or whether it decreases to the extent that it becomes unobservably small at g_3.

The present study uses ^{13}C ENDOR to identify signals from three distinct, C_2H_2-derived ^{13}C nuclei that are bound to an EPR-active FeMo-cofactor that gives rise to the S_{EPR1} species observed for the α-Gln[195] MoFe protein. Given that the

FeMo-cofactor is the site of C_2H_2 reduction, and that the above section showed that CO binds to this EPR-active site during enzymatic turnover, these three signals can be assigned to at least two C_2H_2-derived molecules that are interacting with the FeMo-cofactor. This conclusion is in line with other results that indicate the presence of multiple C_2H_2 binding sites within the MoFe protein (28,30,84). There are also signals arising from rather strongly coupled protons that are derived from C_2H_2 that are not exchangeable with solvent (data not shown). Finally, there are no strongly-coupled signals that arise from a solvent proton (37).

The data presented here can be used to consider possible binding mode(s) for the C_2H_2-derived species. Firstly, because C_2H_4 is the only detectable product of C_2H_2 reduction for both the wild-type- and α-Gln[195] MoFe protein, we conclude that the ^{13}C we measure arises from cluster-bound C_2H_x species. The ^{13}C ENDOR hyperfine coupling tensors estimated for the S_{EPR1}-associated C_2H_x species are summarized in **Table 1**. Orientation-selective ^{13}C ENDOR experiments of CO-inhibited MoFe protein in the above section and of the MoFe protein incubated under turnover conditions in the presence of $^{13}CS_2$ yielded the ^{13}C hyperfine coupling tensors of bound inhibitor and CS_2-related species (**Table 1**) (33,35). The hyperfine interactions of C1 and C2 for S_{EPR1} are comparable to those previously reported for hi-^{13}CO (C1) and $^{13}CS_2$ adducts in that the tensors are mostly isotropic, with similar magnitudes for the isotropic couplings (**Table 1**). In these previous cases, we interpreted this type of ^{13}C hyperfine interaction as arising from terminally bound ^{13}C compounds (33,35). For this current discussion, we likewise assign C1 and C2 of S_{EPR1} to acetylene-derived species terminally bound to the cofactor. Such binding can occur either with coordination by a single carbon or by both carbons in a bridging arrangement with two Fe atoms as shown in **Figure 7**.

EXAFS (69,85) and crystallographic (86) data have shown that the adjacent Fe-Fe distances are ca. 2.5-2.6 Å and the diagonal, cross-face, Fe-Fe distances are ca. 3.6~3.8 Å. A simple energy minimization calculation for the model of cis-$(SH)_3FeCH=HCFe(SH)_3$ with tetrahedral Fe geometries gave an Fe-Fe distance of 3.9 Å (37). Considering the potential flexibility of the FeMo-cofactor, this model appears acceptable. However, acetylene has been shown to bind to, and bridge, transition metal atoms in a wide variety of different geometries, thus making this but one of many possibilities that could be consistent with the data presented here. At first glance, it would appear that one must consider separately the possibilities that C1 and C2 are associated with the same C_2H_x fragment, or that each represents a bound carbon from a separate fragment. There are many scenarios, depending on the spin-coupling scheme of the cluster and whether one considers the single-point or bridging modes of attachment. However, any interaction with the cluster that gives rise to appreciable coupling to one carbon likely would give rise to a coupling of comparable magnitude to the other. For example, the spin of the C_2H_1 vinyl radical, which is to first approximation localized in a σ-orbital on one carbon, gives rise to *larger* hyperfine couplings to the two protons on the other carbon than to the single proton on the "spin-bearing" carbon (*87,88*). Hence, we suggest that C1 and C2 are associated with the same C_2H_x fragment, and that C3 is associated with another whose couplings are small. Note, however, that small couplings need not imply weak binding.

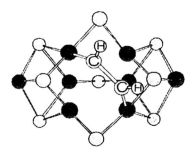

Figure 7. Proposed binding mode of acetylene bound to the FeMo-cofactor of the nitrogenase α-Gln^{195} MoFe protein under turnover in the presence of acetylene (0.1 atm) (37).

Reduction of C_2H_2 in D_2O by both wild type and α-Gln^{195} nitrogenase MoFe proteins produces predominantly *cis*-CHDCHD (*89-91*). Therefore, the C_2H_x fragment being examined must have $x \geq 2$, because loss of hydrogen upon binding, to form an acetylide (C_2H_1) species, almost certainly would lead to enzymatic formation of CD_2CHD. To elucidate the value for x, we incorporated the results of the ^1H ENDOR measurements (*37*), and examined the various possibilities offered by considering that the S_{EPR1} state is associated with a C_2H_x fragment, where $x = 2$, 3, or 4. Overall, it seemed to us most plausible to suggest that $x = 2$, and that C_2H_2 binds to two Fe atoms in a bridging fashion (formally, as bridging dianion), as in **Figure 7**. Such a structure explains ^1H ENDOR results (*37*) in that there is no solvent-derived proton to give rise to the strongly-coupled signal. It is also consistent with an earlier proposal (*92*) of a mechanism for the addition of D atoms to C_2H_2, so as to produce *cis*-CHDCHD.

Conclusions

Our ^{13}C and ^{57}Fe ENDOR studies of CO bound states of nitrogenase gave the *first* direct observations of a small-molecule (CO)binding to the FeMo-cofactor and clearly identified the properties of the bound diatomic inhibitor (CO): [FeMo-co][CO]$_n$, is the origin of the EPR signals from both lo-CO (n = 1) and hi-CO (n

= 2, or possibly 3) (*31,34*). The complete ^{13}C hyperfine tensors of bound ^{13}CO show substantial differences in the bonding characteristics of the single CO bound to the cofactor of lo-CO and the two characterized CO molecules bound to hi-CO. We have suggested that the CO of lo-CO may bridge, or semibridge, two iron ions, while each of the two CO bound to hi-CO is a terminal ligand as presented in **Figure 4** (*35*).

The ^{57}Fe ENDOR signals presented here for the seven Fe ions of the FeMo-cofactor of lo-CO can be completely assigned and interpreted in terms of four magnetically distinct ^{57}Fe signals (*36*). These signals have been analyzed within a conceptual framework provided by the deep understanding of FeS clusters generated through the joint efforts of Mössbauer (*75*) (and ENDOR (*62*)) spectroscopy and theory (*63-65*). The interpretation of the ENDOR data for the iron-molybdenum cofactor of the lo-CO nitrogenase MoFe protein has led us to propose valence assignments for the inorganic part of the cofactor cluster: [Mo, Fe_7, S_9]$^+$ = [Mo^{4+}, $Fe^{3+}{}_1$, $Fe^{2+}{}_6$, $S^{2-}{}_9$]$^+$, with a formal d-electron count of 43. It is further proposed that these ions are organized into one $Fe^{2.5+}$ pair and five Fe^{+2} ions: [Mo^{4+}, $(2Fe^{2.5+})_1$, $Fe^{2+}{}_5$, $S^{2-}{}_9$]$^+$. A variety of arguments indicate that the lo-CO, hi-CO, and resting states of the FeMo-cofactor are all at the same oxidation level. Hence, the proposed valency assignments apply to *all* three states as depicted in **Figure 4**.

Q-band ^{13}C ENDOR of the S_{EPR1} turnover state of α-Gln^{195} MoFe protein formed in the presence of $^{13}C_2H_2$ revealed the first direct evidence of the molecular interaction between the FeMo-cofactor and C_2H_2. At least two C_2H_x species are bound to the cofactor in the S_{EPR1} state. ^{13}C ENDOR, combined with ^1H ENDOR data obtained from the α-Gln^{195} MoFe protein with $C_2{}^{1,2}H_2$ in $^{1,2}H_2O$ buffer (*37*), led us to propose that one of the species is C_2H_2 bound in the bridging mode to two Fe ions of the FeMo-cofactor, thereby stabilizing the $S = 1/2$ cluster state (**Figure 7**).

Prospective

The current results establish a methodology for analysis of ENDOR data from a complicated system, the FeMo-cofactor of nitrogenase, so as to extract detailed information on the interactions between the cofactor and bound small molecules. This encourages one to continue searching for an altered (or wild-type) MoFe protein that would exhibit an EPR signal that is specifically associated with binding of the physiologically relevant substrate, N_2, under turnover conditions. As shown here for CO and C_2H_2, such studies would provide considerable insight about where and how N_2 becomes bound to the FeMo-cofactor during fixation by nitrogenase.

Recent high-resolution crystallographic studies of the nitrogenase MoFe protein surprisingly revealed electron density arising from a previously unrecognized light species (X) located in the center of the cofactor as shown in **Figure 8** (*93*). This ligand is coordinated to six central 'trigonal prismatic' Fe ions of the cofactor with Fe-X distances of 1.95 ~ 2.07 Å. It has been suggested that X is a nitrogen atom, based on a combination of electron density analysis of the x-ray crystallographic data (*93*) and on electron spin echo envelope modulation (ESEEM) studies of resting-state MoFe protein, which observed several nitrogen signals (*94-96*). Although this assignment is tentative, the presence of X has significant potential for understanding the catalytic function of nitrogenase. There are two most urgent questions related to the finding of X: its identity, and the relationship between X and the catalytic mechanism. Because the resting and many turnover states of nitrogenase FeMo-cofactor are EPR-active, ENDOR and ESEEM, which have the ability of detecting magnetic coupling between the nucleus of X and the electron spin in the cofactor, will be in a central place to solve these questions.

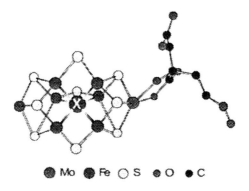

● Mo ● Fe ○ S ● O ● C

Figure 8. Schematic drawing of the FeMo-cofactor structure containing the central ligand (X), revealed by 1.16 Å resolution crystallographic analysis (93).

Acknowledgments

This work has been supported by the Korea Research Foundation (KRF-2001-015-DP0251, HIL), USDA (99-35305-8643, BMH and 1999-3695, BJH), and NSF (MCB-9904018, BMH and MCB-0211384, DRD).

174

References

1. Burgess, B. K.; Lowe, D. L *Chem. Rev.* **1996**, *96*, 2983-3011.
2. Rees, D. C. ; Howard, J. B. *Curr. Opin. Chem. Biol.* **2000**, *4*, 559-566.
3. Newton, W. E., In *Biological Nitrogen Fixation*; Stacey, G.; Burris, R. H.; Evans, H. J., Eds.; Chapman and Hall: New York, 1992; pp. 877-929.
4. Seefeldt, L. C. ; Dean, R.D. *Acc. Chem. Res.* **1997**, *30*, 260-266.
5. Howard, J. B.; Rees, D. C. *Chem. Rev.* **1996**, *96*, 2965-2982.
6. Georgiadis, M. M.; Komiya, H.; Woo, D.; Kornuc, J. J.; Rees, D. C. *Science* **1992**, *257*, 1653-1659.
7. Kim, J.; Rees, D. C. *Science* **1992**, *257*, 1677-1682.
8. Kim, J.; Rees, D. C. *Nature* **1992**, *360*, 553-560.
9. Chan, M. K.; Kim, J.; Rees, D. C. *Science* **1993**, *260*, 792-794.
10. Mayer, S. M.; Lawson, D. M.; Gormal, C. A.; Roe, S. M.; Smith, B. E. *J. Mol. Biol.* **1999**, *292*, 871-891.
11. Kim, J.; Woo, D.; Rees, D. C. *Biochemistry* **1993**, *32*, 7104-7115.
12. Münck, E.; Rhodes, H.; Orme-Johnson, W. H.; Davis, L. C.; Brill, W. J.; Shah, V. K. *Biochim. Biophys. Acta* **1975**, *400*, 32-53.
13. Zimmermann, R.; Münck, E.; Brill, W. J.; Shah, V. K.; Henzl, M. T.; Rawlings, J.; Orme-Johnson, W. H. *Biochim. Biophys. Acta* **1978**, *536*, 185-207.
14. Rawlings, J.; Shah, V. K.; Chisnell, J. R.; Brill, W. J.; Zimmermann, R.; Münck, E.; Orme-Johnson, W. H. *J. Biol. Chem.* **1978**, *253*, 1001-1004.
15. Huynh, B. H.; Münck, E.; Orme-Johnson, W. H. *Biochim. Biophys. Acta* **1979**, *527*, 192-203.
16. Davis, L. C.; Shah, V. K.; Brill, W. J.; Orme-Johnson, W. H. *Biochim. Biophys. Acta* **1972**, *256*, 512-523.
17. Huynh, B. H.; Henzl, M. T.; Christner, J. A.; Zimmerman, R.; Orme-Johnson, W. H. *Biochim. Biophys. Acta* **1980**, *623*, 124-138.
18. Smith, B. E.; Lowe, D. J.; Bray, R. C. *Biochem. J.* **1972**, *130*, 641-643.
19. True, A. E.; Nelson, M. J.; Venters, R. A.; Orme-Johnson, W. H.; Hoffman, B. M. *J. Am. Chem. Soc.* **1998**, *110*, 1935-1943.
20. True, A. E.; McLean, P. A.; Nelson, M. J.; Orme-Johnson, W. H.; Hoffman, B. M. *J. Am. Chem. Soc.* **1990**, *112*, 651-657.
21. Venters, R. A.; Nelson, M. J.; McLean, P. A.; True, A. E.; Levy, M. A.; Hoffman, B. M.; Orme-Johnson, W. H. .*J. Am. Chem. Soc.* **1986**, *108*, 3487-3498.
22. Stiefel, E. I. In *ACS Symposium Series 535*; Coucouvanis, D.; Newton, W. E., Eds.; American Chemical Society: Washington, DC, 1993; pp. Chapters 10-23.

23. Demadis, K. D.; Malinak, S. M.; Coucouvanis, D. *Inorg. Chem.* **1996**, *35*, 4038-4046.

24. Deng, H.; Hoffmann, R. *Angew. Chem. Int. Ed. Engl.* **1993**, *32*, 1062-1065.

25. Dance, I. *J. Biol. Inorg. Chem.* **1996**, *1*, 581-586.

26. Stavre, K. K.; Zerner, M. C. *Chem. Euro. J.* **1996**, *2*, 83-87.

27. Yates, M. G.; Lowe, D. J. *FEBS Lett.* **1976**, *72*, 121-126.

28. Lowe, D. J.; Eady, R. R.; Thorneley, R. N. F. *Biochem. J.* **1978**, *173*, 277-290.

29. Orme-Johnson, W. H.; Davis, L. C. In *Iron-Sulfur Proteins*; Lovenberg, W., Ed.; Academic: New York, 1978; pp. 15-60.

30. Davis, L. C.; Henzl, M. T.; Burris, R. H.; Orme-Johnson, W. H. *Biochem.* **1979**, *18*, 4860-4869.

31. Pollock, R. C.; Lee, H.-I.; Cameron, L. M.; DeRose, V. J.; Hales, B. J.; Orme-Johnson, W. H.; Hoffman, B. M. *J. Am. Chem. Soc.* **1995**, *117*, 8686-8687.

32. Sørlie, M.; Christiansen, J.; Dean, D. R.; Hales, B. J. *J. Am. Chem. Soc.* **1999**, 121, 9457-9458.

33. Ryle, M. J.; Lee, H.-I.; Seefeldt, L. C.; Hoffman, B. M. *Biochemistry* **2000**, 39, 1114-1119.

34. Christie, P. D.; Lee, H.-I.; Cameron, L. M.; Hales, B. J.; Orme-Johnson, W. H.; Hoffman, B. M. *J. Am. Chem. Soc.* **1996**, *118*, 8707-8709.

35. Lee, H.-I.; Cameron, L. M.; Hales, B. J.; Hoffman, B. M. *J. Am. Chem. Soc.* **1997**, 119, 10121-10126.

36. Lee, H.-I.; Hales, B. J.; Hoffman, B. M. *J. Am. Chem. Soc.* **1997**, 119, 11359-11400.

37. Lee, H.-I.; Sørlie, M.; Christiansen, J.; Song, R.; Dean, D. R.; Hales, B. J.; Hoffman, B. M. *J. Am. Chem. Soc.* **2000**, 122, 5582-5587.

38. Burgess, B. K.; Jacobs, D. B.; Stiefel, E. I. *Biochim. Biophys. Acta* **1980**, *614*, 196-209.

39. Christie, P.D. Ph.D. Thesis, MIT, Cambridge; **1996**.

40. Scott, D. J.; Dean, D. R.; Newton, W. E. *J. Biol. Chem* **1992**, *267*, 20002-20010.

41. Christiansen, J.; Goodwin, P. J.; Lanzilotta, W. N.; Seefeldt, L. C.; Dean, D. R. *Biochemistry* **1998**, *37*, 12611-12623.

42 Chromy, V.; Fischer, J.; Kulhanek, V. *Clin. Chem.* **1974**, *20*, 1362-1363.

43. Laemmli, U. K. *Nature* **1970**, *227*, 680-685.

44. Kim, C.-H.; Zheng, L.; Newton, W. E.; Dean, D. R. In *New Horizons in Nitrogen Fixation*; Palacios, R., Mora, J., Newton, W. E., Eds.; Kluwer Academic Publishers: Norwell, MA 1992; pp 105-110.

45. Peters, J. W.; Fisher, K.; Dean, D. R. *J. Biol. Chem.* **1994**, *269*, 28076-28083.

46. Werst, M. M.; Davoust, C. E.; Hoffman, B. M. *J. Am. Chem. Soc.* **1991**, *113*, 1533-1538.

47. Mailer, C.; Taylor, C. P. S. *Biochim. Biophys. Acta* **1973**, *322*, 195-203.

48. Feher, G. *Phys. Rev.* **1959**, *114*, 1219-1244.

49. Hoffman, B. M.; DeRose, V. J.; Ong, J. L.; Davoust, C. E. *J. Magn. Reson.* **1994**, *110*, 52-57.

50. Davoust, C. E.; Doan, P., E; Hoffman, B. M. *J. Mag. Res., A* **1996**, *119*, 38-44.

51. Hoff, A.J. *Advanced EPR: Applications in Biology and Biochemistry*, Ed.; Elsevier: Amsterdam, 1989.

52. Schweiger, A. *Angew. Chem. Int. Ed. Engl.* **1991**, *30*, 265-292.

53. Gemperle, C.; Schweiger, A. *Chem. Rev.* **1991**, *91*, 1481-1505.

54. Doan, P. E.; Hoffman, B. M. *Chem. Phys. Lett.* **1997**, *269*, 208-214.

55. Abragam, A.; Bleaney, B. *Electron Paramagnetic Resonance of Transition Metal Ions*, 2nd Ed.; Clarendon Press: Oxford, 1970.

56. Hoffman, B. M. *Acc. Chem. Res.* **1991**, *24*, 164-170.

57. Hoffman, B. M.; DeRose, V. J.; Doan, P. E.; Gurbiel, R. J.; Houseman, A. L. P.; Telser, J. In *Biological Magnet Resonance*; Berliner, L. J., Reuben, J., Eds.; Plenum Press: New York and London 1993; Vol. 13, pp 151-218.

58. Hoffman, B. M.; Martinsen, J.; Venters, R. A. *J. Magn. Reson.* **1984**, *59*, 110-123.

59. Hoffman, B. M.; Venters, R. A.; Martinsen, J. *J. Magn. Reson.* **1985**, *62*, 537-542.

60. Gordy, W. *Theory and Applications of Electron Spin Resonance*; John Wiley & Sons: New York, 1980.

61. Telser, J.; Benecky, M. J.; Adams, M. W. W.; Mortenson, L. E.; Hoffman, B. M. *J. Biol. Chem.* **1986**, *261*, 13536-13541.

62. Telser, J.; Benecky, M. J.; Adams, M. W. W.; Mortenson, L. E.; Hoffman, B. M. *J. Biol. Chem.* **1987**, *262*, 6589-6594.

63. Mouesca, J.-M.; Noodleman, L.; Case, D. A.; Lamotte, B. *Inorg. Chem.* **1995**, *34*, 4347-4359.

64. Noodleman, L.; Peng, C. Y.; Case, D. A.; Mouesca, J.-M. *Coord. Chem. Rev.* **1995**, *144*, 199-244.

65. Lovell, T.; Li, J.; Liu, T.; Case, D. A.; Noodleman, L. *J. Am. Chem. Soc.* **2001**, *123*, 12392-12410.

66. Derose, V. J.; Liu, K. E.; Lippard, S. J.; Hoffman, B. M. *J. Am. Chem. Soc.* **1996**, *118*, 121-134.

67. Willems, J.-P.; Lee, H.-I.; Burdi, D.; Doan, P. E.; Stubbe, J.; Hoffman, B. M. *J. Am. Chem. Soc.* **1997**, *119*, 9816-9824.

68. Weiss, B. Ph.D Thesis, University of California, Davis, 1997.

69. Chen, J.; Christiansen, J.; Tittsworth, R. C.; Hales, B. J.; George, S. J.; Coucouvanis, D.; Cramer, S. P. *J. Am. Chem. Soc.* **1993**, *115*, 5509.
70. Greenwood, N. N.; Earnshaw, A. *Chemistry of the Elements*; Pergamon: Oxford, 1984.
71. Cameron, L. M.; Hales, B. J. *Biochemistry* **1998**, *37*, 9449-9456.
72. Noodleman, L.; Baerends, E. J. *J. Am. Chem. Soc.* **1984**, *106*, 2316-3227.
73. Blondin, G.; Girerd, J. J. *Chem. Rev.* **1990**, *90*, 1359-1376.
74. Girerd, J. J. *J. Chem. Phys.* **1983**, *79*, 1766-1775.
75. Papaefthymiou, V.; Girerd, J.-J.; Moura, I.; Moura, J. J. G.; Münck, E. *J. Am. Chem. Soc.* **1987**, *109*, 4703-4710.
76. Burgess, B. K. *Chem. Rev.* **1990**, *90*, 1377-1406.
77. Huang, H. Q.; Kofford, M.; Simpson, F. B.; Watt, G. D. *J. Inorg. Biochem.* **1993**, *218*, 59-75.
78. Wink, D. A.; McLean, P. A.; Hickman, A. B.; Orme-Johnson, W. H. *Biochem.* **1989**, *28*, 9407-9412.
79. Yang, S.-S.; Pan, W.-H.; Friesen, G. D.; Burgess, B. K.; Corbin, J. L.; Stiefel, E. I.; Newton, W. E. *J. Biol. Chem.* **1982**, *257*, 8042-8048.
80. Yoo, S. J.; Angove, H. C.; Papaefthymiou, V.; Burgess, B. K.; Münck, E. *J. Am. Chem. Soc.* **2000**, *122*, 4926-4936.
81. Burris, R. H. *J. Biol. Chem.* **1991**, *266*, 9339-9342.
82. Henderson, R. A. *Angew. Chem., Int. Ed. Engl.* **1996**, *35*, 947-967.
83. Kim, C.-H.; Newton, W. E.; Dean, D. R. *Biochemistry* **1995**, *34*, 2798-2808.
84. Shen, D. J.; Dean, D. R.; Newton, W. E. *Biochemistry* **1997**, *36*, 4884-4894.
85. Christiansen, J.; Tittsworth, R. C.; Hales, B. J.; Cramer, S. P. *J. Am. Chem. Soc.* **1995**, *117*, 10017-10024.
86. Peters, J. W.; Stowell, M. H. B.; Soltis, S. M.; Finnegan, M. G.; Johnson, M. K.; Rees, D. C. *Biochemistry* **1997**, *36*, 1181-1187.
87. Fessenden, R. W.; Schuler, R. H. *J. Chem. Phys.* **1963**, *39*, 2147.
88. Adrian, F. J.; Cochran, E. L.; Bowers, V. A. *Free Radicals in Inorganic Chemistry*; American Chemical Society: Washington, DC, 1962.
89. Dilworth, M. J. *Biochim. Biophys. Acta* **1966**, *127*, 285-294.
90. Hardy, R. W. F.; Holsten, R. D.; Jackson, E. K.; Burns, R. C. *Plant Physiol.* **1968**, *43*, 1185-1207.
91. Newton, W. E.; Fisher, K.; Kim, C.-H.; Shen, J.; Cantwell, J. S.; Thrasher, K. S.; Dean, D. R. In *Nitrogen Fixation: Fundamentals and Applications*; Tekhenovich, I. A., Provorov, N. A., Romanov, V. I., Newton, W. E., Eds.; Kuwer Academic Publisher: Dordrecht, The Netherlands, 1995; pp 121-127,
92. Stiefel, E. I. *Proc. Natl. Acad. Sci. U.S.A.* **1973**, *70*, 988-992.
93. Einsle, O.; Tezcan, F. A.; Andrade, L. A.; Schmid, B.; Yoshida, M.; Howard, J. B.; Rees, D. C. *Science* **2002**, *297*, 1696-1700.

94. Thomann, H.; Bernado, M.; Newton, W. E.; Dean, D. R. *Proc. Natl. Acad. Sci. U.S.A.* **1991**, *88*, 6620-6623.
95. DeRose, V. J.; Kim C.-H.; Newton, W. E.; Dean, D. R.; Hoffman, B. M. *Biochemistry*, **1995**, *34*, 2809-2814.
96. Lee, H.-I.; Thrasher, K. S.; Dean, D. R.; Newton, W. E.; Hoffman, B. M. *Biochemistry*, **1998**, *37*, 13370-13378.
97. Telser, J.; Smith, E. T.; Adams, M. W. W.; Conover, R. C.; Johnson, M. K.; Hoffman, B. M. *J. Am. Chem. Soc.* **1995**, *117*, 5133-5140.
98. Telser, J.; Huang, H.; Lee, H.-I.; Adams, M.W.W.; Hoffman, B.M. *J. Am. Chem. Soc.* **1998**, *120*, 861-870.

Chapter 9

Variable Frequency Pulsed EPR Studies of Molybdenum Enzymes: Structure of Molybdenum Enzymes

John H. Enemark, Andrei V. Astashkin, and Arnold M. Raitsimring

Department of Chemistry, University of Arizona, 1306 East University, Tucson, AZ 85721–0041

Variable frequency (C- to Ku-band) pulsed EPR reveals the detailed coordination structures of the Mo(V) sites in SO and DMSO reductase. A distributed orientation of the exchangeable ligand is a common feature of the Mo center in these enzymes. For SO, the conformation of the coordinated cysteinyl residue varies slightly with pH. The similarity of the pulsed ENDOR spectra of chicken and human SO indicates that the structures of the Mo(V) sites in these enzymes are nearly identical. Further refinement of the Mo(V) site structure can be achieved by means of Ka-band pulsed EPR investigations of the enzymes with the ligands containing ^{17}O.

Introduction

In this Chapter we describe our pulsed EPR investigations of two molybdenum enzymes: sulfite oxidase (SO) and DMSO reductase. SO is a physiologically vital enzyme that catalyzes the oxidation of sulfite to sulfate. The crystal structure of chicken liver SO shows a five-coordinate molybdenum center with a distorted square pyramidal geometry (*1*). One of the equatorial ligands is the thiolate side chain of a cysteinyl residue that is conserved in all species and essential for catalytic activity (*2,3*). The proposed mechanism for SO involves oxygen atom transfer followed by successive coupled electron/proton transfers that pass through the Mo(V) oxidation state, which is detectable by electron paramagnetic resonance (EPR) (*4-6*).

At least three spectroscopically distinct forms of SO have been identified by CW EPR (*7,8*). The high pH (*hpH*) form is obtained at pH 9-9.5 and low concentrations of anions. The form observed at low pH (6.5-7) depends upon the anion in the buffer. The phosphate inhibited (*Pi*) form is observed in the presence of phosphate, whereas the low pH (*lpH*) form occurs in the absence of phosphate. The EPR spectrum of *lpH* SO shows a hyperfine interaction (*hfi*) with a single strongly coupled exchangeable proton, but the *hpH* and *Pi* forms do not show any resolved *hfi* in H_2O solution. The differences in the CW EPR spectra of SO imply structural variations in its Mo(V) site that are dependent upon pH and anions. However, the resolution of CW EPR is too low to determine the details of these structural variations. Therefore, we have employed variable frequency pulsed EPR spectroscopy to investigate the structures of the Mo(V) sites of all three forms of SO.

Insufficient resolution in the CW EPR spectrum was also a problem in our investigation of another Mo enzyme, dimethyl sulfoxide (DMSO) reductase. The periplasmic DMSO reductases of the photosynthetic bacteria *Rhodobacter capsulatus* (*R.c.*) and *Rhodobacter sphaeroides* (*R.s.*) function in a respiratory chain with DMSO as the terminal electron acceptor, producing dimethyl sulfide (DMS). The high resolution (1.3 Å) crystal structure of *R.s.* DMSO reductase (*9*) revealed two forms that have five- or six-coordinate Mo centers at the active site. In the six-coordinate form, four ligands are the sulfurs from the two dithiolate groups. A terminal oxo ligand and the side chain oxygen of Ser147 complete the coordination sphere. This view is supported by recent biochemical and resonance Raman data for *R.c.* DMSO reductase (*10,11*). Although the structure of the Mo(V) site in DMSO reductase is unknown, the EXAFS (*12*) and resonance Raman (*13*) spectroscopic studies suggest that the Mo(V) center remains coordinated by two dithiolate moieties and the oxygen atom from Ser147; however, a hydroxyl group replaces the terminal oxo group.

The CW EPR spectrum of Mo(V) in *R.s.* DMSO reductase at pH 7.5 showed a strong *hfi* with the proton of the OH ligand (*14*). Increasing the pH resulted in gradual variation of the EPR spectrum shape, unlike the abrupt change observed in SO. As in SO to understand the details of the OH ligation, we applied high resolution pulsed EPR spectroscopy.

Experimental

Preparation of highly purified chicken liver SO, His-tagged recombinant human SO and the recombinant wild type *R.s.* DMSO reductase is described elsewhere *(15,16)*. Pulsed EPR measurements were mainly performed on home-built variable frequency pulsed EPR spectrometers covering the range of the microwave (mw) frequencies v_{mw} from 2 to 18 GHz. However, for the optimal observation of the ^{17}O ligand in SO by ESEEM, a significantly higher $v_{mw} \sim 35$ - 40 GHz was required. As such frequencies were not available with our instrumentation, the ^{17}O ESEEM measurements were performed at ETH, Zürich, in collaboration with Prof. A. Schweiger.

Results and Discussion

General remarks.

The pulsed EPR measurements were performed at several magnetic fields B_0 across the g-anisotropic ($\Delta g \sim 5 \times 10^{-2}$) EPR spectra of the Mo(V) centers in SO and DMSO reductase. Such measurements allowed us to selectively study the Mo(V) complexes with different orientations relative to the $\mathbf{B_0}$ vector, and to determine the orientation of the *hfi* and quadrupole interaction (*nqi*) tensors of the studied magnetic nuclei with respect to the g-tensors.

The success of many of our ESEEM studies critically depended on the multifrequency capability of our pulsed EPR instrumentation. Indeed, to optimize the ESEEM amplitude, we chose in each experiment the magnetic field B_0 and the corresponding v_{mw} in such a way, as to provide the conditions as close as possible to the mutual cancellation of the nuclear Zeeman interaction and the *hfi* in one of the electron spin manifolds. For a given magnetic nucleus, the cancellation condition requires that $v_I \approx A/2$, where v_I is the nuclear Zeeman frequency and A is the secular part of the total *hfi* constant contributed by the isotropic *hfi* constant a_{iso} and the anisotropic *hfi* tensor T_{ij}. In addition, to simplify the analysis of the ESEEM spectra, it was also advantageous to provide the so-called weak interaction condition ($v_I > A/2$), while still keeping the system close to the Zeeman/*hfi* cancellation.

ESEEM spectra of Pi SO (17).

A typical primary ESEEM spectrum of *Pi* SO in D_2O buffer is shown in Fig. 1. The labels near the spectral lines indicate their assignment to fundamental (v_H, v_D) or sum combination ($v_{H\sigma}$, $v_{D\sigma}$, $v_{P\sigma}$) lines of ^{1}H, D and ^{31}P.

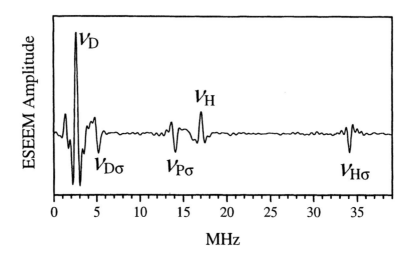

Fig. 1. Cosine Fourier transform (FT) of the primary ESEEM of Pi SO in deuterated phosphate buffer, recorded at $v_{mw} = 11.001$ GHz, $B_o = 4000$ G near $g_X = 1.965$).
(Reproduced from reference 17. Copyright 1996 American Chemical Society.)

The various ^{1}H and D lines are located at Zeeman (v_H, v_D) or at twice the Zeeman frequencies ($v_{H\sigma}=2v_H$, $v_{D\sigma}=2v_D$) of these nuclei, and the respective spectra are typical of distant and very weakly coupled protons and deuterons. However, the situation is different for ^{31}P. The unequivocal assignment of the $v_{P\sigma}$ line was based on measurements in a wide range of v_{mw} (~ 9-15.5 GHz) and corresponding B_o values. In all these experiments the $v_{P\sigma}$ line was always located near twice the Zeeman frequency of ^{31}P (v_P). On the other hand, there was no trace of the ^{31}P Zeeman line (v_P) in the ESEEM spectra. This symptomatic feature immediately tells us that the $v_{P\sigma}$ line originates from the ^{31}P nucleus that has an appreciable *hfi* with the Mo(V) ion. Such a *hfi* can only be realized if Mo(V) is coordinated by the PO$_4$ group, and therefore these experiments unambiguously demonstrated that in the *Pi* form of SO the PO$_4$ group is ligated to Mo(V).

The numerical simulations of the ESEEM spectra showed that a_{iso} of ^{31}P in the phosphate ligand is statistically distributed in very wide limits, $a_{iso} \in [0, 20]$

MHz. Such a distribution implies a structural inhomogeneity of the Mo-O-PO$_3$ fragment that results in a distribution of the Mo-O-P plane orientation with respect the Mo(V) d$_{xy}$ orbital plane. Interestingly, for the corresponding AsO$_4$ derivative of SO, a well-resolved [75]As hfi constant of about 17 MHz was directly detectable by CW EPR (*18*), implying little freedom of the AsO$_4$ ligand relative to the Mo(V) d$_{xy}$ orbital. Therefore, the ESEEM data obtained for the *Pi* form of SO were the first indication that the Mo(V) site of SO does not have a unique geometry, but allows a certain orientational freedom in the binding of ligands.

ESEEM and ENDOR spectra of lpH SO (19,20).

The presence of a single strongly coupled exchangeable proton at the Mo(V) site in the *lpH* form of SO was established already in early publications, from the splittings in CW EPR spectra (*7,8*). The proton *hfi* was up to 30 MHz, and based on information obtained from model Mo(V) compounds (*21*), it could be concluded that the OH proton is close to the equatorial position of the complex. The purpose of the pulsed EPR measurements in this system was to achieve a more detailed description of the Mo(V) site structure. However, the ESEEM experiments on *lpH* SO in buffered H$_2$O solution, performed at X and Ku bands, showed no features attributable to the strongly coupled proton. This finding is not surprising, given the magnitude of the *hfi* that resulted in unfavorable ESEEM excitation conditions and fast damping of the modulation.

Therefore, the ESEEM experiments were performed with samples in buffered D$_2$O solution because for a deuteron the *hfi* is reduced by a factor of ~6.5 and the excitation conditions are considerably improved. The *hfi* parameters obtained from the ESEEM spectra were in agreement with those found earlier by CW EPR. The main axis of the deuterium *nqi* tensor (that is aligned along the OD bond) was found to be lying very close to the XY plane of the *g*-frame, making an angle of ~ 25° with the X-axis of the *g*-frame.

The ESEEM spectra in that early work were interpreted assuming fixed *hfi* parameters because the EPR and ESEEM data provided no direct indication of possible distributions. In comparison with the *Pi* SO and *hpH* SO (see below) the situation was thus puzzling because the lack of the distribution in *hfi* parameters implied a higher degree of order in the binding geometry of the OH group in *lpH* SO. Recently we found; however, that the ^{1}H pulsed ENDOR spectra of the *lpH* SO in a protonated buffer (see Fig. 2) could not be simulated using the *hfi* data obtained from CW EPR and ESEEM. A successful fit with fixed *hfi* parameters required that the anisotropic *hfi* values be 60 - 80% greater than those used in the ESEEM simulations (see Fig. 2). With such *hfi* parameters; however, the simulated EPR splitting at g_X was considerably greater than that observed in experiment. Further simulations have shown that this

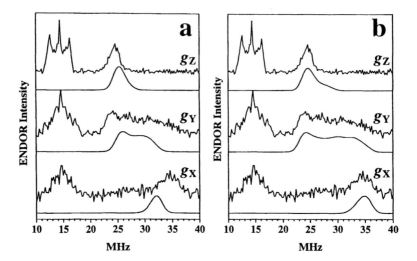

Fig. 2. Noisy traces, Davies ENDOR spectra of the chicken liver lpH SO in H_2O, recorded at $g_X \approx 1.966$, $g_Y \approx 1.972$, $g_Z \approx 2.004$. Experimental conditions: $v_{mw} = 9.446$ GHz; mw pulse durations, 40, 20 and 40 ns. Smooth lines in panel (a) are simulations with the hfi parameters obtained earlier by ESEEM ($a_{iso} = 26$ MHz; $T_{11} = T_{22} = -5.12$ MHz; $T_{33} = 10.24$ MHz; $\theta = 82\,°$; $\varphi = 8\,°$). Smooth lines in panel (b) are simulations with $a_{iso} = 26.5$ MHz; $T_{11} = -9.2$ MHz; $T_{22} = -7$ MHz; $T_{33} = 16.2$ MHz; $\theta = 80\,°$, $\varphi = 0\,°$.

disagreement could only be reconciled by introducing broad static distributions of the *hfi* and/or the orientations of the proton *hfi* tensor relative to the g-frame. The best fit was with ψ being uniformly distributed between 40° and 120° (*20*). Since the *hfi* distribution is most probably caused by the distribution of the OH binding geometry, these distributions are expected to be correlated.

Thus, also in *lpH* SO the geometry of the OH coordination to Mo(V) has a certain degree of freedom, which may be smaller than that of the phosphate ligand in *Pi* SO, but is nevertheless detectable by pulsed EPR.

ESEEM and ENDOR spectra of hpH SO (6,19,22).

CW EPR spectra of *hpH* SO do not show any splittings indicative of the presence of nearby exchangeable protons, although it was suggested (*23*) that certain features of the spectrum can be explained as forbidden transitions associated with such protons. In order to establish the presence or absence of these protons with certainty, ESEEM experiments were performed. The first X-band ESEEM spectra of *hpH* SO in buffered H_2O solution still showed no evidence for nearby exchangeable protons. However, the ESEEM of *hpH* SO in

D_2O was consistent with at least one nearby deuteron (19). Assuming a single exchangeable deuteron, the estimated *hfi* parameters of $a_{iso} \leq 2$ MHz and $T_\perp \sim -1.23$ MHz were obtained. This anisotropic *hfi* value translated into $T_\perp \sim -8$ MHz for ^1H, and it became clear that the proton ESEEM was not directly observable primarily because of a large width of spectral lines and, correspondingly, very short damping time of the ESEEM harmonics (shorter than the dead time of $\sim$200 ns). The *nqi* constant of the close deuteron determined in these experiments was found to be $\sim$2-2.5 times smaller than that normally expected for a hydroxyl deuteron, and we still do not have any plausible explanation for this observation. Nonetheless these first experiments clearly demonstrated the presence of exchangeable proton(s) close to the Mo(V) center in *hpH* SO.

A breakthrough in direct observation of the close exchangeable proton(s) was made after invention of the refocused primary (RP) ESEEM technique, a two-dimensional technique with a zero dead time in one of the time dimensions (6). A simulation of the RP ESEEM spectra (Fig. 3) has confirmed the small isotropic and large anisotropic *hfi* constants, but has established that *there are*

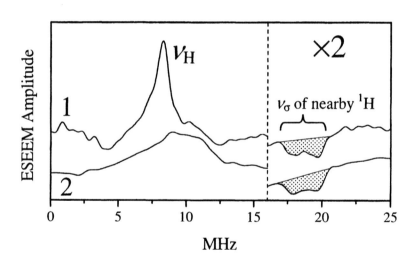

Fig. 3. Trace 1, cosine FT of integrated RP ESEEM of hpH SO in H_2O detected at $v_{mw} = 5.404$ GHz, $B_o = 1970$ G (near g_Y). Trace 2, simulated for two nearby protons with $a_{iso} = 0$ and distributed anisotropic hfi: $T_\perp = -4.1-3.2cos^2\gamma$, where γ is an angle running from 0 to 90°.

actually two close protons with such parameters. The anisotropic *hfi* was found to be distributed in very broad limits, with two distribution maxima at $T_\perp \sim -4.1$ MHz and $T_\perp \sim -7.3$ MHz. However, it was not possible to tell if the two protons contribute to different maxima, or if each maximum is contributed by both protons. Compared with the *lpH* form, the decrease in a_{iso} was more than an order of magnitude, implying that the ligand protons are substantially out of the d_{xy} orbital plane. To clarify the origin of the proton-containing group coordinated to Mo(V) that would have two protons (H_2O, an OH group associated with another proton donor, or some other species) and the unusually small *nqi* constant for the deuterons, pulsed EPR studies of ^{17}O-enriched samples are necessary.

Non-exchangeable protons (22).

The non-exchangeable protons surrounding the Mo(V) center in different forms of SO can be monitored using pulsed ENDOR spectroscopy. Based on the X-ray crystal structure (*1*), the largest splittings in these spectra are expected to arise from C_α and C_β–protons of Cys 185 (chicken liver SO numbering); the C_γ–proton of Arg 138 and the C_ϵ–proton of Tyr 322. In a broad perspective, the shape of the ENDOR spectra, even without any interpretation of separate lines, allows one to compare the structures of the Mo(V) centers from wild-type and mutated SO from different organisms. Of more specific interest may be the conformation of the coordinated cysteine residue because it could provide important insight for understanding the catalytic cycle of the enzyme. Examples of the Davies ENDOR spectra of non-exchangeable protons of *lpH* and *hpH* SO are shown in Fig. 4. One can see that in *lpH* SO the largest splitting between the ENDOR lines is ~6 MHz, while in *hpH* SO it is only ~4.4 MHz. Using the X-ray crystallography data for chicken SO and independent determination of the largest anisotropic *hfi* value by 4-pulsed ESEEM at $\nu_{mw} \approx 4.75$ GHz, the largest splitting in the ENDOR spectra is assigned to the C_α-proton of the coordinated cysteine residue. The conformation of this residue in *lpH* SO is found to be virtually the same as that in the single crystal samples studied by X-ray crystallography (*1*). The decrease of the maximal splitting in the ENDOR spectra of *hpH* SO is then attributed to the slight variation of the conformation of the cysteine residue. The required change in the ENDOR spectra can be accomplished by two consecutive 15° rotations around the C_α-C_β and C_β-S_γ bonds. This conformational change might reflect the overall change of the protein geometry as a result of different pH.

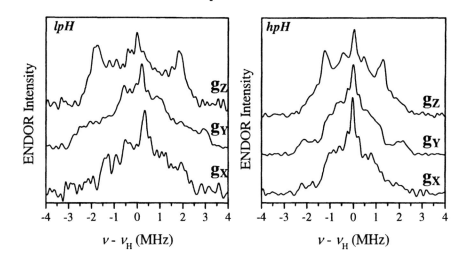

Fig. 4. Davies ^1H ENDOR spectra of lpH SO (left) and hpH SO (right) in D_2O recorded at the EPR turning points. Experimental conditions, $v_{mw} = 9.449$ GHz; mw pulse durations, 120, 60 and 120 ns.

Similarity of EPR results for chicken and human SO.

The results shown above for *lpH* and *hpH* SO were obtained for the SO from chicken liver. Similar experiments with the recombinant human SO have shown a remarkable similarity of all EPR, ESEEM and ENDOR spectra with those obtained for the chicken liver SO. This indicates that the structures of the Mo(V) center in these enzymes are nearly identical, as could be expected from the strong sequence homology for the two enzymes and the conserved sequence near the essential coordinating cysteine residue (Cys 185 in chicken; Cys 207 in human).

DMSO reductase: ESEEM experiments (24).

Previous CW investigations of DMSO reductase (*12*) showed the ligation of Mo(V) by an OH group, with the average *hfi* of the proton being in the range of 35 MHz. In this respect, DMSO reductase was very similar to the *lpH* form of SO. To obtain detailed structural information about this OH group, Ku band ESEEM experiments with DMSO reductase in buffered D_2O solution were performed. As an example, Fig. 5a shows a primary ESEEM spectrum obtained at the middle turning point of the EPR spectrum.

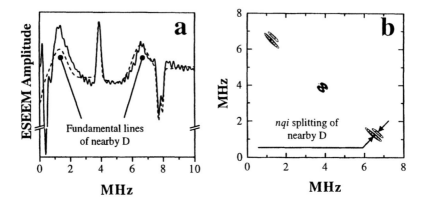

Fig. 5. (a) Solid line, cosine FT of the primary ESEEM of DMSO reductase in D_2O obtained at $v_{mw} = 16.44$ GHz and $B_o = 5921$ G (near g_Y). Dashed line, simulation with a_{iso} Gaussian-distributed around 4.8 MHz, with the width between the maximum slope points of 1.6 MHz; $T_\perp = -0.6$ MHz; $\theta = 70°$; $\varphi = 20°$; $e^2Qq/h = 0.25$ MHz; $\theta_q = 83°$; $\varphi_q = 30°$. The simulation also included the distant matrix deuterons that were assumed to be distributed with a density of 0.015 $Å^{-3}$, starting from a distance of 3.4 Å from Mo. (b) HYSCORE spectrum ((++) quadrant) of DMSO reductase in D_2O obtained at the same conditions.

The attempts to simulate the ESEEM spectra with fixed *hfi* parameters failed, as they did not result in smooth featureless fundamental lines similar to those seen in Fig. 5a. To reproduce the experimental lineshapes, we had to introduce a distribution of the *hfi* parameters. These simulations showed that it was mostly the isotropic *hfi* constant which was distributed, whereas the anisotropic *hfi* constant was restricted to very narrow limits, and could be considered as fixed. The *hfi* tensor axis was found to be directed at $\theta \approx 70°$ and $\varphi \approx 20°$ relative to the g-frame.

The direction of the *nqi* axis ($\theta_q = 90° \pm 10°$; $\varphi_q = 30° \pm 10°$) that coincides with the direction of the OD bond was obtained from the quadrupole splittings observed in the HYSCORE spectra (Fig. 5b). The angle between the axes of the *hfi* and *nqi* tensors is thus only $\Delta\theta_{hq} \approx 16°$. This angle is substantially smaller than that between the Mo-H and O-H directions (52°), which indicates that the anisotropic *hfi* of the proton (deuteron) is substantially affected by the spin density on the hydroxyl oxygen. Thus, the anisotropic *hfi* cannot be used directly to determine the Mo-H distance, nor the orientation of the Mo-H vector. Such structural information could only be derived after taking the contribution of the spin density on oxygen into account. Since no quantum chemical analysis of the *hfi* in Mo(V) complexes was available in the literature, our analysis was based on an analogy with CH_2OH radical (25), where the OH proton *hfi* is also determined by the π-bonding interactions between the oxygen orbitals with the unpaired electron orbital. Using this analogy, we were able to estimate the spin density $\rho_O \approx 0.05$ on the hydroxyl oxygen from the isotropic hfi constant of the OH-proton. This spin density agrees with that found from the hfi of ^{17}O determined by CW EPR (12). The anisotropic hfi tensor of the hydroxyl proton can be then be constructed as a sum of two main contributions, from the large (~0.7) spin density on Mo(V) and from the small (~0.05) spin density on the hydroxyl oxygen. For the tensor evaluated in such a way, the angle between the main hfi axis and the OH (OD) bond direction is about 24°, close to $\Delta\theta_{hq} \approx 16°$ obtained in experiment.

The isotropic hfi constant of the OH proton is expected to depend on the geometry of the Mo-OH fragment as follows:

$$a_{iso} \approx a_{max}\cos^2\chi \qquad (1)$$

where a_{max} is in the range of 36 - 45 MHz, and χ is the dihedral angle between the Mo-O-H plane and the plane corresponding to the orientation of the Mo-OH fragment giving the maximal isotropic hfi constant (in the ideal case, the plane of the d_{xy} orbital).

Using Eq. (1) with a_{max} = 36 MHz (lower limit of possible a_{max} values), we can find that the observed variation of a_{iso} from ~26 MHz to ~36 MHz translates to a distribution of χ within the limits from 0° to 32°. This range of χ is possibly somewhat overestimated, and it decreases if a greater a_{max} value is assumed (e.g., for a_{max} = 45 MHz, $\chi \in [27°, 41°]$). Based on the X-ray crystallographic data, it is difficult to assume an internal hydrogen bond between the exchangeable hydrogen and any other ligand because the distance from this hydrogen to the closest of these ligands, the oxygen of the Ser147 residue, is about 2.8 Å. The rather well-defined orientation of the OH bond relative to the g-frame is possibly caused by a hydrogen bond to some other amino acid residue or to a buffer or solvent molecule situated nearby.

^{17}O HYSCORE experiment.

The above investigations show that the structural interpretation of the hfi parameters of protons of $OH_{(2)}$ ligands critically depends on our knowledge of the amount of spin density delocalized to the oxygen atom, and of how this spin density is distributed over various oxygen orbitals. It is therefore obvious that the next step in refining the structural information for the Mo(V) sites in Mo-containing enzymes (including, but not limited to, SO and DMSO reductase) is to directly determine the oxygen hfi for the $OH_{(2)}$ ligands, and, combining it with the proton hfi data, to assess the spin density distribution over the $OH_{(2)}$ fragment. For this, the experiments should be performed with samples enriched with ^{17}O. CW EPR spectra of samples of SO and DMSO reductase in solutions containing ^{17}O-enriched water (26) indicate an appreciable spin density transfer from Mo(V) to ^{17}O. However, the inhomogeneous broadening of CW EPR spectra often prevents the accurate determination of the isotropic *and* anisotropic *hfi* parameters of ^{17}O. Moreover, CW EPR cannot resolve the weak *nqi* of ^{17}O, which may provide a valuable information on the nature of the oxygen-containing ligand.

Investigation of ^{17}O *hfi* by ESEEM is difficult because the *hfi* constant is rather large (30-40 MHz), while the nuclear Zeeman frequency, even at the upper limit of the Ku-band, is only ~ 4 MHz. Therefore, in all forms of SO and in DMSO reductase, ^{17}O is in the so-called "strong interaction limit", and the ESEEM amplitude is very small. The ESEEM experiments, therefore, should be performed, at least, at Ka-band, where ν_O increases to ~8 MHz. With ν_0 ~ 8 MHz the system is still in the strong interaction limit, but the ESEEM amplitude substantially increases and should become measurable. To verify this point, preliminary ^{17}O experiments on SO were performed at the ETH, Zurich, in collaboration with Prof. A. Schweiger. The spectra shown in Fig. 6 demonstrate

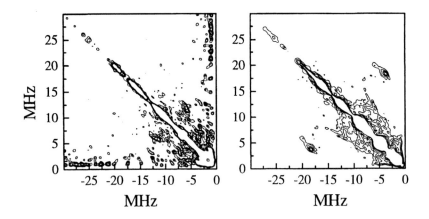

Fig. 6. HYSCORE spectra ((-+) quadrants) of lpH SO in H_2O enriched with 40% ^{17}O. Experimental conditions: (a) $\nu_{mw} = 16.011$ GHz, $B_o = 5796$ G, all mw pulses of 15 ns duration, 120 accumulations per point; (b) $\nu_{mw} = 35.285$ GHz, $B_o = 12794$ G, mw pulses of 20, 20, 16 and 20 ns duration, 20 accumulations per point.

the expected increase of intensity of the ^{17}O lines in the Ka-band experiment. Additional thorough experiments will be required to obtain the complete *hfi* tensor and to resolve the *nqi* of ^{17}O. However, these preliminary results already show that solving this problem is quite a feasible task.

Conclusions

Pulsed EPR spectroscopy enables the structures of the key Mo(V) states of the catalytic cycles of enzymes to be investigated in detail through analysis of the *hfi* and *nqi* parameters of nearby nuclei. Since these *hfi* parameters range over very broad limits, their optimal observation and analysis required the full use of the multifrequency capability of our pulsed EPR instrumentation. Our work has shown that a common feature of Mo(V) sites is a distributed orientation of exchangeable ligands. Pulsed ENDOR signals from non-exchangeable protons can be used to elucidate the conformations of Mo(V) active sites. Finally, the preliminary Ka-band ^{17}O HYSCORE experiments clearly indicate the promise of the technique to gain further insight into the active site structures of molybdenum enzymes and the trafficking of oxygen atoms during catalysis.

Acknowledgements

We are grateful to R. Codd, C. Feng, J. N. Johnson, K. J. Nelson, A. Pacheco and K. V. Rajagopalan for samples of SO and DMSO reductase. We thank Prof. A. Schweiger and Dr. I. Gromov of the ETH for the Ka band HYSCORE data for SO. This research was supported by the NIH (GM-37773 to J.H.E) and the NSF (DBI 9604939 and BIR 9224431 for construction the pulsed EPR spectrometers).

References

1. Kisker, C.; Schindelin, H.; Pacheco, A.; Wehbi, W. A.; Garrett, R. M.; Rajagopalan, K. V.; Enemark, J. H.; Rees, D. C. *Cell* **1997**, *91*, 973-983.
2. Garrett, R. M.; Rajagopalan, K. V. *J. Biol. Chem.* **1996**, *271*, 7387-7391.
3. George, G. N.; Garrett, R. M.; Prince, R. C.; Rajagopalan, K. V. *J. Am. Chem. Soc.* **1996**, *118*, 8588-8592.
4. Rajagopalan, K. V. *Molybdenum and Molybdenum Containing Enzymes*; Coughlan, M., Ed.; Pergamon Press: New York, 1980; pp 243-272.
5. Brody, M.; Hille, R. *Biochemistry* **1999**, *38*, 6668-6677.
6. Astashkin, A. V.; Mader, M. L.; Pacheco, A.; Enemark, J. H.; Raitsimring, A. M. *J. Am. Chem. Soc.* **2000**, *122*, 5294-5302.

7. Bray, R. C.; Gutteridge, S.; Lamy, M. T.; Wilkinson, T. *Biochem. J.* **1983**, *211*, 227-236.
8. Lamy, M. T.; Gutteridge, S.; Bray, R. C. *Biochem. J.* **1980**, *185*, 397-403.
9. Li, H-K.; Temple, C. A.; Rajagopalan, K. V.; Schindelin, H. *J. Am. Chem. Soc.* **2000**, *122*, 7673-7680.
10. Bray, R. C.; Adams, B.; Smith, A. T.; Bennett, B.; Bailey, S. *Biochemistry* **2000**, *39*, 11258-11269.
11. Bell, A. F.; He, X.; Ridge, J. P.; Hanson, G. R.; McEwan, A. G.; Tonge, P. J. *Biochemistry* **2001**, *40*, 440-448.
12. George, G. N.; Hilton, J.; Temple, C.; Prince, R. C.; Rajagopalan, K. V. *J. Am. Chem. Soc.* **1999**, *121*, 1256-1266.
13. Garton, S. D.; Hilton, J.; Oku, H.; Crouse, B. R.; Rajagopalan, K. V.; Johnson, M. K. *J. Am. Chem. Soc.* **1997**, *119*, 12906-12916.
14. Bastian, N. R.; Kay, C. J.; Barber, M. J.; Rajagopalan, K. V. *J. Biol. Chem.* **1991**, *266*, 45-51.
15. Bennett, B.; Benson, N.; McEwan, A. G.; Bray, R. C. *Eur. J. Biochem.* **1994**, *225*, 321-331.
16. Hilton, J.; Temple, C.; Rajagopalan, K. V. *J. Biol. Chem.* **1999**, *274*, 8428-8436.
17. Pacheco, A.; Basu, P.; Borbat, P.; Raitsimring, A.; Enemark, J. H. *Inorg. Chem.* **1996**, *35*, 7001-7008.
18. George, G. N.; Garrett, R. M.; Graf, T; Prince, R. C.; Rajagopalan K. V. *J. Am. Chem. Soc.* **1998**, *120*, 4522-4523.
19. Raitsimring, A. M.; Pacheco, A.; Enemark, J. H. *J. Am. Chem. Soc.* **1998**, *120*, 11263-11278.
20. Astashkin, A. V.; Raitsimring, A. M.; Feng , C.; Johnson, J. L.; Rajagopalan, K. V.; and Enemark, J. H. *Appl. Magn. Reson.* **2002**, *22*, 421-430.
21. Wilson, G. L.; Greenwood, R. J.; Pilbrow, J, R,; Spence, J. T.; Wedd, A. G. *J. Am. Chem. Soc.* **1991**, *113*, 6803-6812.
22. Astashkin, A. V.; Raitsimring, A. M.; Feng, C.; Johnson, J. L.; Rajagopalan, K. V.; Enemark, J. H. *J. Am. Chem. Soc.* **2002**, *124*, 6109-6118.
23. George, G. N. *J. Magn. Reson.* **1985**, *64*, 384-394.
24. Raitsimring, A. M.; Astashkin, A. V.; Feng, C.; Enemark, J. H.; Nelson, K. J.; Rajagopalan, K. V. *J. Biol. Inorg. Chem.* **2002**, in press.
25. Schastnev, P. V.; Likhosherstov, V. M.; Musin, R. N. *J. Struct. Chem.* **1974**, *15*, 474-479.
26. Cramer, S. P.; Johnson, J. L.; Rajagopalan, K. V.; Sorrell, T. N. *Biochem. Biophys. Res. Commun.* **1979**, *91*, 434-439.

Chapter 10

Mn^{2+} Sites Investigated by Advanced EPR Techniques: In-Depth Study of Mn^{2+} Ion-Binding Sites in the Hammerhead Ribozyme

Matthew Vogt and Victoria J. DeRose

Department of Chemistry, Texas A&M University, 3255 TAMU, College Station, TX 77842

EPR techniques have been employed to investigate the role that metal ion binding plays in the function of RNA molecules. The hammerhead ribozyme is an RNA capable of performing a catalytic phosphodiester bond cleavage in the presence of metal ions. EPR binding isotherms provided evidence of two classes of Mn^{2+} ion binding sites in the hammerhead. A high affinity Mn^{2+} ion binding site in the hammerhead ribozyme has been characterized by EPR, ENDOR, and ESEEM spectroscopies. These techniques have aided in determining the coordinating ligands as a phosphate oxygen and the N7 of the guanine at the A9/G10.1 site.

RNA plays many roles in cellular processes, with its major functions involving regulation of gene expression (*1*). Many types of cellular RNA are known, the most familiar being mRNA, tRNA, and rRNA with roles in protein synthesis. The study of RNA splicing led to the discovery that some RNA molecules can perform chemical reactions, and it now appears that the RNA components of the ribosome and the spliceosome assist in the reactivity of these complexes (*2*). While RNA is often predicted as a single-stranded molecule, it folds into complex structures with various tertiary motifs. Cations are important in the function of RNA molecules in stabilizing the structure of RNA and potentially playing a role in catalysis (*3,4*).

Catalytic RNAs, or ribozymes, can perform a variety of chemical reactions including promoting the cleavage or ligation of a phosphodiester bond (*5*). These catalytic RNA motifs can range in size from 40 nucleotides to well over 400 nucleotides in size. Examples of smaller, naturally-occurring ribozymes that catalyze phosphoryl transfer reactions include the hammerhead, hairpin, and the hepatitis delta virus, while the group I and group II introns are examples of larger ribozymes (*5*). At moderate ionic strength, divalent metal ions are required for most ribozyme activities. Metal ions have been proposed to play several roles in the catalytic reaction, including acting as a general base to aid in deprotonating the reactive sugar hydroxyl groups, stabilizing the negative charge on phosphate oxygen leaving groups, or aligning the phosphate for nucleophilic attack (*5*).

The hammerhead ribozyme is an RNA molecule capable of performing a site-specific phosphodiester bond cleavage reaction (Figure 1). The hammerhead contains a conserved core of nucleotides, alteration of which may either diminish or even prohibit catalytic activity. Although high concentrations of monovalent ions can support catalytic activity, divalent metal ions are required for efficient catalysis. X-ray crystal structures of the hammerhead have shown several potential divalent metal ion binding sites. Possible metal binding sites include A9/G10.1, near the G5 residue, as well as residuesnear the cleavage site. Mn^{2+}, Mg^{2+}, Cd^{2+}, Co^{2+}, Zn^{2+}, and Tb^{3+} have all been observed bound to the hammerhead in X-ray crystal structures (reviewed in (*2*)).

Although the high cellular concentration of Mg^{2+} makes it the most likely metal ion to be found in the hammerhead in Nature, other divalent metal ions, including Mn^{2+}, can substitute for Mg^{2+} and retain catalytic activity. The Mn^{2+}-substituted hammerhead ribozyme has a 2 to 5 fold increase in maximum rate when compared to the Mg^{2+} substituted hammerhead. A possible explanation for this is the difference of the pK_a value of H_2O bound to Mn^{2+} (10.6) versus Mg^{2+} (11.4), so that hydrated Mn^{2+} has a 6 fold higher M-(OH) concentration than Mg^{2+} at the same pH (*6*).

Since Mn^{2+} supports high levels of hammerhead activity, it is a convenient metal substitute that can also be used as an EPR spectroscopic probe to study the metal ion binding sites in this ribozyme. Mn^{2+} is d^5 transition metal that is virtually always high spin, having $S = 5/2$, and is 100% abundant as ^{55}Mn, with $I = 5/2$. Detailed analysis of Mn^{2+} EPR spectra is complicated by the superposition of the 5 sets of transitions between the six M_s levels. For Mn^{2+} ions in asymmetrical environments, zero-field splitting of the M_s levels produces even more complex EPR spectra with contributions of spin-forbidden transitions becoming possible (7).

RNA provides several potential sites where metal ions can bind. Common coordination sites include the phosphodiester or nucleobase keto oxygen, and purine N7 ligands. The presence of monovalent and divalent cations, such as Na^+, K^+, and Mg^{2+}, is required to neutralize the negative charge from the phosphate backbone and often to stabilize the correctly folded tertiary structure of the RNA. Metal ions can be bound to a specific RNA site either through hydrogen bonding to a metal aqua ligand, or directly coordinated to a ligand provided by the RNA, and these are referred to as outer and inner sphere coordination modes respectively. The goal of the present work is to use EPR, ENDOR, and ESEEM spectroscopic methods to identify the site(s) at which Mn^{2+} ions bind to the hammerhead ribozyme. EPR spectroscopy is used in measurements of global metal binding properties to this RNA, whereas ENDOR and ESEEM spectroscopies have allowed identification of a high-affinity site that is critical to catalysis in the hammerhead ribozyme.

Mn^{2+}-RNA Binding Isotherms Using EPR Spectroscopy

Room temperature X-band EPR has been used to obtain binding affinities of Mn^{2+} ions for the hammerhead ribozyme (8,9). In the presence of RNA, the room temperature EPR signal of Mn^{2+} ions bound to the RNA is broadened beyond detection leaving only "unbound" Mn^{2+} contributing to the EPR signal. In this technique, the assumption is made that all ions bound to the RNA have undetectable EPR signals at room temperature. The loss of the EPR signal is due to extreme line broadening caused by an increase in the zero-field splitting parameter D, which affects the Mn^{2+} electron spin relaxation rate as $T_2 \sim |D|^2$. Replacement of a water ligand in $Mn(H_2O)_6^{2+}$ by another ligand is thought to increase D sufficiently to broaden the room temperature Mn^{2+} signal beyond detection (10). By comparing the EPR signal intensities of Mn^{2+} standards of the same concentration with samples containing RNA, the amount of free Mn^{2+} in the RNA sample can be determined. The difference in signal intensity is directly related to the number of bound Mn^{2+} ions per RNA molecule and an apparent K_d value can be obtained from fitting the data with equation 1. Equation 1 assumes j classes of n_i noninteracting sites with dissociation constants $K_{d(i)}$.

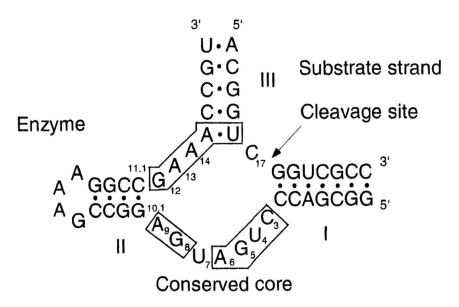

Figure 1: Hammerhead ribozyme consisting of enzyme and substrate strands.
The conserved core of nucleotides are boxed.

$$\frac{\left[Mn_b^{2+}\right]}{\left[ribozyme\right]} = \sum_{i=1}^{j} \frac{n_i\left[Mn_f^{2+}\right]}{\left(K_{d(i)} + \left[Mn_f^{2+}\right]\right)} \qquad (1)$$

For the hammerhead in 100 mM NaCl, two classes of Mn^{2+} sites are found. A set of high affinity sites of 3-4 metals with a K_d of 3.3 ± 2.7 μM, as well as a set of weaker sites with ~ 7 metals with a much higher K_d of 122 ± 48 μM are observed (8,9). Variation in Na^+ concentration influences the number and affinity of Mn^{2+} ions bound to the RNA by competing for metal binding sites and changing the electrostatic environment around the RNA. In the presence of 1 M NaCl a single high affinity Mn^{2+} site with a K_d of less than 10 μM was observed (Figure 2) (8). In the absence of Na^+, the hammerhead binds a greater number of Mn^{2+} ions, ~16, compared to the ~8 total sites for the 100 mM NaCl hammerhead samples.

The number of bound Mn^{2+} ions obtained by EPR has been correlated to cleavage activity in order to determine the effect the metal ions have on the rate of hammerhead activity. An increase in activity is observed when 4-8 Mn^{2+} ions are bound to the RNA with maximum activity attained at 5 mM Mn^{2+}. Maximum activity occurs under conditions in which all the metal sites in the hammerhead are populated. Importantly, this result does not mean that activity requires all of these sites, but does demonstrate that a weakly bound ion or set of ions is required. As described below, at least one of the high-affinity Mn^{2+} sites also is critical for hammerhead ribozyme activity.

Binding competition studies have also been performed in order to obtain a relative affinity for a series of metal ions to the hammerhead in comparison with Mn^{2+} (9). Keeping the Mn^{2+} concentration constant in the presence of the hammerhead, and increasing the concentration of the competing divalent metal ion, such as Mg^{2+}, Cd^{2+}, Co^{2+}, and Zn^{2+}, the number of Mn^{2+} ions displaced from the RNA can be determined. The results from this experiment show relative affinities for this series of metals to the hammerhead as $Mn^{2+} \sim Co^{2+} \sim Zn^{2+} > Cd^{2+} >> Mg^{2+}$. The maximum rate of cleavage of the hammerhead reaction follows a trend as $Mn^{2+} > Co^{2+} > Cd^{2+} > Mg^{2+}$ and is similar to the trend for the metal affinity to the hammerhead determined by EPR. Overall conclusions from these experiments were that metal ions with lower pK_a values, Mn^{2+} and Co^{2+}, show an increased rate in cleavage and higher apparent overall affinity to the RNA in comparison with metal ions with higher pK_a values, Cd^{2+} and Mg^{2+}.

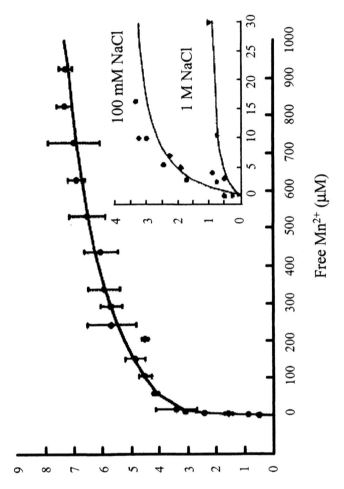

Figure 2: Mn²⁺ EPR room temperature binding isotherm for the hammerhead ribozyme. Data in the main plot are fit to two types of Mn²⁺ sites as described in the text. X-band EPR, 1-10 μM RNA concentrations, 5 mM TEA, pH = 7.8, and 100 mM NaCl for main plot. Taken from reference (8). (Reproduced from reference 8. Copyright 1998 American Chemical Society.)

Low Temperature EPR Studies of Mn^{2+} Bound to the Hammerhead Ribozyme

Low temperature X-band EPR has been used to examine Mn^{2+} ions bound to the hammerhead (*11,12*). Careful lineshape analysis can yield information about slight changes in the electronic environment of the Mn^{2+} ion when bound to the RNA. The studies described above demonstrated that in 1 M NaCl, the hammerhead has a single high-affinity Mn^{2+} site with an apparent K_d of ~10 μM (*8*), meaning that this site should be populated in a 1:1 ratio of Mn^{2+} to RNA at concentrations >100 μM (>10 times the K_d value). Comparison of Mn^{2+} in buffer and Mn^{2+} nucleotide model complexes to a hammerhead sample containing a 1:1 ratio of Mn^{2+} to RNA in 1 M NaCl showed a slight, but reproducible, difference in the Mn^{2+} EPR spectrum of the hammerhead bound metal (Figure 3) (*11,12*). The Mn^{2+}-hammerhead EPR spectrum shows this difference most notably on the sixth hyperfine line, and observation of this change has been linked to the Mn^{2+} ion bound to the hammerhead in a specific environment. This site is referred to as the hammerhead ribozyme *high affinity* Mn^{2+} *ion site*.

To further characterize the source of the EPR spectral feature for Mn^{2+} in the hammerhead high affinity site, addition of substoichiometric Mn^{2+} to hammerhead RNA was investigated. The results from this experiment showed a more defined lineshape and sharper features for the Mn^{2+} EPR signal with a stoichiometry of 0.7 or less. The substoichiometric amount of Mn^{2+} apparently ensures that all ions would be bound to the hammerhead at the same site, producing a more pronounced change in the lineshape. It is interesting to note that in activity studies, it is commonly found that only ~70% of the hammerhead ribozyme molecules react to completion (*8,9*). In combination with the activity studies, the EPR data suggest that up to 30% of the RNA molecules are misfolded in solution.

EPR microwave power saturation studies with successive addition of Mn^{2+} ions to the RNA have been used to determine the proximity of multiple metal ions occupying sites in the hammerhead (*12*). Holding the RNA concentration constant, the equivalents of Mn^{2+} ions to RNA were increased from 1:1 to 4:1 Mn^{2+} to RNA ratios, in order to observe any changes in the relaxation behavior of the metal ions. At microwave powers below saturation, the EPR signal amplitude increases linearly with the incident microwave power. At microwave powers above saturation, the Boltzmann population difference of the ground and excited spin states become reduced and the EPR signal amplitude decreases. When the Mn^{2+} ions are in close proximity, there will be additional relaxation pathways that will cause an increase in the microwave power required to saturate the EPR signal. Results for the hammerhead ribozyme

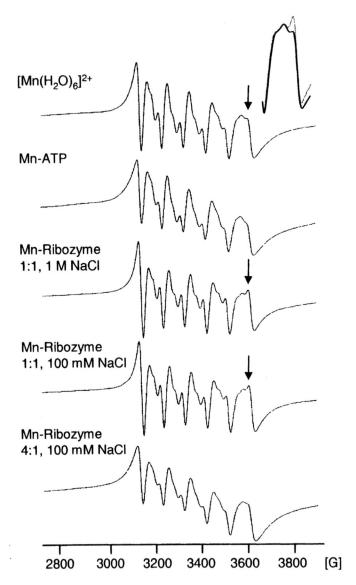

Figure 3: Comparison of Mn²⁺ EPR spectra showing lineshape changes indicative of Mn²⁺ binding in hammerhead high affinity site. Inset shows overlay of the 6ᵗʰ line feature of [Mn(H₂O)₆]²⁺ (black) and Mn-Ribozyme (gray). Conditions: 10 K, 15 G Mod. Amp., 0.063 mW microwave power, 9.44 GHz, 1 scan. Taken from reference (11).

(Reproduced from reference 11. Copyright 2000 American Chemical Society.)

showed an increase in the saturation power upon addition of the first three equivalents of Mn^{2+} to hammerhead. In comparison, in an RNA duplex, there was very little change in the power required to saturate the EPR signal when more ions were added to the RNA (Figure 4). Thus, in the presence of the hammerhead RNA, the Mn^{2+} ions are in close enough proximity that they can affect the average measured saturation behavior. However, the Mn^{2+} are not close enough to produce an EPR spectrum of coupled Mn^{2+} ions. Although the relaxation data for multiple Mn^{2+} ions are too complex to quantitatively analyze, this result suggests a picture in which the three highest-affinity Mn^{2+} ions are within approximately 15-20 Å of each other in the hammerhead ribozyme. This result is consistent with distances between Mg^{2+} ions observed in crystal structures of the hammerhead (13).

The EPR-detected Mn^{2+} binding isotherm for the 13-nucleotide RNA duplex showed 3-4 weakly coordinated Mn^{2+} ions with a K_d of ~150 μM, indicating nonspecific metal ion interactions (data not shown). As predicted, the power saturation study showed very little changes in the relaxation properties from the addition of Mn^{2+} ions to the duplex, indicating very little magnetic interactions between the metal ions when binding is nonspecific.

The Hammerhead High Affinity Site Investigated by Q-band ENDOR Spectroscopy

To gain further understanding and identification of the ligand environment around the high affinity Mn^{2+} ion in the hammerhead, ENDOR and ESEEM spectroscopies have been employed. These techniques are extremely useful in detecting weak hyperfine interactions not observed in conventional EPR spectroscopy, as well as in assigning the identity of the coordinating ligands to a metal ion (14,15). Nuclei that were predicted to be observed from RNA ligands include natural abundance ^{31}P from phosphodiester groups, $^{14}N/^{15}N$ from nucleobase ligands, and $^{1}H/^{2}H$ from exchangeable and nonexchangeable nearby protons.

Q-band ENDOR has been used to obtain information on the coordinating ligands to the Mn^{2+} ion bound to the hammerhead. To first order, features in ENDOR spectra are described by equation 2,

$$v_{\pm} = |v_N \pm A/2| \qquad (2)$$

where v_N is the Larmor frequency and A is the observed hyperfine coupling constant. ENDOR of Mn^{2+} ions is complicated due to contributions from the different m_s levels within the $S = 5/2$ spin manifold (11,15,16,17). The main contributions to the central six-line pattern are from the $M_s = +1/2 \leftrightarrow -1/2$ transitions. Lying under the central pattern are broad, unresolved contributions

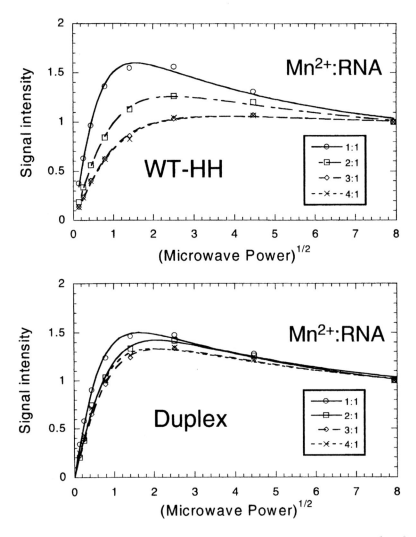

Figure 4: Microwave power saturation comparison of the hammerhead and a 13nt RNA duplex. As observed for the hammerhead ribozyme but not the duplex sample, changes in relaxation parameters occur if Mn^{2+} ions added to the RNA are in close proximity. Conditions: 20 K, 9.4 GHz, 15 G Mod. Amp., RNA conc. 250 μM, 250 μM to 1 mM Mn^{2+} in 100 mM NaCl, 5 mM TEA, pH = 7.8.

from higher M_s states, which extend into the wings of the spectrum. ENDOR obtained at any position on the Mn^{2+} EPR signal will be associated with signals from all $\Delta M_s = \pm 1$ transitions of the $S = 5/2$ Mn^{2+} ion that contribute at that magnetic field position. Hyperfine splitting from the $M_s = \pm 3/2$ and $\pm 5/2$ sublevels are multiplied by factors 3 and 5, respectively, over those observed for the EPR transitions between the $M_s = \pm 1/2$ sublevels.

The ^1H ENDOR spectra of $Mn(H_2O)_6^{2+}$ obtained near $g = 2$ show two sets of ^1H signals separated by 0.5 and 2.0 MHz centered around the ^1H Larmor frequency of ~ 50 MHz (15,11). The ^1H ENDOR spectrum of the high affinity Mn^{2+} site in the hammerhead has features similar to those observed for the $Mn(H_2O)_6^{2+}$ sample (Figure 5) (11). Following exchange into deuterated solvent, these features in the ENDOR spectrum were confirmed as arising from aqueous ligands to the Mn^{2+} ion bound to the hammerhead. However, additional features in the ^1H ENDOR spectrum were still present due to non-exchangeable protons. The ^1H ENDOR spectrum of nonexchangable protons in the hammerhead high affinity Mn^{2+} site is assigned to the C8-H of a purine residue based on comparison studies of Mn-GMP (guanosine 5'-monophosphate) model complexes with H and D at the C8 position in D_2O.

Data for the field dependence of the ^1H ENDOR signals of the nonexchangable protons in the hammerhead were obtained (11). By sampling points on the Mn^{2+} EPR envelope at different magnetic fields, contributions from the different M_s levels can be observed. As the magnetic field is lowered from $g \sim 2$, new features in the ^1H ENDOR spectrum grow in at 3 and 5 times the A value of 1.7 MHz seen for the ^1H signals due to the $+1/2 \leftrightarrow -1/2$ EPR transition. The field-dependent spectra aided in assigning the nonexchangable ^1H ENDOR features and estimated distances were obtained for two sets of ^1H at distances of ~ 3.6 Å and ~ 4.8 Å, based on a point-dipole approximation between the ^1H and Mn^{2+}. The 3.6 Å distance is consistent with assignment to the proton at the C8 position of guanine (~ 3.4 Å) while the further proton, at 4.8 Å, may be from another base or sugar in the RNA.

^{31}P ENDOR has been used to examine phosphate coordination to the high affinity Mn^{2+} site in the hammerhead ribozyme (11). Model complexes of GMP and ATP (adenosine 5'-triphosphate) were used to observe outer versus inner sphere interaction of Mn^{2+} with the phosphate. The Mn-GMP complex has a ^{31}P ENDOR spectrum consisting of only a single feature centered at the ^{31}P Larmor frequency (~ 20 MHz) and a breadth of 0.9 MHz, consistent with weak interaction. This is consistent with the crystal structure of Mn-GMP, in which the Mn^{2+} ion is coordinated to the N7 position with five H_2O molecules and a through-water contact to the phosphate. Outer-sphere phosphate coordination in Mn-GMP and a Mn-tRNA site have also been measured using Mims electron-spin-echo (ESE)-detected ENDOR methods (18). The ^{31}P

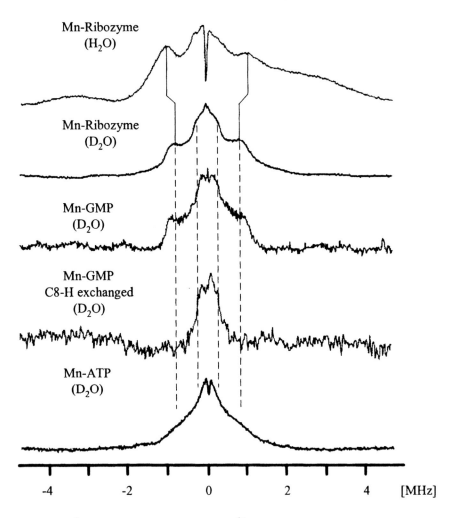

Figure 5: ^1H ENDOR comparison of Mn^{2+} hammerhead and GMP samples in H₂O and D₂O solvents. Conditions: 34 GHz, 2 K. Taken from reference (11). (Reproduced from reference 11. Copyright 2000 American Chemical Society.)

ENDOR of the Mn-ATP complex had additional hyperfine features at $\sim$ 4 MHz due to strong interaction of the Mn^{2+} ion directly coordinated to the phosphate oxygen (Figure 6). This is consistent with the Mn-ATP crystal structure showing coordination of the Mn^{2+} ion with the phosphate oxygens. This hyperfine interaction is similar to that observed by Tan et al. for a Mn^{2+} oxalate ATP complex in pyruvate kinase (16a). ENDOR results for the hammerhead high affinity Mn^{2+} site also showed features separated with a hyperfine value of $\sim$ 4 MHz. This is consistent with direct phosphate coordination to the Mn^{2+} ion. The observed value of $A(^{31}P) \sim$ 4 MHz is large in comparison with what would be expected for a simple point-dipole interaction between Mn and ^{31}P, suggesting significant polarization transfer through the Mn-O-P bond (16a).

Mn^{2+} RNA Ligands Determined by ESEEM Spectroscopy

X-band ESEEM spectroscopy, in collaboration with Dr. R. D. Britt's laboratory, was performed to further determine the coordination environment of the high affinity Mn^{2+} ion in the hammerhead (19). As described elsewhere in this volume, the 3-pulse ESEEM sequence consists of a $\pi/2$-τ-$\pi/2$-T-$\pi/2$-echo that measures the T dependence of the resulting electron spin echo amplitude. Fourier transform of the time domain data results in a frequency domain spectrum whose features are a function of hyperfine, Larmor, and quadrupole interactions. To aid in assignment of the ESEEM features, spectra of natural abundance ^{14}N (I = 1) samples of Mn-GMP in addition to Mn-hammerhead samples were obtained. The resulting ESEEM spectra contained several features < 6 MHz that are typical of Mn^{2+} coupled to a ^{14}N ligand, and the overall lineshape is similar to a Mn^{2+} ion coordinated to a histidine or imidazole ligand. This similarity would be consistent with the Mn^{2+} ion coordinated to the N7 of a purine. To determine the RNA residue responsible for this signal, ^{15}N (I = 1/2) labeled GTPs were used for *in vitro* transcription to isotopically label all the G residues in the enzyme strand of the hammerhead ribozyme. The ESEEM spectra of Mn^{2+} with ^{15}N G-labeled hammerhead and, for comparison, a sample of Mn^{2+} with ^{15}N GMP, both contained a single feature at 3.4 MHz that is consistent with ^{15}N coordination to the Mn^{2+} ion (Figure 7). This result confirms that Mn^{2+} coordinates to guanine from the hammerhead enzyme strand.

The EPR, ENDOR, and ESEEM data all are consistent with identification of the high-affinity Mn^{2+} site in the hammerhead ribozyme to a site that comprises inner-sphere coordination to Mn^{2+} by the A9 phosphate nonbridging oxygen and by the N7 of guanine G10.1. This site has been located in several crystal structures (13). In order to identify the exact location of the RNA

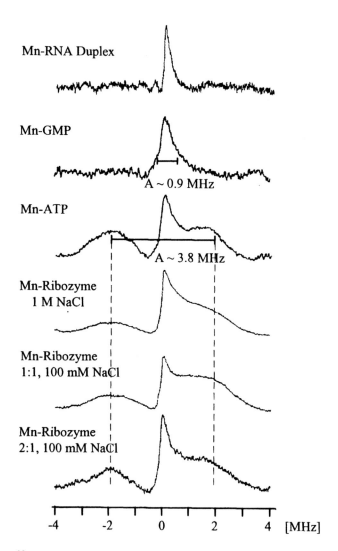

Mn-RNA Duplex

Mn-GMP

A ~ 0.9 MHz

Mn-ATP

A ~ 3.8 MHz

Mn-Ribozyme
1 M NaCl

Mn-Ribozyme
1:1, 100 mM NaCl

Mn-Ribozyme
2:1, 100 mM NaCl

-4   -2   0   2   4   [MHz]

Figure 6: ^{31}P ENDOR comparison of model complexes and RNA samples for measuring inner- and outer-sphere Mn^{2+} phosphate interactions. Conditions: 34 GHz, 2 K. Taken from reference (11).
(Reproduced from reference 11. Copyright 2000 American Chemical Society.)

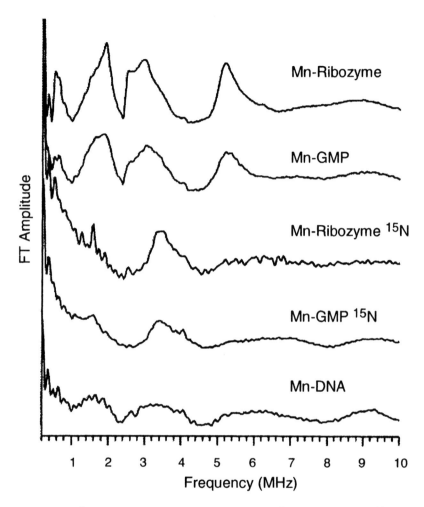

Figure 7: Mn²⁺ ESEEM spectra comparison of Mn²⁺ coordinated to ¹⁴N and
¹⁵N labeled hammerhead ribozyme and nucleotide samples. The hammerhead
ribozyme contains ¹⁵N-labeled G's in the enzyme strand. Conditions: 4.2 K,
microwave frequency 10.23 GHz, 3600 G magnetic field, 11 ns pulse length,
3.2 W microwave power. Taken from reference (19).
(Reproduced from reference 19. Copyright 1999 American Chemical Society.)

residue bound to the high affinity Mn^{2+} ion, preparation of a site-specific ^{15}N-labeled guanine residue in the hammerhead enzyme strand was pursued. While ^{15}N-labeled 2'-deoxyribose guanosine phosphoroamidites are currently commercially available, the equivalent ribose derivatives are not easily obtained for chemical synthesis of the site-specific labeled hammerhead. Deoxy modification of G10.1 was found to have little impact on Mn^{2+}-dependent activity or the low-temperature EPR signal from the high-affinity Mn^{2+} site in the hammerhead ribozyme (20). Thus, a hammerhead enzyme strand was chemically synthesized using ^{15}N-labeled 2'-deoxyribose guanine phosphoroamidites to specifically place a labeled guanine residue *only* at the G10.1 site. The resulting ESEEM spectrum is consistent with ^{15}N coordination to the Mn^{2+} ion, indicating that the G10.1 site supplies the N coordination to the high affinity Mn^{2+} site and also that this site is uniquely populated in a 47-nucleotide hammerhead RNA molecule (20).

Hoogstraten and Britt have taken advantage of the quantitative nature of ESEEM spectroscopy to obtain the number of aqua ligands in Mn-nucleotide model compounds as well as in a Mn^{2+} site in tRNA (21). This technique should be useful in further developing the Mn^{2+} coordination sphere in the hammerhead and other high-affinity Mn^{2+}-RNA sites.

Conclusions

The use of EPR techniques has provided great insight into the metal ion binding properties of RNA molecules. Although X-ray crystal structures of complex RNA molecules have demonstrated many types of potential metal sites, and studies of RNA function including both folding and catalysis have demonstrated a need for divalent cations, the lack of direct measurements of metal binding to RNA under solution conditions has been a troubling lacuna for the field of RNA structural biology. Binding affinities and number of bound Mn^{2+} ions obtained by solution EPR spectroscopy provide differentiation between tightly and weakly bound metal ions in the hammerhead ribozyme. Low temperature EPR showed a specific environment for the 1:1 Mn^{2+}: hammerhead sample, while multiple bound Mn^{2+} ions showed evidence for being in close proximity to each. The assignment of the ligands to the high affinity Mn^{2+} ion in the hammerhead based on ENDOR and ESEEM spectroscopy leads to confirmation that this site is the A9/G10.1 site. This site, along with others, is observed populated in several X-ray crystal structures of the hammerhead ribozyme; only the types of studies described above would allow the apparent affinity of this metal site to be determined under solution conditions. A model for this site is seen in Figure 8 wherein ENDOR provided confirmation of an aqueous metal ion bound to a phosphate oxygen, and

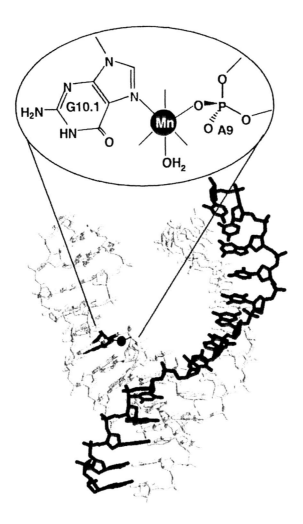

Figure 8. Hammerhead ribozyme with 'enzyme' strand in grey and 'substrate' strand in black. The Mn^{2+} site critical to activity, as deduced from ^{1}H, ^{31}P, and $^{14/15}$N ENDOR/ESEEM, is displayed as the inset (Based on reference 19). (Reproduced from reference 19. Copyright 1999 American Chemical Society.)

ESEEM clarified nitrogen coordination from the G10.1 residue in the hammerhead enzyme strand. The A9/G10.1 site in the hammerhead, although ~ 20 Å from the cleavage site, is essential for catalytic activity and even the slightest alteration at this site can have a dramatic effect on catalysis. Paramagnetic resonance techniques applied to metals bound to this catalytically important site is a valuable new examining metal-based catalysis in the hammerhead ribozyme.

Acknowledgments

This work is supported by the NIH (RO1 GM58096) to VJD and GM 61211 to RDB, the NSF (CHE-0111696), the Robert A. Welch Foundation, the Cottrell Scholars Program of the Research Corporation, and the Texas Advanced Research Program. EPR facilities at Texas A&M University are supported by the NSF (CHE-0092010).

References:

1. Cech, T. R. In *The RNA World*; Gesteland, R., Atkins, J., Eds.; Cold Spring Harbor Laboratory Press: Cold Spring Harbor, NY, 1993, pp 239-270.
2. DeRose, V. J. *Chemistry & Biology* **2002**, *9*, 961-969.
3. Misra V. K.; Draper D. E. *J. Mol. Biol.* **1999**, 294, 1135-47.
4. Pyle, A. M. *Journal of Biological Inorganic Chemistry* **2002**, *7*, 679-690.
5. Takagi, Y.; Warashina, M.; Stec, W. J.; Yoshinari, K.; Taira, K. *Nucleic Acid Research* **2001**, *29*, 1815-1834.
6. Dahm, S. C.; Derrick, W. B.; Uhlenbeck, O. C. *Biochemistry* **1993**, *32*, 13040-13045.
7. Reed, G. H.; Markham, G. D. *Biol. Magn. Reson.* **1984**, *6*, 73-142.
8. Horton, T. E.; Clardy, R.; DeRose, V. J. *Biochemistry* **1998**, *37*, 18094-18101.
9. Hunsicker, L. M.; DeRose, V. J. *Journal of Inorganic Biochemistry* **2000**, *80*, 271-281.
10. Ottaviani, M. F.; Montalti, F.; Romanelli, M.; Turro, N. J.; Tomalia, D. A. *J. Phys. Chem.* **1996**, *100*, 11033-11042.
11. Morrissey, S. R.; Horton, T. E.; DeRose, V. J. *J. Am. Chem. Soc.* **2000**, *122*, 3473-3481.
12. Hunsicker, L. M.; Vogt, M.; Morrissey, S. R.; DeRose, V. J. manuscript in preparation.

13. a) Scott W. G.; Finch J. T.; Klug A. *Cell* **1995**, *81*, 991-1002. b) Scott W. G.; Murray J. B.; Arnold J. R.; Stoddard B. L.; Klug A. *Science* **1996**, *274*, 2065-2069.

14. Hoffman, B.M. *PNAS*, in press.

15. DeRose, V. J.; Hoffman, B. M. *Methods in Enzymol.* **1995**, *246*, 554-589.

16. a) Tan, X.; Poyner, R.; Reed, G. H.; Scholes, C. P. *Biochemistry* **1993**, *32*, 7799-7810; (b) Tan, X.; Bernardo, M.; Thomann, H.; Scholes, C. P. *J. Chem. Phys.* **1993**, *98*, 5147-5157.

17. a) Sturgeon, B. E.; Ball, J. A.; Randall, D. W.; Britt, R. D. *J. Phys. Chem.* **1994**, *98*, 12871-12883. b) Arieli, D.; Vaughan, D. E. W.; Strohmaier, K. G.; Goldfarb, D. *J. Am. Chem. Soc.* **1999**, *121*, 6028-6032.

18. Hoogstraten, C. G.; Grant, C. V.; Horton, T. E.; DeRose, V. J.; Britt, R. D.; *J. Am. Chem. Soc.* **2002**, *124*, 834-842.

19. Morrissey, S. R.; Horton, T. E.; Grant, C. V.; Hoogstraten, C. G.; Britt, R. D.; DeRose, V. J. *J. Am. Chem. Soc.* **1999**, *121*, 9215-9218.

20. Vogt, M.; Hoogstraten, C. G.; Lahiri, S.; Aznar, C.; Britt, R. D.; DeRose, V. J. manuscript in preparation.

21. Hoogstraten, C.G.; Britt, R.D. *RNA* **2002**, *8*, 252-260.

Nuclear Paramagnetic Resonance

Chapter 11

Structure and Dynamics of Paramagnetic Proteins by NMR

Milka Kostic and Thomas C. Pochapsky[*]

Department of Chemistry, Brandeis University, 415 South Street, Waltham, MA 02454–9110

Many proteins of interest for NMR studies contain paramagnetic centers such as metal ions and radical species. Such proteins pose special challenges to NMR characterization. This article describes approaches to characterization of paramagnetic protein structure and dynamics using NMR. Two types of paramagnetic proteins will be discussed. First, NMR studies of a class of $Cys_4Fe_2S_2$ ferredoxins that transfer electrons to cytochromes P450 will be described. In particular, the ferredoxin putidaredoxin (Pdx) has been extensively characterized by NMR in terms of structure, dynamics and function. Pdx exhibits specific redox-dependent binding to its redox partner, cytochrome $P450_{cam}$ (CYP101). We have examined the redox dependence of protein dynamics in Pdx, redox-dependent conformational selection and have performed a detailed comparison of functional domains of Pdx with other ferredoxins. Our findings underline the role of dynamics in protein-protein interactions, linking it to the redox state of protein and implicate the specific hydrogen bonding network as a way of transmitting redox-dependent behavior. A second protein under investigation is the metal-containing acireductone dioxygenase (ARD). ARD is capable of binding a single metal ion, the nature of which determines the chemistry that is catalyzed by the enzyme. Using ^{13}C-detected 2D NMR methods, we are now able to observe resonances in the vicinity of the high-spin paramagnetic center of Ni^{+2}-bound ARD.

Putidaredoxin as a model for redox-dependent dynamics in electron transfer proteins.

The $Cys_4Fe_2S_2$ ferredoxins are a class of small (80-130 residue) acidic electron-transfer proteins that are found in all living organisms. These ferredoxins function as single electron carriers in many biologically relevant processes, including photosynthesis, aerobic and anaerobic respiration, oxidative metabolism, gene regulation and expression. They cycle between Fe(II)/Fe(III) and Fe(III)/Fe(III) oxidation states in their metal clusters. Most of the $Cys_4Fe_2S_2$ ferredoxins that have been structurally characterized so far have the same folding topology. Their structure can be viewed as consisting of two "lobes", N-terminal and C-terminal, with the iron-sulfur cluster binding site situated between them (see Figure 1). The N-terminal lobe contains the conserved hydrophobic core, and assumes the so-called ubiquitin superfold or β-grasp motif (1), with a single α-helix crossing a mixed four-stranded β-sheet approximately perpendicularly. Unlike the N-terminal lobes, the C-terminal lobes of $Cys_4Fe_2S_2$ ferredoxins show considerable differences, especially if plant-type are compared with vertebrate-type ferredoxins. In vertebrate-type ferredoxins, C-terminal lobes are typically longer (with 50-70 residues) with more regular secondary structure. Vertebrate-type ferredoxins with known structures include putidaredoxin (Pdx) (2, 3), terpredoxin (Tdx) (4), adrenodoxin (Adx) (5, 6) and *Escherichia coli* ferredoxin (Fdx) (7).

All of the structurally characterized vertebrate-type ferredoxins, except the *E. coli* ferredoxin (for which a function is not yet known (9)), play a role in oxidative metabolism acting as electron shuttles between NAD(P)H-dependent reductases and cytochromes P450. Adx functions as a reductant for mammalian P450 isozymes, CYP11A1, CYP11B1 and CYP11B2, catalyzing selective oxidations of cholesterol (10). Tdx shuttles electrons to cytochrome $P450_{terp}$ (CYP108), initiating terpinol metabolism (11) in *Pseudomonas*.

Pdx, a 106-residue ferredoxin, is the physiological reductant and effector for cytochrome $P450_{cam}$ (CYP101) in the camphor hydroxylase system from *Pseudomonas putida*. CYP101 catalyzes the 5-*exo*-hydroxylation of camphor by molecular oxygen in the first step of camphor metabolism by *P. putida*. The hydroxylation requires two electrons, shuttled sequentially, in two discrete events per turnover (for review, see ref. 12). The interactions between Pdx and CYP101 are quite specific. CYP101 exhibits low or no cross-reactivity with mammalian Adx or bacterial Tdx, two ferredoxins that are highly functionally and structurally homologous to Pdx.

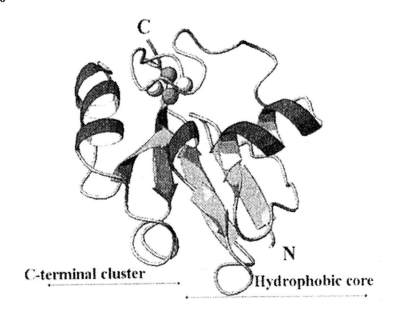

*Figure 1. Ribbon representation of the solution NMR structure of Pdx (2, 3).
Two domains of $Cys_4Fe_2S_2$ ferredoxins, the hydrophobic core (N-terminal lobe)
and the C-terminal cluster are shown, with the Fe_2S_2 cluster positioned between
them. All figures of protein structure were prepared using MOLMOL (8).*

The reduced form of Pdx (Pdx^r) binds to CYP101 two orders of magnitude more tightly than the oxidized form (Pdx^o). Examination of redox-dependent differences in ^1H and ^{15}N chemical shifts and nuclear Overhauser effects (NOE) between Pdx^o and Pdx^r revealed that the largest changes in the diamagnetic region of protein are localized within C-terminal cluster, specifically at residues His 49, Ala 76, Glu 77, and Trp 106 (13). Comparing redox dependence of amide proton exchange rates, we found that, in general, Pdx^o was more dynamic than Pdx^r, on millisecond to kilosecond time-scale (14). Also, probing local dynamics of the diamagnetic regions of Pdx as a function of oxidation state by measuring ^{15}N relaxation rates (picosecond to nanosecond time scale), showed that order parameters for Pdx^r were higher than for Pdx^o (15), pointing out that Pdx dynamics exhibit similar redox effects over the whole range of experimentally accessible protein motions. Therefore, Pdx^o appears to sample a larger region of conformational space, on time average, than Pdx^r, with Pdx^r occupying only a fraction of conformational substates occupied by Pdx^o (Figure 2). Based on this analysis, it is expected that Pdx^r would be entropically disfavored relative to Pdx^o (16). This was confirmed by observed temperature effects upon the reduction potential of Pdx (17).

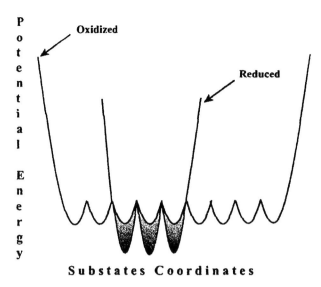

Figure 2. Simplified energy diagram illustrating proposed effects of redox-induced conformational selection in Pdx, where reduced form occupies a subset (highlighted by shading) of the conformations occupied by oxidized form. (Reproduced from reference 20. Copyright 2001 American Chemical Society.)

Origin of Redox Dependent Dynamics in Pdx. It is reasonable to assume that the redox-state dependence of the dynamics of Pdx is triggered by structural changes taking place within the Fe_2S_2 binding loop. This surface-exposed loop (G_{37}-D_{38}-$\underline{C_{39}}$-G_{40}-G_{41}-S_{42}-A_{43}-S_{44}-$\underline{C_{45}}$-A_{46}-T_{47}-$\underline{C_{48}}$-H_{49}) contains three of the four cysteinyl ligands of the metal cluster (underlined). Considering the amino acid composition of the loop, considerable flexibility of the polypeptide backbone might be expected. Several experimental results suggest the existence of considerable differences in local dynamics between Pdx^o and Pdx^r in this region. The largest changes in NMR-accessible protein dynamics occur for residues either sequentially or spatially adjacent to the metal cluster binding loop (14, 15). Also, we have recently reported redox-dependent hyperfine-shifts and nuclear spin relaxation rates for amide ^{15}N and $^{13}C'$ resonances in both oxidation states of Pdx (18, 19). While these effects are strongly dependent on differences in unpaired electron spin density between the two oxidation states, differential effects across the polypeptide chain could be due to changes in time average conformational populations as well.

Patterns of hyperfine shifts, temperature effects and nuclear spin relaxation rates suggest localized redox-dependent conformational selection in the Pdx metal cluster binding loop. It was noted that ^{15}N resonances of several residues

(Gly 41, Ala 43 and Ala 46) relax more slowly and shift upfield upon reduction, and show smaller temperature dependence in Pdx^r than Pdx^o. On the other hand, for the adjacent residues, Ser 42, Ser 44 and Thr 47, ^{15}N relaxation rates increase, resonances move downfield and become more temperature sensitive going from Pdx^o to Pdx^r (Table 1). These observations imply a conformational selection that, on a time average, moves the H^N groups of Gly 41, Ala 43 and Ala 46 further away from, and the H^N groups of Ser 42, Ser 44 and Thr 47 closer to the metal cluster upon reduction. Based on comparison with the crystal structure of oxidized Adx (5), the H^N protons of Ser 44 and Thr 47 are expected to hydrogen bond to the cysteinyl S$^\gamma$ sulfur atoms of Cys 39 and Cys 45, respectively. The observed behavior upon reduction of Ser 44 and Thr 47 ^{15}N shifts suggests increased unpaired electron spin delocalization, which could reflect the strengthening of the hydrogen bonds to the cluster from those amides, as expected due to the increase of negative charge at the metal cluster. This alternant pattern of redox-dependent changes in hyperfine shifts is also seen for ^{13}C' shifts in the cluster binding loop. In this case, since carbonyl groups are not implicated in direct interactions with metal cluster, the changes in ^{13}C' hyperfine may indicate real structural perturbations upon reduction.

Table 1. Comparison of Pdx^o and Pdx^r: ^{15}N T$_1$ Relaxation Times and Chemical Shifts for Residues in the Metal Cluster Binding Loop.

Residue	$^{15}N\ T_1$ (Pdx^o) [ms]	$^{15}N\ T_1$ (Pdx^r) [ms]	$^{15}N\ \delta$ (Pdx^o) [ppm]	$^{15}N\ \delta$ (Pdx^r) [ppm]	$\Delta\delta\ ^{15}N$ (Pdx^o-Pdx^r) [ppm]
D38	42	30	121.7	120.8	0.9
C39	29	8	107.2*	182.5*	-75.3
G40	56	25	117.6	124.2	-6.6
G41	19	50	156.2	124.8	31.4
S42	47	12	139.5	176.6	-37.1
A43	30	80	131.0	120.7	10.3
S44	32	10	132.1	87.9	44.2
C45	22	9	136.2	140.8	-4.6
A46	34	60	137.5	121.0	16.5
T47	37	20	146.6	85.6	61
C48	58	4	126.7*	272.7*	-146
L84	55	65	120.6	122.8	-2.2
C85	52	30	135.4	139.8	-4.4
C86	37	2	151.0	260.5	-109.5
Q87	53	15	129.8	160.2	-30.4

Assignments marked with asterisk are tentative , refs. 18 and 19.

Based on the pattern of redox-dependent changes in amide resonances in the cluster binding loop, it can be concluded that the polypeptide loop from Cys 39 to Ala 46 spends more time in a "puckered" conformation upon reduction. That kind of motion would bring NH groups of residues Ser 42, Ser 44 and Thr 47 closer to the metal cluster, while moving Gly 41, Ala 43 and Ala 46 further away and it can be regarded as the consequence of shortening hydrogen bonds between metal cluster and binding loop amides in Pdx^r, resulting in a contraction of the binding loop around metal center.

Redox-Dependent Conformational Selection in Pdx (20). The Roles of Gly 40 and Gly 41 in Controlling Local Dynamics. The proposed contraction of the metal binding loop upon reduction requires that some parts of the loop polypeptide chain take up the slack that would be generated by the loop contraction. That portion of the chain should not be critical for metal cluster ligation and should be sterically unhindered and flexible. Two adjacent loop residues in Pdx, Gly 40 and Gly 41, could potentially act as the hinges for multiresidue conformational fluctuations. We mutated these residues, changing Gly to Asn. We expected that the side chain of Asn would increase the barrier to proposed hinge motions, causing an overall slowing down of backbone dynamics in the mutant protein. One mutant in particular, G40N, showed high stability and achieved the desired slowing.

Comparison of WT and G40N Pdx. To compare WT and G40N Pdx in both oxidation states, uniformly ^{15}N labeled protein samples were prepared, and ^1H-^{15}N HSQC (21) experiments were applied to map differences in the proteins structures and dynamics. The primary differences between WT and G40N Pdx are seen in the ^1H-^{15}N HSQC spectra of oxidized forms. Differences between WT and G40N Pdx^o are observed at the NH correlations of Val 6, Ser 7, His 8, Gly 10, Arg 12, Arg 13, Ala 27, Ser 29, Asn 30, Ile 32, Tyr 33, Asp 34, Ile 35, Tyr 51, Thr 75, Lys 79, Asp 103, Arg 104, Gln 105, Trp 106, and in side chain resonances, namely $H^{\delta 1}$ of His 8 and His 49, as well as H^{ε} of Arg 13 and $H^{\varepsilon 1}$ of Trp 106. Some of the resonances are broadened in G40N Pdx^o, sometimes so much so as to be unobservable (Ser 29, Asp 34 and Arg 104). In some cases, peaks in G40N Pdx^o are split into groups of well-defined resonances, that we interpret as representing conformational substates at slow exchange on the chemical shift time scale (Thr 75 and Gly 10), as shown in Figure 3. These perturbed resonances also showed an expected temperature dependence, going from a defined multiplet structure (slow exchange) at low temperatures to a singlet at higher temperature when the fast exchange limit is reached. Remarkably, reduction of G40N Pdx resulted in the collapse of broadened and multiplied peaks observed in G40N Pdx^o into peaks that are essentially identical to those in WT Pdx^r. This behavior can be taken as an indication that fewer conformational substates are occupied in Pdx^r than in Pdx^o, and that the same conformational substates are occupied in both WT and G40N Pdx^r.

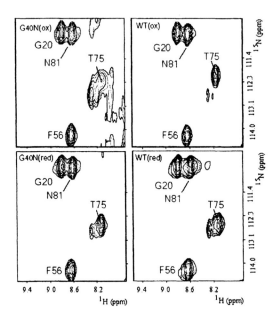

Figure 3. Comparison of selected region of the 11.74 T (500 MHz ^{1}H) ^{1}H-^{15}N HSQC spectra of oxidized and reduced forms of WT and G40N Pdx. Multiplet structure of Thr75 resonance in G40N Pdx^o can be clearly seen. Phe 56 and Gly 20, unperturbed resonances provide a reference. Figure adopted from Pochapsky et al. (20).

Time Scales of Affected Motion. The temperature dependence of the G40N Pdx^o spectra confirms that the spectral perturbations introduced by mutation are the result of dynamic processes that are slow or intermediate on the chemical shift time scale. In order to unambiguously establish the time scale of affected motions in WT Pdx, we performed an analysis of protein backbone motion based on ^{15}N relaxation rates measurements. ^{15}N T_1 and T_2 values were determined on a per residue basis for WT Pdx^o and Pdx^r, at two field strengths, 11.74 T (500 MHz ^{1}H) and 14 T (600 MHz ^{1}H). To judge the importance of the exchange contribution to the transverse relaxation rate, we applied a selection criterion based on the analysis of ^{15}N T_1/T_2 ratios (22), that is, the R_{ex} term must be included in transverse relaxation rate calculations if, for a certain residue, ^{15}N T_1/T_2 ratio is more than one standard deviation above the mean. In Pdx°, the R_{ex} term has to be taken into account for Val 6, His 8, Arg 12, Tyr 33, Asp 34 and Lys 79 (Plate 1). All these residues are affected by G40N mutation in Pdx°. The static magnetic field dependence of ^{15}N T_1 and T_2 values can be used to evaluate the time scale of the exchange contribution (23). In our case, applying this method confirmed that the conformational changes observed in WT Pdx° are in the fast exchange limit and that the observed G40N Pdx° behavior is indeed due to slowing of protein dynamics.

Coupling of the Metal Binding Loop Conformational Changes to the C-Terminal Cluster. One obvious conclusion of our NMR studies is that mutations within the metal cluster binding loop can strongly affect Pdx dynamics. Three main regions where these effects are seen include residues immediately preceding the metal binding loop (Asn 30-Ile 35) and residues in contact with them (Val 6, Ser 7, His 8, Gly 10, Ala 27 and Ser 29), residues that contact metal binding loop near residues Gly 40 and Gly 41 (Ser 22 and Leu 23) and residues in the C-terminal cluster region. Perturbations in the C-terminal cluster region are the most far-reaching, indicating that there must be a way to transmit conformational changes from the metal binding loop to the C-terminal cluster.

The conserved basic residue His 49 occupies a position immediately following the third cysteinyl ligand of the Fe_2S_2 cluster in the metal binding loop of Pdx, making it ideally situated to respond to any redox-dependent conformational and/or dynamic changes in metal binding loop. The CXTCH (X = A, S) motif is common to many ferredoxins. In some cases, like in Tdx, the His residue is replaced by another basic residue, Arg. The imidazole of His 49 has some unusual properties. It has an unusually low pK_a and does not protonate to form the imidazolium ion over the pH range of Pdx stability (9 > pH > 6.5). The unprotonated $N^{\epsilon 2}$ nitrogen exhibits a chemical shift of 247 ppm, near the extreme for any diamagnetic His $N^{\epsilon 2}$ nitrogen (24). The behavior of His 56 (corresponding to His 49 in Pdx) shows nearly identical behavior. Xia *et al.* (25) indicated that His $H^{\delta 1}$ has a pK_a < 5 and we have assigned the $H^{\delta 1}$ at 11.67 ppm and $N^{\epsilon 2}$ at 240 ppm (26). These observations can be explained by assuming that His 49 in Pdx and His 56 in Adx are forming strong hydrogen bonds with residues in the C-terminal cluster region in both proteins. In the crystal structure of truncated Adx° (5), the $H^{\delta 1}$ of His 56 is hydrogen bonded to the backbone carbonyl oxygen of Tyr 82. This crystal structure also shows a hydrogen bond between Ser 88 H^γ and His 56 $N^{\epsilon 2}$. Ser 88 H^γ gives a signal in the ^{1}H NMR spectrum of Adx° at 11.2 ppm, with well-defined NOE's with His 56 imidazole protons that confirm this proton is involved in a strong hydrogen-bonding interaction, since it is extremely downfield shifted (OH protons usually show up between 0.5 and 5 ppm) and not exchanging with the solvent on NMR time scale. The situation in Pdx is similar but there exists an important difference. The $H^{\delta 1}$ of His 49 shows NOEs to the side chain resonances of Val 74 and the H^N of Ala 76. This is consistent with the hydrogen bond between carbonyl of Ala 76 and $H^{\delta 1}$ of His 49 which is precisely homologous to His 56 $H^{\delta 1}$ – Tyr 82 carbonyl hydrogen bond in Adx. However, the distance between Ser 82 H^γ and the His 49 $N^{\epsilon 2}$ is too great for a direct hydrogen-bonding interaction. Instead, in Pdx, Tyr 51 H^η lies approximately between the Ser 82 H^γ and the His 49 $N^{\epsilon 2}$ and may act as a part of hydrogen bonding relay (Plate 2).

A number of NOEs between these residues in Pdx support this possibility. In addition, we prepared a number of site-directed mutations to probe the importance of His 49 and Tyr 51. We found that both H49F and H49R Pdx mutants were expressed as apoproteins (without Fe_2S_2 cluster incorporation) and could not be reconstituted under normal conditions (26). From these results, it is clear that the stability of Pdx is seriously compromised by the replacement of His 49. It is interesting to note that Adx can tolerate H56R, H56N and H56Q mutations, although these are less stable than the WT protein (27).

The other member of proposed hydrogen-bonding network in Pdx, Tyr 51, was mutated to Phe. This mutation should be able to maintain any hydrophobic interactions found in WT Pdx and perturb only the hydrogen bonding ability. Y51F Pdx was expressed primarily as apoprotein and was reconstituted with Fe_2S_2 cluster in vitro. The optical properties of reconstituted Y51F Pdx° are similar to those of WT. However, the thermal and acid stability of the mutant protein is greatly reduced. Y51F Pdx retains only 33% activity of WT Pdx in the fully reconstituted CYP101 camphor hydroxylase system. To compare WT and Y51F Pdx in more detail, homonuclear ^1H 2D NOESY and TOCSY spectra of the two proteins were compared. NMR spectral data showed that no major structural differences exist between the two proteins. The largest chemical shift changes were observed, as expected, for residue 51. However, one of the signature peaks of complete folding of Pdx, the Ser 82 H^γ resonance at 10.10 ppm, was completely lost in Y51F Pdx. This serves as direct evidence that the hydrogen bonding network that holds the Ser 82 H^γ in place in WT Pdx, is disrupted or perturbed in Y51F Pdx mutant. However, the relatively small perturbations of the His 49 imidazole resonances suggest that this residue still forms multiple hydrogen bonds. It is possible that in Y51F Pdx, the hydrogen bonding network is reconstituted with a solvent water molecule replacing the H^η of Tyr 51. NOEs from bulk water to aromatic protons of Phe 51 support this notion.

Comparison of Local Dynamics in Pdx, Adx and Tdx via H/D Exchange Measurements. Amide proton exchange rates provide a measure of a local solvent accessibility as well as local backbone dynamics in protein. It is interesting to compare the differences between protein dynamics of oxidized and reduced forms of Adx, Pdx, and Tdx, keeping in mind the postulated role of a hydrogen bonding network as a mechanical transmitter of redox dependent changes from the metal binding loop to the remote sites in the C-terminal cluster. In Adx, His 56 is hydrogen bonded to Tyr 82 and Ser 88, in Pdx His 49 is bonded to Ala 76 and Tyr 51, which is in turn bonded to Ser 82 and in Tdx Arg 49 doesn't have a well-defined hydrogen bonding pattern (Plate 2). As seen from the H/D exchange study results, all three proteins contain two redox-responsive regions. One is the region just preceding the metal binding loop and the other region is the C-terminal cluster. Of the three proteins, Tdx exhibits the smallest effects on changing redox state; this may result from the greater flexibility afforded by Arg 49. The overall larger effects of reduction on amide exchange rates observed in Adx relative to Pdx suggest that the shorter hydrogen bonding network in Adx transmits conformational and dynamic information more effectively.

^{13}C and ^{15}N Sequence-Specific Resonance Assignments in Paramagnetic Proteins.

The presence of unpaired electron spin density in a protein has NMR relevant consequences, including an increase of nuclear spin relaxation rates and changes in chemical shifts of neighboring spin active nuclei. This often makes standard ^{1}H-detected methods for sequence specific assignments inapplicable. For identifying nuclei near paramagnetic centers, special approaches are needed, especially those based on direct detection of ^{13}C and ^{15}N. For example, our investigations of redox-dependent behavior within the metal binding loop of Pdx have been made possible by specifically assigning hyperfine shifted ^{15}N and ^{13}C' resonances. These assignments were made using combination of doubly and multiply labeled samples and selective ^{13}C/^{15}N difference decoupling NMR experiments, optimized for detection of fast-relaxing nuclei. We have used this method for obtaining unambiguous assignments for most of the amide ^{15}N and ^{13}C' resonances in the paramagnetic region of Pdx in both oxidation states (18). Since these difference decoupling experiments are based on detecting small changes in linewidths due to the coupling, they cannot be applied if the linewidth exceeds the relevant coupling constant by a factor greater than 10. Also, due to the requirement for selective labeling, this method is fairly laborious.

Knowledge of the exact assignments within metal binding region is, in many instances, crucial for understanding the functional features of a given metalloprotein. Therefore, we are developing rapid recycle ^{13}C detected methods based on the heteronuclear multiple quantum coherence (HMQC) experiment that can be applied to uniformly ^{13}C and ^{15}N samples in order to yield information about connectivities and assignments.

For testing our experimental approach, we are using a 20 kDa Ni^{2+}-containing enzyme, acireductone dioxygenase (ARD) (28), from the methionine salvage pathway of *Klebsiella pneumoniae*. Due to the presence of a high-spin Ni^{2+}, resonances could not be assigned for 34 of the 179 residues in the protein (29). We have designed and applied two experiments to date. First, we used a pair of $^{13}C^{\alpha}$ {$^{13}C'$} multiple quantum coherence experiments to obtain correlation between $^{13}C^{\alpha}$, directly detected, and $^{13}C'$, indirectly detected. The pulse sequence used is a standard four pulse HMQC experiment (30). Selectivity of ^{13}C pulses was achieved using soft pulses set at the center of either $^{13}C^{\alpha}$ or $^{13}C'$ region and null at $^{13}C'$ or $^{13}C^{\alpha}$ region, respectively. Results are shown in Plate 3. Cross-peaks shown in green result from using experimental conditions optimized for detection of nuclei that are not affected by paramagnetism, and those shown in red are obtained by optimizing pulse sequence delays to allow for detection of fast relaxing resonances. A new doublet appears in the chemical shift region corresponding to a glycine C^{α} in the paramagnetic spectrum. Since ARD has a single glycine in paramagnetic region of the protein, we assigned Gly99 $^{13}C^{\alpha}$ and $^{13}C'$ resonances to be 41.1 and 176.8 ppm, respectively (31).

The second pair of experiments was designed to detect correlations between $^{13}C'$ of a given residue and ^{15}N of the following residue. Again, a pair of experiments was performed with two sets of experimental conditions, one optimized for detection of diamagnetic and the other for detection of paramagnetic signals. Obtained spectral results are presented in Plate 4. The most pronounced differences between the two datasets are the appearance of several signals outside of standard ^{15}N chemical shift region (well resolved doublets above 140 ppm). Also, there is a new peak is observed in the paramagnetic spectrum, corresponding to Gly 99 $^{13}C'$-Glu 100 ^{15}N correlation (31).

Previously, Machonkin et al. (32) have reported the application of ^{13}C{^{13}C} CT-COSY for identifying connectivities between fast relaxing ^{13}C resonances, but to best of our knowledge our work represents the first report on applying ^{13}C detected heteronuclear and homonuclear multiple quantum coherence for resonance assignments in paramagnetic proteins. Also, assignments obtained using these methods represent the first assignments of resonances within the paramagnetic region of ARD.

Conclusion

Understanding the details of structure and function of proteins is one of the fundamental aims of biochemical research. Here, we described some recent efforts made by the researchers in our laboratory to understand behavior of several paramagnetic vertebrate-type ferredoxins, adrenodoxin, putidaredoxin and terpredoxin, and to develop new strategies for resonance assignments in paramagnetic proteins, using Ni^{2+} acireductone dioxygenase as a test system. Our results have demonstrated the existence of redox-dependent behavior in Pdx, have pointed out the possible origin of this dependency and implicated a specific hydrogen-bonding network as a mechanical linker between metal cluster binding loop and C-terminal domain, responsible for specific interactions with redox partners. Also, we have showed that rapid recycle ^{13}C detected HMQC experiments, specially tailored for detection of fast relaxing nuclei, if applied systematically, can provide us with backbone assignments within the regions of metalloproteins that are invisible to standard ^1H-detected protein NMR experiments.

Acknowledgments

This work has been supported by NIH grant R01-GM44191 (T.C.P.). Parts of manuscript have been adopted from work referenced (20, 26, 31).

References

1. Overington, J. P. *Curr. Opin. Struct. Biol.* **1992**, *2*, 394.
2. Pochapsky, T. C.; Ye, X. M.; Ratnaswamy, G.; Lyons, T. A. *Biochemistry* **1994**, *33*, 6424.
3. Pochapsky, T. C.; Jain, N. U.; Kuti, M.; Lyons, T. A.; Heymont, J. *Biochemistry* **1999**, *38*, 4681.
4. Mo, H.; Pochapsky, S. S.; Pochapsky, T. C. *Biochemistry* **1999**, *38*, 5666.
5. Muller, A.; Muller, J. J.; Muller, Y. A.; Uhlmann, H.; Berhardt, R.; Heinenmann, U. *Structure* **1998**, *6*, 269.
6. Pikuleva, I. A.; Tesh, K.; Waterman, M. R.; Kim, Y. C. *Arch. Biochem. Biophys.* **2000**, *373*, 44.
7. Kakuta, Y.; Horio, T.; Takahashi, Y.; Fukuyama, K. *Biochemistry* **2001**, *40*, 11007.
8. Koradi, R.; Billeter, M.; Wüthrich, K. *J. Mol. Graphics* **1996**, *14*, 51.
9. Takashi, Y.; Nakamura, M. *J. Biochem.* **1999**, *126*, 917.
10. Bernhardt, R.; Muller, A.; Uhlmann, H.; Grinberg, A.; Muller, J. J.; Heinenmann, U. *Endocr. Res.* **1998**, *24*, 531.
11. Peterson, J. A.; Lu, J. Y.; Geisselsoder, J.; Graham-Lorence, S.; Carmona, C.; Witney, F.; Lorence, M. C. *J. Biol. Chem.* **1992**, *267*, 14193.

12. Mueller, E. J.; Loida, P, J.; Sligar, S. G. in *Cytochrome P450: Structure, Function and Biochemistry* (Ortiz de Montellano, P., Ed.); Plenum Press, New York, **1995**, pp 83.
13. Pochapsky, T. C.; Ratnaswamy, G.; Patera, A. Biochemistry 1994, 33, 6433.
14. Lyons, T. A.; Ratnaswamy, G.; Pochapsky, T. C. *Protein Sci.* **1996**, *5*, 627.
15. Sari, N.; Holden, M. J.; Mayhew, M. P.; Vilker, V. L.; Coxon, B. *Biochemistry* **1999**, *38*, 9862.
16. Pochapsky, T. C.; Arakaki, T.; Jain. N.; Kazanis, S.; Lyons, T. A.; Mo, H.; Patera, A.; Ratnaswamy, G.; Ye, X. in *Structure, Motion, Interaction and Expression of Biological Macromolecules,* Vol. 2, Adenine Press, New York, **1997**, pp. 79.
17. Reipa, V.; Holden, M. J.; Mayhew, M. P.; Vilker, V. L. *Biochim. Biophys. Acta* **2000**, *1459*, 1.
18. Jain, N. U.; Pochapsky, T. C. *J. Am. Chem. Soc.* **1998**, *120*, 12984.
19. Jain, N. U.; Pochapsky, T. C. *Biochem. Biophys. Res. Commun.* **1999**, *258*, 54.
20. Pochapsky, T. C.; Kostic, M.; Jain, N.; Pejchal, R. *Biochemistry* **2001**, *40*, 5602.
21. Kay, L. E.; Keifer, P.; Saarinen, T. *J. Am. Chem. Soc.* **1992**, *114*, 10663.
22. Clore, G. M.; Driscoll, P. C.; Wingfield, P. T.; Gronenborn, A. M. *Biochemistry* **1990**, *29*, 7387.
23. Millet, O.; Loria, P. J.; Kroenke, C. D.; Pons, M.; Palmer. A. G., III *J. Am. Chem. Soc.* **2000**, *122,* 2867.
24. Farr, E. A.; Seavey, B. R.; Conti, A. M.; Westler, W. M.; Markley, J. L. *J. Cell. Biochem.* **1993**, 252.
25. Xia, B.; Cheng, H.; Skjeldal, L.; Coghlan, V. M.; Vickery, L. E.; Markley, J. L. Biochemistry 1995, 34, 180.
26. Kostic, M.; Pochapsky, S. S.; Obenauer, J.; Mo, H.; Pagani, G. M.; Pejchal, R.; Pochapsky, T. C. *Biochemistry* **2002**, *41*, 5978.
27. Beckert, V.; Schrauber, H.; Bernhardt, R., Vandijk, A. A.; Kakoschke, C.; Wray, V. *Eur. J. Biochem.* **1995**, *231*, 226.
28. Dai, Y.; Pochapsky, T. C.; Abeles, R. H. *Biochemistry* **2001**, *40*, 6379.
29. Mo, H.; Dai, Y.; Pochapsky, S. S.; Pochapsky, T. C. *J. Biomol. NMR* **1999**, *14*, 287.
30. Müller, L. *J. Am. Chem. Soc.* **1979**, *101*, 4481.
31. Kostic, M.; Pochapsky, S. S.; Pochapsky, T. C. *J. Am. Chem. Soc.* **2002**, *124*, 9054.
32. Machonkin, T. E., Westler, W. M., Markley, J. L. *J. Am. Chem. Soc.* **2002**, *124*, 3204.

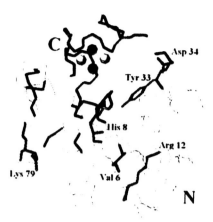

Plate 1. Backbone representation of Pdx. Shown in red are residues from the C-terminal cluster (His 49, Tyr 51, Leu 71, Val 74, Thr 75, Ala 76, Leu 78, Lys 79, Ser 82, Pro102 and Trp 106), in green is metal cluster binding loop and in purple are residues with R_{ex} contribution to their T_2. Iron-sulfur cluster is presented as 4 spheres, green = Fe and yellow = S.

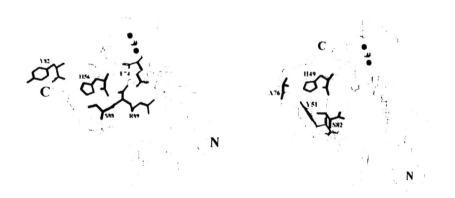

Plate 2. Backbone representation of Adx structure (left), with sidechains of residues involved in the conserved hydrogen bonding network,(His 56, Tyr 82 and Ser 88) and conserved salt bridge (Glu 74 and Arg 89) shown in. Also, residues involved in H-bond network in Pdx (right) (His 49, Tyr 51, Ala 76 and Ser 82) are shown. The iron-sulfur cluster is shown as four spheres. In Tdx hydrogen bonding network is not well defined.

(Reproduced from reference 26. Copyright 2002 American Chemical Society.)

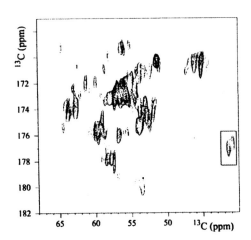

Plate 3. $^{13}C^\alpha\{^{13}C'\}$ *Homonuclear multiple quantum coherence data for 1 mM [U-^{13}C, ^{15}N] ARD obtained with experimental conditions optimized for detection of slow relaxing (green) and fast relaxing resonances (red). Note the appearance of Gly 99 C$^\alpha$-C' correlation under fast pulsing conditions. (Reproduced from reference 31. Copyright 2002 American Chemical Society.)*

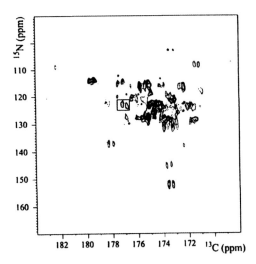

Plate 4. $^{13}C'\{^{15}N\}$ *Heteronuclear multiple quantum coherence data for 1 mM [U-^{13}C, ^{15}N] ARD obtained with experimental conditions optimized for detection of slow relaxing (green) and fast relaxing resonances (red). Note the appearance of Gly 99 C'-Glu 100 ^{15}N correlation under fast pulsing conditions, as indicated by the box. (Reproduced from reference 31. Copyright 2002 American Chemical Society.)*

Chapter 12

An Approach to NMR Treatment of Structural Perturbations in Paramagnetic Proteins too Big for Solution Structure Determination

Judith M. Nocek[1], Kai Huang[2], and Brian M. Hoffman[1]

Departments of [1]Chemistry and [2]Structural Biology, NMR Facility, Northwestern University, 2145 North Sheridan Road, Evanston, IL 60208

The classic description of the quaternary structure change accompanying cooperative ligand binding by hemoglobin (Hb) assumes the existence of two stable tetrameric, quaternary structural classes: the fully-oxygenated (R-state) and the completely deoxygenated (T-state) structures. To characterize the structures that actually occur in solution, to improve the ability of the NMR experiment to distinguish members within a structural class, and ultimately to use this improved ability to assign the structures of intermediate ligation states to members within such families, we have been developing a new approach in which experimental pseudocontact shifts are analyzed to obtain a site-specific description of structural changes produced by mutations in the intradimer interface.

Historically, the structure changes accompanying cooperative ligand binding by the hemoglobin tetramer (Hb) have been taken to involve two stable quaternary structures: the R-structure assumed by fully oxygenated Hb and the T-structure assumed by completely deoxygenated protein. Although details of the mechanism by which these structures interconvert is not yet fully understood, T → R conversion involves a change in the orientation of the two heterodimers ($\alpha_1\beta_1$ and $\alpha_2\beta_2$) of Hb with respect to each other. Characterization of the intermediate ligation states is a critical component of a full understanding of the mechanism whereby the quaternary structure of Hb changes upon uptake and release of oxygen. Although these intermediates are present only in small amounts in the erythrocyte, making them difficult to prepare and characterize, studies with them(1-3) indicate that the mechanism of oxygen uptake/release is far more complicated than the simple analysis implies. Recent crystallographic studies have shown that more than two quaternary structures (designated R and R_2) are, in fact, energetically accessible to the *ligated* state of tetrameric human hemoglobin(4,5) while examination of the x-ray structures for a set of mammalian Hbs has shown that the ligated state can actually adopt any of an *ensemble* of structures that encompasses both the R-structure and the R_2-structure.(6) Together, these results have prodded a re-examination of the structural relationships and the thermodynamic linkages that lead to cooperative ligand binding.(7)

To characterize the structures that actually occur in solution, to improve the ability of the NMR experiment to distinguish members within a structural class, and ultimately to use this improved ability to assign the structures of intermediate ligation states to members within such families, measurements that yield geometric information are needed with site-specific resolution. While significant progress has been made toward obtaining a complete NMR structure of Hb,(8-14) the four probes assigned to exchangeable protons on aromatic residues in the $\alpha_1\beta_2$ and $\alpha_1\beta_1$ interfaces are still the primary NMR signatures for assessing the quaternary state of the protein in solution.(11) Clearly, more extensive assignments are needed.

In a first effort to identify new probes, we recently incorporated ^{15}N into the β-chain Val residues of R-structure carboxy-Hb(βV1M) and fluoromet-Hb(βV1M), and used a novel assignment strategy, based on comparison of predicted and measured pseudocontact shifts, to identify the amide backbone signals in ^{15}N-val labeled Hb.(14) The 17 valine residues in the β-chain are distributed throughout the chain and several are close enough to the heme that they exhibit large pseudocontact shift increments. We,(14,15) like others,(12,13,16-24) have found that such experimental pseudocontact chemical shift increments can be utilized to obtain site-specific structural information for paramagnetic metalloproteins. In particular, we have shown that these probes in Hb are sensitive to the quaternary structure changes induced by the addition of allosteric effectors.(14)

We now discuss the use of these resonances in characterizing the structure of the Hb(βW37E) interface mutant, whose functional(25,26) and structural properties(27) indicate that it exemplifies a new class of high-affinity, unligated tetramers, denoted T_{hi}. We find that this mutation causes widespread changes in the β chain, supporting the idea that the structure of the mutant is different from that of native R-structure carboxy-Hb. As the chemical shift differences ($\Delta\delta_i$) cannot at present be used to derive quantitative information about the changes in atomic position of the nuclei, we are developing a protocol for deriving a site-specific geometric description of the mutation-induced structural changes from the pseudocontact shift increments ($\Delta\delta_i^P$). In this progress report, we present results that support the view that the liganded state of the Hb(βW37E) mutant adopts a structure that indeed differs from the native R-structure.

General Approach

Imagine modifications, made either by chemical means or by mutagenesis, to a large paramagnetic protein of known structure. If that protein is too large to obtain a complete NMR structure, can one nevertheless make use of a limited set of NMR assignments to assess the effects of the modification on that structure? Here we describe an approach in which the presence of a paramagnetic center bound to a protein is exploited firstly to compare the overall quaternary state of the modified protein with the native protein, and secondly to quantify the structural perturbations resulting from the modification. This requires that a suitable set of NMR probes have been assigned, and that the chemical shifts of these probes can be measured in both the paramagnetic state and a suitable diamagnetic state having a similar structure.

Pseudocontact Shifts

The total observed chemical shift (δ_i^{obs}) for a paramagnetic system can be partitioned into diamagnetic (δ_i^{dia}) and paramagnetic contributions, the latter being separable into pseudocontact (δ_i^P) and Fermi contact (δ_i^{FC}) components.(**Equation 1**)

$$\delta_i^{obs} = \delta_i^{dia} + \delta_i^{P} + \delta_i^{FC} \tag{1}$$

For protons that are *not* covalently-linked to the paramagnetic center, the Fermi contact term is negligible ($\delta_i^{FC} = 0$), leaving only the pseudocontact contribution to the paramagnetic shift. In this case, experimental pseudocontact shift increments for any given atom are calculated by subtracting the chemical shift for a suitable diamagnetic reference sample from those of the paramagnetic sample ($\delta_i^P = \delta_i^{obs} - \delta_i^{dia}$).

The pseudocontact shifts for the i-th proton can also be calculated (*17,23*) from a known x-ray structure through use of **Equation 2**.

$$\delta_i^P = \frac{\Delta\chi_{AX}\,(3\cos^2\theta_i - 1) + \Delta\chi_{RH}\,(\cos 2\theta_i\,\sin\varphi_i)}{3NR_i^3} \tag{2}$$

Here $\Delta\chi_{AX}$ and $\Delta\chi_{RH}$ are the axial and rhombic contributions to the magnetic susceptibility and N is Avogadro's number. The distance (R_i) between the paramagnetic center and the i-th proton, and the polar and azimuthal angles (θ_i and φ_i) describe the location of the i-th proton within the framework of the susceptibility tensor.(**Figure 1**)

In its most general form, the pseudocontact shift for the i-th proton (δ_i^P) includes contributions from all the paramagnetic centers in a protein (**Equation 3**),

$$\delta_i^P = \sum \delta_{\gamma i}^P \tag{3}$$

with the pseudocontact shift for each magnetic center being determined by **Equation 2**. In the case of Hb, there are four magnetic centers, each within a different subunit. As large changes in the quaternary state of Hb are linked to changes in the tertiary structure of a given subunit, the contribution to the pseudocontact shift from the heme within the subunit where the probe resides usually provides a good approximation of the total shift. The exception to this would be probes that are situated between two (or more) of the paramagnetic centers (e.g., Val 98 that is near the $\alpha_1\beta_2$ intrasubunit interface).

Structural perturbations which cause a nucleus to move relative to the susceptibility tensor frame give rise to an increment in the pseudocontact shift ($\Delta\delta_i^P$). The sensitivity of the pseudocontact shift to such a structural modification varies with the location of the probe nucleus. **Figure 2** illustrates the well-known angular dependence for the calculated pseudocontact shift increment (δ_i^P, **Equation 2**) for a probe nucleus on the surface of a sphere at a distance of 10 Å from a paramagnetic center with an axial **g-tensor** ($\Delta\chi_{AX} = 1.0 \times 10^{-9}$ m^3/mol; $\Delta\chi_{RH} = 0$ m^3/mol). The magnitude of the shift varies periodically with the angular (θ) position of the probe on the surface of the sphere. The maximum upfield shift occurs at $\theta = \pi/2$, while the maximum downfield shift occurs at $\theta = 0, \pi$.

With a rhombic susceptibility tensor, the pseudocontact shift varies with both the polar (θ) and azimuthal (φ) angles. **Figure 3A** presents a set of contour plots describing the angular dependencies for δ_i^P for a probe nucleus at 10 Å from a paramagnetic center whose tensor properties are chosen to be comparable to those reported for cyanomet-Hb ($\Delta\chi_{AX} = 1.5 \times 10^{-9}$ m^3/mol; $\Delta\chi_{RH} = -0.5 \times 10^{-9}$ m^3/mol). For all values of φ, the maximum upfield shift occurs at $\theta = \pi/2$ and the maximum downfield shift occurs at $\theta = 0, \pi$, as is observed for the axial limit ($\varphi = 0$). The pseudocontact shift shows a strong dependence on the azimuthal angle only when θ is near $\pi/2$.

Site-Sensitivity Factors

If the mutation-induced structural change in an R_{M-i} vector is small in comparison to the R_{M-i} vector, the mutation-induced increment in the pseudocontact shift of the i-th atom, $\Delta\delta_i^P = [\delta_i^P(\text{mutant}) - \delta_i^P(\text{wild type})]$, can be written as the

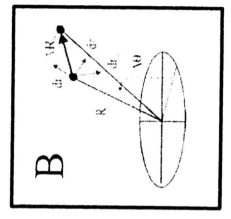

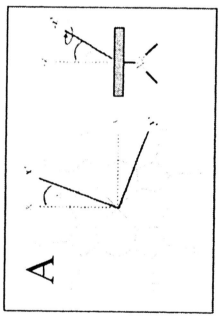

Figure 1: *A. Coordinate system describing the magnetic axes (x', y', z') and the iron-centered pseudoaxes system derived from the x-ray coordinates of oxyHb. The two coordinate systems are related by the Euler transformation angles (α, β, and γ). B. Components describing ΔR$_{fr}$ the mutation-induced structural change in the R$_0$ vector.*

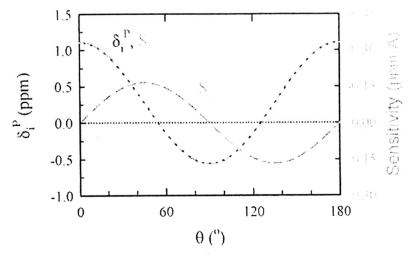

Figure 2: Dependence of the predicted pseudocontact shift and sensitivity factors on θ in the axial limit. Parameters: $\Delta\chi_{AX} = 1.0 \times 10^{-9} \, m^3/mol$; $R = 10 \, Å$.

$$\Delta\delta_i^P = S_R dr_{\parallel} + S_\theta dr\perp + S_\varphi dr_{tan} \tag{4}$$

weighted sum of terms arising from motions of the i-th atom along three mutually-orthogonal directions,(24) as shown in **Figure 1B**: (i) along the R_{M-i} vector (dR$_1$); (ii) perpendicular to this vector and within the plane defined by χ_3 and R_{M-i} (dR$_{M-i}$); (iii) perpendicular to this plane (dR$_{tan}$). We refer to the coefficients (S_R, S_θ, and $S\varphi$) in **Equation 4** as sensitivity factors because they reflect the sensitivity of δ_i^P to motions along each of the three orthogonal directions; taking δ_i^P from **Equation 2**,

$$S_R = \frac{\partial(\Delta\delta_i^P)}{\partial R} = - \frac{[\Delta\chi_{AX}(3\cos^2\theta - 1) + 1.5\Delta\chi_{RH}(\sin^2\theta\cos2\varphi)]}{NR^4} \tag{5}$$

$$S_\theta = \frac{1}{R} \cdot \frac{\partial(\Delta\delta_i^P)}{\partial\theta} = \frac{(\Delta\chi_{RH}\cos2\varphi - 2\Delta\chi_{AX})\cos\theta\sin\theta}{NR^4} \tag{6}$$

$$S_\varphi = \frac{1}{R} \cdot \frac{\partial(\Delta\delta_i^P)}{\partial\varphi} = - \frac{\Delta\chi_{RH}\sin^2\theta\sin2\varphi}{NR^4} \tag{7}$$

these factors are given by **Equations 5-7**:

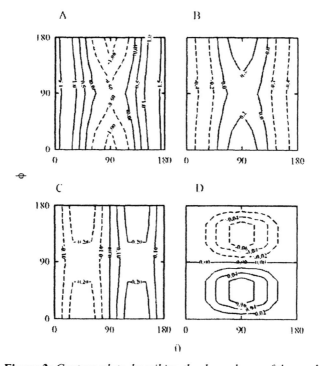

A   B

C   D

φ

θ

Figure 3: *Contour plots describing the dependence of the predicted pseudocontact shift (A) and sensitivity factors (S_R, B; S_θ, C; S_φ, D) on θ and φ. Parameters: $\Delta\chi_{AX} = 1.5 \times 10^{-9}$ m³/mol; $\Delta\chi_{RH} = -0.5 \times 10^{-9}$ m³/mol; R = 10 Å.*

The sensitivity factors, like the pseudocontact shift, depend on the proximity of the probe nucleus to the paramagnetic center as well as on the spatial orientation of the probe nucleus within the framework of the $\Delta\chi$-tensor. **Figure 2** shows the angular dependence for the sensitivity factors (S_R and S_θ, **Equations 5, 6**) for a probe nucleus at a distance of 10 Å from the paramagnetic center, with an axial susceptibility tensor ($\Delta\chi_{AX} = 1.0 \times 10^{-9}$ m³/mol; $\Delta\chi_{RH} = 0$ m³/mol). S_R has the same orientation dependence as δ_i^P (maximum positive value at $\theta = \pi/2$; maximum negative value at $\theta = \pi$), whereas S_θ is shifted in phase (largest positive value at $\theta = 3\pi/4$; largest negative value at $\theta = \pi/4$). The consequence of this phase offset is that positions where δ_i^P is most sensitive to structural perturbations *along* the R_{M-i} vector are those where it is also least sensitive to structural changes *perpendicular* to the R_{M-i} vector. Contributions to $\Delta\delta_i^P$ from the two orthogonal components of the motion of the *i*-th atom can either add or subtract, depending on the location of the probe within the magnetic framework.

With a rhombic susceptibility tensor, δ_i^P is sensitive to motions in all three orthogonal directions. **Figure 3B-D** present a set of contour plots describing the angular dependencies for the three sensitivity factors for a probe nucleus on the surface of a sphere that is 10 Å from a paramagnetic atom, with tensor properties that are comparable to those reported for cyanomet-Hb ($\Delta\chi_{AX} = 1.5 \times 10^{-9}$ m³/mol; $\Delta\chi_{RH} = -0.5 \times 10^{-9}$ m³/mol). As in the axial case, nuclei at positions that are highly sensitive to structural shifts *along* the R_{M-i} vector are quite insensitive to changes that are *perpendicular* to it. In general, $S_R \sim S_\theta \gg S_\varphi$, indicating that $\Delta\delta_i^P$ is least sensitive to changes tangential to the R_M vector. The orientations where S_R and S_θ change signs are slightly shifted in the rhombic limit ($\varphi > 0$) relative to the axial limit ($\varphi = 0$).

Quantifying the Structural Changes

We now discuss how mutation-induced changes in the pseudocontact shift increment, $\Delta\delta_i^P$, can be converted into quantitative estimates of the mutation-induced changes in position of a probe nucleus for a molecule whose unmodified structure is known. In the simplest case of a protein-bound paramagnetic center having an axial susceptibility tensor, **Equation 4** shows that the pseudocontact shift increment ($\Delta\delta_i^P$) is a linear function of the changes perpendicular and parallel to the R_{M-i} vector ($dr_\perp$ and $dr_\parallel$). Paired values of the correlated changes, ($dr_\perp$ $dr_\parallel$), that are consistent with an observed $\Delta\delta_i^P$ can be visualized by plotting $dr_\perp$ as a function of $dr_\parallel$. In such a plot, the slope ($M = -S_R/S_\theta$) is the ratio of the sensitivities in the two orthogonal directions. Inspection of **Figure 2** shows that the slope of such a plot is negative for $0° < \theta < 55°$ and $90 < \theta < 125°$. Although one cannot use **Equation 4** to uniquely determine the changes along the individual axes, the equation does define the minimum mutation-induced change in distance (ΔR_{min}^{AX}) that is compatible with the experimental value of $\Delta\delta_i^P$, **Equation 8**.

$$\Delta R_{min}^{AX} = \sqrt{dr_\parallel^2 + dr_\perp^2} = \sqrt{\frac{(\Delta\delta_i^P)^2}{S_\theta^2 + S_R^2}}$$

(8)

Because the pseudocontact shift measured for a sample with an axial tensor is insensitive to motions tangential to the R_{M-i} vector, ΔR_{min}^{AX} represents a lower bound to the total structural change: $\Delta R_{min}^{AX} \leq \Delta R = (dr_\perp^2 + dr_\parallel^2 + dR_{tan}^2)^{1/2}$.

Application to Hemoglobin

Is the pseudocontact shift a sufficiently sensitive tool for determining the quaternary structure of a mutant of Hb? To test this, we used **Equation 2** to calculate the pseudocontact shifts for the 17 backbone amide protons of the β-chain valine residues of human Hb using coordinates obtained for the R-structure (1ird.pdb), the R_2-structure (1bbb.pdb), the T-structure (1dxu.pdb), and the T_{hi}-quaternary structure (1aoy.pdb). In this calculation, the paramagnetic center was taken to have a g-tensor representative of a fluoromet-Hb with $\Delta\chi_{AX} = 1.0 \times 10^{-9}$ m^3/mol, $\Delta\chi_{RH} = 0$ m^3/mol positioned within the crystallographic framework using the Euler rotation angles, $\alpha = 15°$; $\beta = 10°$. **Figure 4** shows the pseudocontact shift

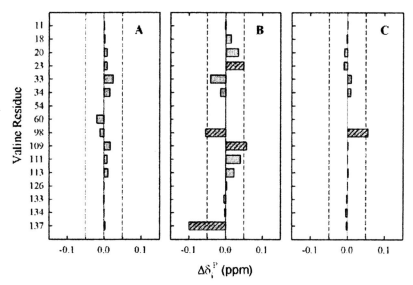

Figure 4: *Predicted pseudocontact shift increments for the backbone amide signals of the β-Val probes in the axial limit derived for pairs of x-ray structures. (A) two members of the R-ensemble, R_2 - R; (B) two structures describing the classical quaternary structure change, T -R; and (C) two T-state structures,T_{hi} - T. Parameters: $\Delta\chi_{AX} = 1.0 \times 10^{-9}$ m^3/mol; $\alpha = 10°$; $\beta = 15°$.*

increments calculated for three pairs of these x-ray structures. The first two maps are the patterns predicted for pairs with similar quaternary structures (R →R_2 and T → T_{hi}). Not surprisingly, the predicted pseudocontact shift increments for these pairs are negligible. On the other hand, measurable pseudocontact shift increments are predicted when the pairs have different quaternary structures (R → T).

To determine the pseudocontact shifts of a set of assigned NMR probes requires that the chemical shifts of those probes can be reliably measured both for the paramagnetic heme form and for a diamagnetic reference heme form with the same quaternary structure. In general, a liganded diamagnetic heme form of the mutant (oxy-Hb or carboxy-Hb) will have the same quaternary structure as a liganded paramagnetic form (cyanomet-Hb or fluoromet-Hb in the absence of effectors), and can be used as the reference for calculation of pseudocontact shifts (δ_i^P, **Equation 1**). One can then compare the pattern of pseudocontact shifts for the mutant to that for the liganded native Hb which adopts the R structure (**Table I**). Likewise, one can assume that a paramagnetic form of the unliganded mutant (deoxyHb) has the same structure as the diamagnetic unliganded reference form (ZnHb), and calculate a set of δ_i^P which can be compared to those for the unliganded T-structure Hb(βV1M). In principle, one might even compare the δ_i^P for a mutant directly to the δ_i^P calculated from the coordinates of R- and T-structure Hb.

Table I: Magnetic properties of Hb derivatives.

	R	T
diamagnetic	HbCO	ZnHb
	HbO$_2$	
paramagnetic	Hb$^+$F$^-$ (S = 5/2)	Hb$^+$F$^-$ + IHP (S = 5/2)
	Hb$^+$CN$^-$ (S = 1/2)	deoxyHb (S = 2)

Fluoromet-Hb can be readily shifted to the T-structure, and thus need not necessarily have the same quaternary structure as the carboxy-Hb form. In this case, one may test the quaternary structure of a fluoromet-Hb form by first assuming it has the same quaternary structure as the carboxy-Hb and calculating a set of δ_i^P, and next by assuming that the fluoromet-Hb (but not the carboxy-Hb mutant) has tipped its quaternary structure to the T-structure and instead use the ZnHb mutant as the reference for calculating δ_i^P. If one doesn't have both the ZnHb and carboxy-Hb forms of the mutant, one can, in effect, make and test a hypothesis about the quaternary structure of the mutant by seeing whether the assumption of equality with carboxy-Hb succeeds or fails.

The Hb(βW37E) Mutant

We now use these ideas to examine the structure of the Hb(βW37E) mutant in solution by measuring the mutation-induced changes in the chemical shifts ($\Delta\delta_i$) and pseudocontact shifts ($\Delta\delta_i^P$) of the β-chain valine amide protons. We provisionally identified the signals for 16 of the 17 valine amide probes through

Table II: Sensitivity factors and minimal mutation-induced structural changes for the backbone amides of the β-val probes in the axial limit.[a]

Valine[b]	S_R (ppm/Å)	S_θ (ppm/Å)	$\Delta\delta_i^P$ (ppm)	R_{min}^{AX} (Å)
11/K	0.005	-0.000	-	-
18/H	0.004	-0.006	-	-
20/G	0.002	-0.013	-	-
23/E	-0.013	-0.026	0.05	1.72
33/B	-0.040	-0.031	-	-
34/D	-0.018	-0.024	-	-
54/C	-0.029	-0.005	-	-
60/O	-0.110	-0.041	-0.05	0.42
98/A	0.067	0.172	0.12	0.65
109/M	0.020	-0.033	-0.06	1.55
111/L	0.015	-0.019	-	-
113/F	-0.000	-0.016	0.08	4.75
126/J	0.003	-0.000	-	-
133/N	0.013	0.006	-	-
134/P	0.026	0.008	-	-
137/I	0.01	0.042	-	-

a. $\Delta\chi_{AX} = 0.86 \times 10^{-9}$ m^3/mol; $\alpha = 5.9°$; $\beta = 13.8°$; 1ird.pdb.
b. New results with cyanomet-Hb have led us to modify the assignments for several of the valine signals from our previous assignments.(*14*)

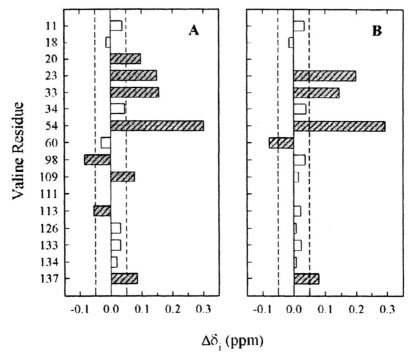

Figure 5: *Effect of the βW37E mutation on the ¹H chemical shifts (Δδᵢ) derived from [¹H, ¹⁵N]-TROSY NMR spectra of ¹⁵N-val fluoromet-Hb (A) and ¹⁵N-val carboxy-Hb (B). Conditions: Varian Inova 600 NMR spectrometer; 37°C; 100 mM KPi buffer, pH 6.*

analysis of their pseudocontact shifts in fluoromet-Hb(βVIM).(*14*) As the [¹H, ¹⁵N]-TROSY spectra of ¹⁵N-β-val-fluoromet-Hb(βW37E) and ¹⁵N-β-val-carboxy-Hb(βW37E) are sufficiently similar to those of fluoromet-Hb(βV1M) and carboxy-Hb(βV1M), assignments for the Hb(βW37E) mutant can be made directly by comparison to the corresponding peaks in the spectrum of Hb(βV1M). Stimulated by new data with cyanomet-Hb (not shown), we are currently re-evaluating our published assignments, and present here a progress report utilizing interim revised assignments for Hb(βV1M) (**Table II**).

Currently, mutation-induced chemical shift differences, $\Delta\delta_i = \delta_i(\beta W37E) - \delta_i(\beta V1M)$, cannot be analyzed to give quantitative information about mutation-induced changes in structure, but they do provide a *qualitative* means of examining whether the conformational state of a particular ligation form of Hb(βW37E) matches that of Hb(βV1M). For the diamagnetic carboxy-Hb, eight signals (assigned to residues 20, 23, 33, 54, 60, 109, 113, and 137) show significant changes in their ^{1}H chemical shifts ($\Delta\delta_i \geq 0.04$ ppm) upon substitution of the Trpβ37 with Glu (**Figure 5**). This set includes four of the five valine residues (valines 33, 34, 54, 109, and 98) that are within 15 Å of the mutation site, as well as four valines (val 23, 60, 113, and 137) that are much further from the mutation site. Likewise, five of these peaks show significant mutation-induced changes for the fluoromet-Hb form (**Figure 5**). If the quaternary structure of carboxy-Hb(βW37E) and fluoromet-Hb(βW37E) were the same as those of the corresponding liganded R-structure Hb(βV1M) proteins, little change should be observed for backbone amides remote from the mutation site. This suggests that the quaternary structures of both carboxy-Hb(βW37E) and fluoromet-Hb(βW37E) differ from the R-structure of the liganded Hb(βV1M) protein.

Deriving Quantitative Mutation-Effects From Pseudocontact Shift Increments, $\Delta\delta_i^P$

We calculated pseudocontact shifts, $\delta_i^P = \delta_i^{HbF} - \delta_i^{HbCO}$, for fluoromet-Hb($\beta$W37E) and fluoromet-Hb($\beta$V1M) using the corresponding carboxy-Hb reference spectra. **Figure 6** plots the differences between the experimental δ_i^P of fluoromet-Hb(βW37E) and R-structure fluoromet-Hb(βV1M), $\Delta\delta_i^P = \delta_i^P(\beta$W37E$) - \delta_i^P(\beta$V1M$)$. Five peaks show small, but significant mutation-induced differences: valines 23, 60, 98, 109, and 113. In this regard, this map resembles the prediction for a tetramer that undergoes a conformational change from the R-structure to the T-structure (**Figure 2**). However, there are significant differences between this map and the predicted map. Thus, it is premature to assert that the differences in the pseudocontact shift maps imply that fluoromet-Hb(βW37E) adopts a T-structure.

We now calculate the site-specific sensitivity parameters (**Equations 5, 6**) for the individual valine amide groups in the β-subunit (**Table II**) for the high-resolution carboxy-Hb crystal structure and use them to estimate the metrical changes in position caused by the βW37E mutation relative to this structure. The site-sensitivity factors vary significantly among the 17 valine residues in the β-subunit. Most of the β-chain valine residues are > 15 Å from the heme, and consequently have small values of S_R and S_θ; those closest to the heme (valines 60 and 98) are the most sensitive structural probes ($|S_i| \geq 0.05$ ppm/Å).

To illustrate the approach, we examine four of the valine residues that show significant ($|\Delta\delta^P| > 0.04$ ppm) mutation-induced pseudocontact shift increments. **Equation 4** relates a measured increment to correlated changes in $dr_\perp$ and $dr_\parallel$, but cannot assign unique values to the two components. **Figure 7** plots the paired values

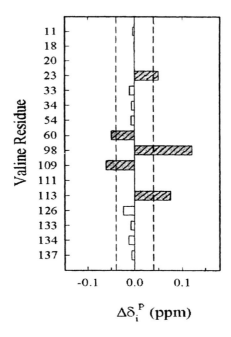

Figure 6: *Effect of the βW37E mutation on the pseudocontact shifts for fluoromet-Hb, $\Delta\delta_i^P = \delta_i^P(\beta W37E) - \delta_i^P(\beta V1M)$. Reference chemical shifts were obtained from [^1H, ^{15}N]-TROSY NMR spectra of ^{15}N-val carboxy-Hb.*

that satisfy **Equation 4**, as discussed above, and the minimum mutation-induced motions, ΔR_{min}^{AX}, as defined by **Equation 8**, are given in **Table II**. For val 60 and val 98, which have the largest sensitivity factors, the measured values of $\Delta\delta_i^P = -0.05$ ppm and +0.12 ppm correspond to much smaller minimal distance changes than those calculated for val 23 and val 109 which have comparable values of $\Delta\delta_i^P$ but are less sensitive to motion (**Table II**). The metrical changes for the latter are especially noteworthy, given that val 23 and val 109 both are over 10 Å from the heme, and that val 109 is nearer to the $\alpha_1\beta_1$ intradimer interface than to the $\alpha_1\beta_2$ interface usually associated with quaternary change.

Summary

We have reported mutation-induced changes in both the chemical shifts and the pseudocontact shifts for the Hb(βW37E) mutant. We find that signals from several of the valine amide probes are sensitive to the mutation, and have described a method for using their mutation-induced pseudocontact shift increments to estimate the metrical changes in position caused by the βW37E mutation. Efforts to identify new probes and to extend this method to other spin states are ongoing.

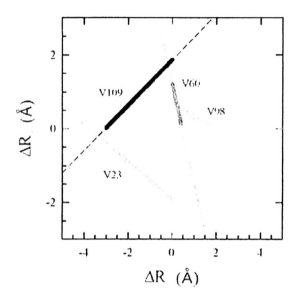

Figure 7: *Sets of structural perturbations for the β-chain Val amides in fluoromet-Hb(βW37E) that show significant experimental pseudocontact shift increments. The heavy lines indicate the limiting ranges where the structural changes are either entirely along the R_{M-i} vector or entirely perpendicular to it.*

Acknowledgements

We thank our colleagues at the University of Iowa (Dr. Jeffrey Kavanaugh, Professor Arthur Arnone, and Mr. Ben Davis) and at the State University of New York in Buffalo (Hilda Hui and Prof. Robert Noble) for preparation of the labeled Hbs. This work has been supported by the National Institute of Health Grant P01 GM58890 (Professor Arthur Arnone, principal investigator).

References

1. Ackers, G. K. *Adv. Protein Chem.* **1998**, *51*, 185-253.
2. Jayaraman, V.; Spiro, T. G. *Biochemistry* **1995**, *34*, 4511-4515.
3. Jayaraman, V.; Rodgers, K. R.; Mukerji, I.; Spiro, T. G. *Science (Washington, D. C.)* **1995**, *269*, 1843-1848.
4. Smith, F. R.; Lattman, E. E.; Carter, C. W. *Proteins-Structure Function and Genetics* **1991**, *10*, 81-91.
5. Silva, M. M.; Rogers, P. H.; Arnone, A. *J. Biol. Chem.* **1992**, *267*, 17248-17256.
6. Mueser, T. C.; Rogers, P. H.; Arnone, A. *Biochemistry* **2000**, *39*, 15353-15364.
7. Tame, J. R. H. *Trends Biochem. Sci.* **1999**, *24*, 372-377.
8. Lukin, J. A.; Simplaceanu, V.; Zou, M.; Ho, N. T.; Ho, C. *Proceedings of the National Academy of Sciences of the United States of America* **2000**, *97*, 10354-10358.
9. Simplaceanu, V.; Lukin, J. A.; Fang, T.-Y.; Zou, M.; Ho, N. T.; Ho, C. *Biophys. J.* **2000**, *79*, 1146-1154.
10. Ho, C. In *Adv. Protein Chem.*; Academic Press, Inc.: New York, 1992; Vol. 43, p 154.
11. Ho, C.; Perussi, J. R. In *Methods Enzymol.*, 1994; Vol. 232, pp 97-139.
12. Kolczak, U.; Han, C.; Sylvia, L. A.; La Mar, G. N. *J. Am. Chem. Soc.* **1997**, *119*, 12643-12654.
13. La Mar, G. N.; Kolczak, U.; Tran, A.-T. T.; Chien, E. Y. T. *J. Am. Chem. Soc.* **2001**, *123*, 4266-4274.
14. Nocek, J. M.; Huang, K.; Hoffman, B. M. *Proc. Nat. Acad. Sci. USA* **2000**, *97*, 2538-2543.
15. Worrall, J. A. R.; Liu, Y.; Crowley, P. B.; Nocek, J.; Hoffman, B. M.; Ubbink, M. *Biochemistry* **2002**, *41*, 11721-11730.
16. Bertini, I.; Luchinat, C.; Piccioli, M. *Methods Enzymol.* **2001**, *339*, 314-340.
17. Bertini, I.; Luchinat, C.; Parigi, G. *Solution NMR of Paramagnetic Molecules: Applications to Metallobiomolecules and Models*; Elsevier Science BV: The Netherlands, 2001; Vol. 2.
18. Worrall, J. A. R.; Kolczak, U.; Canters, G. W.; Ubbink, M. *Biochemistry* **2001**, *40*, 7069-7076.

19. Ubbink, M.; Ejdeback, M.; Karlsson, B. G.; Bendall, D. S. *Structure* **1998**, *6*, 323-335.

20. Morelli, X.; Czjzek, M.; Hatchikian, C. E.; Bornet, O.; Fontecilla-Camps, J. C.; Palma, N. P.; Moura, J. J. G.; Guerlesquin, F. *J. Biol. Chem.* **2000**, *275*, 23204-23210.

21. Sukits, S. F.; Erman, J. E.; Satterlee, J. D. *Biochemistry* **1997**, *36*, 5251-5259.

22. Guiles, R. D.; Sarma, S.; DiGate, R. J.; Banville, D.; Basus, V. J.; Kuntz, I. D.; Waskell, L. *Nat. Struct. Biol.* **1996**, *3*, 333-339.

23. Emerson, S. D.; La Mar, G. N. *Biochemistry* **1990**, *29*, 1545-1556.

24. Feng, Y.; Roder, H.; Englander, S. W. *Biochemistry* **1990**, *29*, 3494-3504.

25. Kwiatkowski, L. D.; Hui, H. L.; Wierzba, A.; Noble, R. W.; Walder, R. Y.; Peterson, E.; Sligar, S.; Sanders, K. *Biochemistry* **1998**, *37*, 4325-4335.

26. Kiger, L.; Klinger, A. L.; Kwiatkowski, L. D.; De Young, A.; Doyle, M. L.; Holt, J. M.; Noble, R. W.; Ackers, G. K. *Biochemistry* **1998**, *37*, 4336-4345.

27. Kavanaugh, J. S.; Weydert, J. A.; Rogers, P. H.; Arnone, A. *Biochemistry* **1998**, *37*, 4358-4373.

Chapter 13

Proton NMR Characterization of Recombinant Ferric Heme Domains of the Oxygen Sensors FixL and Dos: Evidence for Protein Heterogeneity

James D. Satterlee, Christine M. Suquet, Marina I. Savenkova, and Chenyang Lian

Department of Chemistry, Washington State University, Pullman, WA 99164–4630

Proton NMR studies have been carried out on the oxidized (ferric) forms of recombinant FixL heme domains (FixLH) from two species, *Sinorhizobium meliloti* (Sm) and *Bradyrhizobium japonicum* (Bj), and on the recombinant heme domain of the *E. coli* Dos (EcDosH). Hyperfine resonance spectra of the two FixLHs are quite different, despite their structural similarity. Both display high spin proton spectra with shifts extending over 80 ppm to high frequency. In contrast, the EcDosH spectrum is typically low spin, with hyperfine resonance shifts not extending past 40 ppm to high frequency. Careful examination of these spectra reveal line shape heterogeneity in several hyperfine resonances. The spectra depend on the particular preparation and temperature. Spectra of both FixLHs and DosH display this phenomenon. The heterogeneity is due to protein mass loss, which is detectable by mass spectrometry, but is not due to protease activity.

The heme containing proteins FixL and Dos are two examples of a broad group of heme proteins that couple small molecule sensing to biological signaling (1-3). Both of these types of proteins have become objects of NMR studies in our laboratory and here we wish first to describe the similarities in their structure and function, then to present our initial proton NMR results.

In particular, the two FixLs considered here couple heme oxygenation status to expression of the complement of proteins required for nitrogen fixation through a response regulator, FixJ. Dos function in *E. coli* is more ambiguous, but probably involves changes between aerobic and anaerobic metabolism.

FixL was first recognized as integral to control of microaerobic expression of genes required for nitrogen fixation in *Sinorhizobium meliloti* (Sm, formerly *Rhizobium meliloti*)[1](4). Soon thereafter SmFixL was identified as: (a) the specific oxygen sensor protein in this regulatory process (5); (b) a heme protein (6,7); and as (c) a multidomain, combined sensing and signaling, integral membrane protein (7,8). SmFixL is the oxygen sensing and signaling protein that is paired, *in vivo*, with the response regulator protein, FixJ, which acts as the primary transcription factor for the required *nif* and *fix* genes.

In contrast, the FixL from *Bradyrhizobium japonicum* (BjFixL), although similarly heme containing, is not membrane bound. However many properties of BjFixL are closely related to those of SmFixL. Their primary sequence homology is 50% (11). BjFixL also functions as an oxygen sensor *in vivo* by virtue of O_2 binding and release from its heme (12). BjFixl is also a functionally modular protein, containing two of the three domains found in SmFixL: a C-terminal kinase domain and an N-terminal heme domain (7). It also transduces a molecular oxygen based signal to an associated response regulator, which is also called FixJ (11), although the precise regulatory process maybe somewhat different in *S. meliloti* and *B. japonicum* (11,12).

Like the FixLs the *E. coli* Direct Oxygen Sensor (EcDos) contains a heme oxygen sensing PAS (1) domain, but has a phosphodiesterase signaling domain (13,14). Also, like the FixLs (15-17), the heme of EcDos is bound by one histidine in a PAS domain (1,3,18,19). However, unlike the FixLs, whose hemes are natively five-coordinate in the absence of exogenous ligands, the Dos heme is six-coordinate in the presence and absence of heme ligands (13-14). An endogenous methionine has been implicated as the displaceable sixth ligand that coordinates the heme in the absence of exogenous ligands (3).

---

[1]abbreviations used: Bj, *Bradyrhizobium japonicum*; Sm, *Sinorhizobium meliloti*, formerly known as *Rhizobium meliloti* (Rm); Ec, *Escherichia coli*, FixL, full length sequence, including all domains; FixL*, truncated sequence of SmFixL, including only heme and kinase domains; Dos, Direct Oxygen Sensor; DosH, truncated protein consisting only of the Dos Heme domain, defined in Figure 2; FixLH truncated protein including only the FixL heme domain, defined in Figure 2; PAS, abbreviation for the proteins Period, ARNT (Aryl Hydrocarbon Nuclear Transporter), and Simple; IPTG, isopropylthiogalactoside.

Experimental

Protein Production

Plasmids containing RmFixL and BjFixL sequences were obtained from Professor D. Helinski (University of California, San Diego) and Professor Dr. H. Hennecke (ETH, Zurich), respectively. The DosH gene was cloned from *E. coli* strain K12, as described before (13). Standard PCR technology was used to manipulate the DNAs and to create plasmids incorporating either full-length or truncated sequences for expression and protein production. Methods followed those previously published (6,7,20), specific details of which will be described elsewhere (21,22). FixL and DosH genes were subcloned into pET24a+ that were then used to transform *E. coli* strain BL21DE3. Sequences of the inserts were confirmed (Applied Biosystems Model 377 Genetic Analyzer). Expression was induced by 0.1 M IPTG in LB medium containing kanamycin and 0.5% dextrose at 37 °C with 200 rpm agitation for 5 hours. Cells were lysed using a French press (SLM Aminco). Purification also followed published methods (6,7,20). If necessary for NMR spectroscopy, the proteins were oxidized to the ferric state using potassium ferricyanide (Mallinckrodt AR) followed by cycles of washing and concentration with the NMR buffer (next section). Purity of the isolated proteins was assessed using SDS PAGE, optical spectroscopy (Cintra 40; GBC Scientific), and MALDI-TOF mass spectrometry (Voyager DE; PEBiosystems). MALDI-TOF mass spectrometry employed sinapinic acid (3,5-Dimethoxy-4-hydroxy cinnamic acid 98%, Aldrich)/Acetonitrile (Aldrich) matrices. Mass calibrations were carried out externally using horse myoglobin (Sigma) or by using internal protein mass standards (Calibration Mixture 3, PE Biosystems). Protein aging studies were carried out in the presence and absence of the protease inhibitor Pefabloc[R] C hydrochloride (Sigma).

NMR Spectroscopy

Proton NMR spectroscopy was carried out on a Varian Inova spectrometer operating at the nominal proton frequency of 500 MHz. Samples were typically maintained in 99.9% D_2O (Isotec), 0.02M Tris-d_{11} (tris(hydroxymethyl) aminomethane; 99% D, Isotec), 0.1M NaCl (JT Baker). The pH of these solutions was adjusted as necessary with dilute DCl (Isotec, 99% D) or NaOD (Isotec, 99.7% D). Transformation into the deuterated NMR buffer was carried out by at least three consecutive cycles of dilution with the NMR buffer then 10-fold concentration using centricon ultrafiltration (Amicon Corp.). NMR probe temperatures were calibrated using a methanol standard. All spectra presented here were processed without baseline correction.

Results and Discussion

Modularity and Mechanism

The multidomain structure (7) of these oxygen sensors is illustrated using SmFixL as an example in Figure 1. Three separate domains constitute the complete amino acid sequence for SmFixL: membrane binding (N-terminal), heme/oxygen sensing, and the signaling/kinase domain, which is C-terminal and typically has an enzyme function (7). Neither BjFixL, nor Dos has a membrane-binding domain, although both have natural amino acid sequences preceding the heme/sensing domain. The signaling domain also has kinase function in BjFixL, but has phosphodiesterase function in Dos.

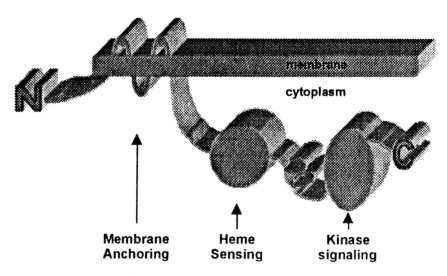

Figure 1. Schematic diagram of the three domains of SmFixL

Figure 2 presents a diagrammatic summary of our current understanding of the SmFixL functional mechanism, focussing only on the sensing and signaling domains. The following description of Figure 2 is also probably relevant to BjFixL, and perhaps even to EcDos. In general the mechanism involves propagation of a signal, triggered by molecular oxygen dissociation from the sensing (heme) domain, to the signaling domain.

Oxygenated FixL is inactive, but when the oxygen partial pressure is reduced O_2 reversibly dissociates from the heme sensing domain (Step 1). This is the initiating signal, for which various details of its propagation have been proposed (15-17,23,24). All of these involve some type of conformational change in the sensing domain (Step 2) that is, in turn, transmitted to the signaling (enzymatic) domain (Step 3). The mode and path of signal propagation are currently not known. Conformational change in the signaling domain (Step 3) apparently has the effect of unmasking a histidine whose role is to facilitate phosphorylation of the response regulator, FixJ. Most recent experimental results (25) indicate that this occurs via what is essentially a multi-body complex (Step 4) involving FixL dimerization. In *S. meliloti* phosphorylation activates FixJ, which subsequently functions as a transcription activator (Step 5) for *nif* A and *fix*N (*fix*K) genes to launch a nitrogen fixation regulatory cascade (4,5,26). In *B. japonicum*, the regulatory status of FixLJ appears to be less absolute than the FixLJ in *S. meliloti* (11), although we speculate that a similar mechanism to that shown in Figure 2 is operable.

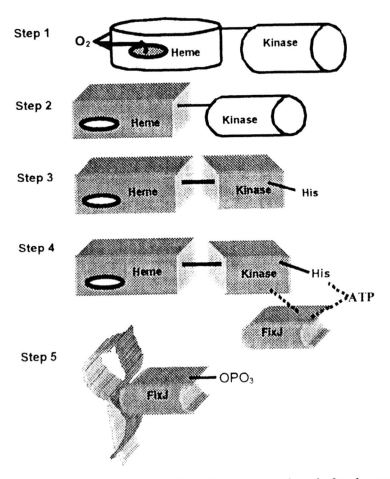

Figure 2. Schematic Description of SmFixL mechanism described in the text.

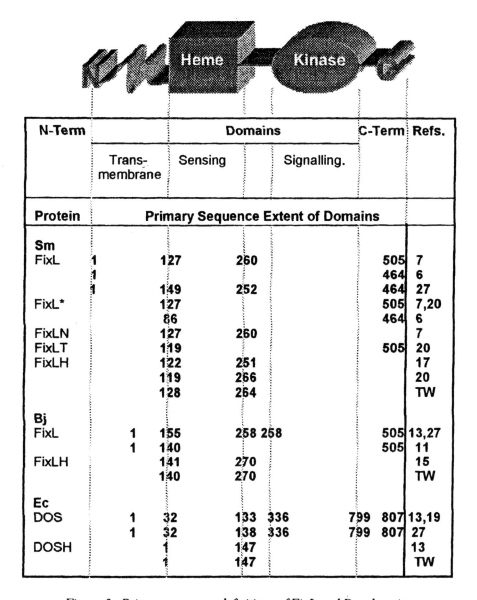

N-Term	Domains			C-Term	Refs.
	Trans-membrane	Sensing	Signalling.		
Protein	Primary Sequence Extent of Domains				
Sm					
FixL	1	127	260	505	7
	1			464	6
	1	149	252	464	27
FixL*		127		505	7,20
		86		464	6
FixLN		127	260		7
FixLT		119		505	20
FixLH		122	251		17
		119	266		20
		128	264		TW
Bj					
FixL	1	155	258 258	505	13,27
	1	140		505	11
FixLH		141	270		15
		140	270		TW
Ec					
DOS	1	32	133 336	799 807	13,19
	1	32	138 336	799 807	27
DOSH		1	147		13
		1	147		TW

Figure 3. Primary sequence definitions of FixL and Dos domains

Expressed Proteins Compared

For all three proteins the extent of each domain, defined by primary sequence number, is given in Figure 3. This figure defines data presented in the literature as well as the heme domain truncations that we used in this work, labeled TW. As shown in Figure 3, several different truncations of the FixL and Dos proteins have been produced. Slight variations in domain definitions also occur, depending on the laboratory of origin. In Figure 3, light vertical lines describe the extent of the domains, referenced to the conceptual protein drawing at the top. Numbers in the body of the figure give the primary sequence positions employed by the referenced workers to define the beginning and ending of the domains. Abbreviations shown in the left column of Figure 3 are defined by the number of domains and sequence length in the figure body. In this work we will focus on the truncations that include only the heme containing oxygen sensing domain (FixLH; DosH) for the Sm, Bj and Ec proteins, which are denoted by TW in the right hand column of Figure 3.

Proton NMR Spectroscopy and Mass Spectrometry

Ferric High-Spin Forms of the FixL Heme Domains (FixLH)

The heme sensing domains of both FixLs are a new type of heme structural motif (15-17). They are heme-binding PAS domains that are structurally dissimilar to globin, cytochrome and peroxidase heme domains (27). While the heme pockets are dissimilar, both FixLs and Dos employ heme b (Fig. 4) as their active site prosthetic group. Previous optical and resonance Raman studies revealed that SmFixLH and BjFixLH are high-spin and five coordinate when the heme iron ion is oxidized to the ferric form (Fe^{3+}) in the absence of potential

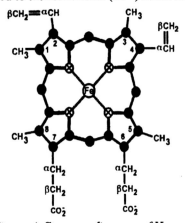

Figure 4. Strucure diagram of Heme b

exogenous ligands (14). This is further illustrated by the ^{1}H NMR spectrum of each of these two proteins shown in Figure 5, which are typical of high-spin ferriheme proteins (28). We focus on the highly shifted hyperfine proton resonances, which result from the paramagnetic moment due to the heme iron ion's five unpaired electrons (28). Experience has shown that in high-spin ferriheme proteins the resonances of protons on carbons directly bonded to the heme macrocycle ring (C_α) appear at high frequency (downfield). For heme-b (Fig. 4) one would expect four individual heme methyl resonances (positions 1,3,5,8) and six individual proton resonances due to the four propionate H_α (positions 6,7) and the two vinyl H_α (positions 2,4) (28).

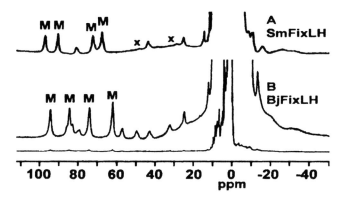

Figure 5. ^{1}H NMR Spectra of the FixL heme domains, as labeled, 21°C, 0.02M Tris, 0.1M NaCl, pH 8.0, 500 MHz. M indicates heme methyl resonances.

In each spectrum shown in Figure 5 the larger four high frequency resonances have been labeled with M to indicate that they are due to the four heme methyl groups. At the present level of analysis it is only necessary to note that the pattern of the heme methyl resonances are dissimilar in the two spectra. In the spectrum of SmFixLH (Fig. 5A) they group into two pairs, similar to the pattern found for the three *Glycera dibranchiata* monomer methemoglobins (29). In Figure 5B, the BjFixLH spectrum, with its evenly dispersed methyl resonances, resembles the pattern shown by metmyoglobin (28,30).

While each spectrum clearly displays four methyl resonances, only the BjFixLH spectrum (Fig. 5B) displays enough single proton resonances to account for the six H_α protons of the two heme propionates and two heme vinyl groups. There are six well-defined single proton resonances in the BjFixLH spectrum. Five of these lie between 20 ppm and 60 ppm. Another occurs at 79 ppm. The methyl resonance at 85 ppm also displays overlap by two smaller peaks. In contrast, for the SmFixLH spectrum (Fig. 5A) there are only three distinguishable single proton resonances (at 25, 42, and 80 ppm). Another may overlap the methyl resonance at 68 ppm, as indicated by asymmetry on the low frequency side of that resonance. The spectrum shown in Figure 5A for

252

SmFixLH is essentially identical to published NMR spectra of two other preparations of SmFixLH (31,32)

Spectral Heterogeneity Indicates Protein Heterogeneity

Careful inspection of the SmFixLH spectrum in Figure 5A reveals small peaks (labeled x) that are less than single proton intensity. As described above, the Bj spectrum (Fig. 5B) contains two more resonances than there are single heme α-substituent protons. There is also preparation-to-preparation variability in the NMR spectra of BjFixLH, as illustrated in Figure 6. Figure 6A and 6B show spectra of two different preparations of BjFixLH. Note the difference in the resonance pattern in the four regions denoted by *. Those resonances are all "doubled" in 6A, but appear as single resonances in the preparation shown in 6B. Figure 6C and 6D are spectral traces of a smaller BjFixLH construct that we also expressed, in which 14 amino acids were truncated from the C-terminal helix (called BjFixLH-h). Its ^1H spectrum is very similar to that of the BjFixLH spectrum in Figure 6B.

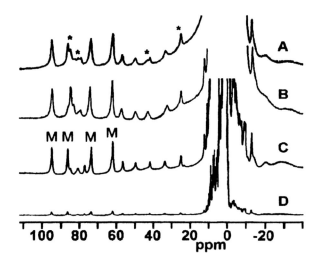

Figure 6. ^1H, 500 MHz spectra of BjFixLH preparations, 0.02M Tris/0.1M NaCl, 21 °C, pH 8.0. A. BjFixLH (prep4); B. BjFixLH (prep5),C, D, BjFixLH-h (prep1).

These three situations are consistent with the idea that varying levels of heterogeneity develop in these NMR solutions of the full-length heme domains (FixLH). The source of that heterogeneity is not impurities in the preparation, for the following reasons. Immediately after purification these preparations show only one line on SDS PAGE and greater than 95% purity by mass spectrometry (not shown). Thus, the heterogeneity must develop on the time frame of preparing NMR samples and performing NMR data acquisitions, from 12 hours to many days. To test this idea we allowed freshly purified samples of both Sm and Bj FixLH to sit at room temperature for four days and obtained mass spectra over this time. Some of our results for SmFixLH are shown in the MALDI-TOF spectra in Figure 7.

The lower trace in Figure 7 is the spectrum of one of our SmFixLH preparations taken approximately 3 hours after the purified protein was removed from the freezer. It is dominated by the peak of the parent compound (peak B) at 15,388.4 D. At higher M/Z is peak A, which is the parent protein that carries an added N-terminal Met. In purified preparations that do not sit at room temperature these are the only two peaks seen in the +1 ion spectrum, indicating the initial high level of purity and homogeneity of our preparations. The peaks C are peaks that develop initially when the sample ages either in the refrigerator or at room temperature. As time passes in unfrozen samples the C peaks show progressively increasing intensity, concomitant with decline in the relative intensity of peaks A and B. After four days at room temperature the sample displayed a spectrum given in the top trace, with multiple peaks (labeled D) in the M/Z range of 12-13.6 kD, and no evidence of the original parent protein peaks (A,B).

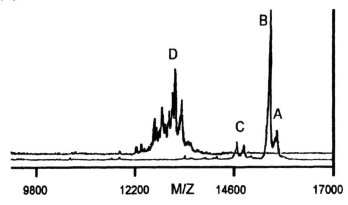

Figure 7. MALDI-TOF mass spectra of SmFixLH mass degradation, as described in the text. Top: aged protein solution; Bottom: less aged.

The combined NMR and mass spectrometry results indicate that time dependent degradation of SmFixLH occurs in solutions that are not frozen. Freezing inhibits this degradation. Similar results have been found for BjFixLH and EcDosH (*not shown*). We do not believe that protease contamination is the source of this phenomenon for the reason that our mass spectrometry/aging studies were carried out in the presence of added protease inhibitors. In particular, the spectra shown in Figure 7 are of a protein sample maintained at

0.1 mM concentration of the protease inhibitor Pefabloc[R]. Further detailed mass spectrometry studies are in progress for the purpose of understanding this phenomenon.

DosH

A ^{1}H NMR spectrum of our preparation of ferric EcDosH is shown in Figure 8. Compared to FixLH spectra the hyperfine resonances of DosH extend over a much narrower range, from only 40 ppm to –9 ppm. This indicates that the electronic structure of EcDosH is different from the FixLHs and is consistent with optical and resonance Raman spectra indicating that this heme domain contains low-spin ferric heme. This spectrum is typical of other low-spin ferriheme proteins (28) in that heme methyl resonances (M) appear in the 15-42 ppm region. However it is also atypical in the fact that four methyl peaks are

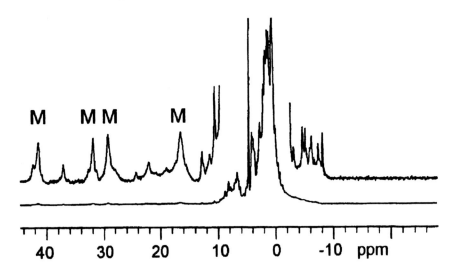

Figure 8. ^{1}H NMR spectrum of recombinant EcDosH at 500 MHz, 21 °C, 0.1M potassium phsophate, pH 6.8. Peaks labeled M are heme methyl resonances.

resolved in the spectrum, since the vast majority of low-spin ferriheme proteins show only two heme methyl resonances in this region. At this level of analysis we also note that despite the rather low resolution of this region and the presence of broad and overlapping peaks, the four heme methyl peaks have approximately equal intensity. There are three possible explanations for this spectrum. 1. The sample degraded during the course of these NMR experiments as determined by mass spectra (not shown) taken at the end of three days of NMR time. Although this was the first spectrum taken in that three-day series, the four heme methyl resonances could represent two different forms of EcDosH, heterogeneous by virtue of mass loss due to degradation. 2. More likely is the possibility that since the spectrum was obtained with the minimum possible "aging time" due to sample preparation, the mass heterogeneity of this sample in this spectrum acquisition was small, and the four heme methyl resonances represent a protein with a unique paramagnetic susceptibility tensor/electronic structure (28). 3. EcDosH crystallizes as a homodimer, with two distinct heme domains per unit cell (33). The NMR spectrum (Fig. 8) is consistent with dimerization in solution. This possibility assigns a typical paramagnetic susceptibility tensor to each monomer EcDosH, with one pair of magnetically inequivalent heme methyls. If the solution dimer were symmetric then only one set of inequivalent heme methy resonances would be predicted. This is not observed (Fig. 8). This situation can only occur if the dimer is asymmetric. Obviously in this scenario the magnetic environment of the two monomer heme domains in such a dimer must be inequivalent so that four methyl hyperfine resonances are distinguished in the dimer spectrum.

Another interesting feature of this spectrum is the absence of a prominent hyperfine-shifted Met-εCH_3 resonance in the low frequency (upfield) part of the spectrum (Fig. 8). Such a methyl peak, appearing near -20 ppm, is characteristic of methionine ligating heme in low-spin, six-coordinate ferriheme proteins like the oxidized cytochromes c (28, 32). Mutagenesis studies (3) have recently shown that EcDosH heme is ligated by both histidine and methionine, just like cytochromes c. In this context the absence of a low frequency hyperfine-shifted Met-εCH_3 resonance is puzzling. Its absence in the EcDosH spectrum can be explained in only two ways. 1. Ferric EcDosH has an unusual paramagnetic susceptibility/electronic structure (see #2 above). 2. The sixth heme ligand in ferric EcDosH is an amino acid other than methionine. In view of the biochemical data (3) we favor the first possibility and have begun studies to resolve this dilemma.

Conclusions

We have demonstrated in this initial study that there is identifiable mass heterogeneity in solutions of Bj and SmFixLH that is detectable by mass spectrometry and manifests itself in the proton hyperfine resonance spectrum of ferric forms of these proteins. This type of heterogeneity also occurs in EcDosH preparations (not shown). Such degradation has obvious implications for past studies, none of which have published mass spectrometry evaluations of their preparations. We note that this degradation does not affect optical spectra of the FixLHs and DosH, but can be seen in NMR and Mass spectra. We have also found that the BjFixLH-h containing the 14 amino acid C-terminal truncation is resistant to mass degradation, indicating that mass loss may come from terminal sequences. Finally, the isolated FixL heme domains display different hyperfine shift patterns indicating that in solution their heme pockets are not structurally identical. We are pursuing comprehensive characterization studies of these proteins.

Acknowledgements

We wish to acknowledge support from NIH grant GM 47645. JDS wishes to acknowledge Harold Goldberg, M.D., Michael Kwasman, M.D., Lawrence Hammond, M.D. and Stacey Dean, M.D., without whose excellent expertise this work would not have been possible. The WSU NMR Center was supported by NIH grants RR0631401 and RR12948, NSF grants CHE-9115282 and DBI-9604689 and the Murdock Charitable Trust.

References

1. Tylor, B. L.; Zhulin, I. B. *Microbiol. Mol. Biol. Rev.* **1999**, *63*, 479-506.
2. Rodgers, K. R. *Curr. Opin. Chem. Biol.* **1999**, *3*, 158-167.
3. Gonzalez, G.; Dioum, E. M.; Bertolucci, C.; Tomita, T.; Ikeda-Saito, M.; Cheesman, M.R.; Watmough, N.J; Gilles-Gonzalez, M.-A. *Biochemistry*, **2002**, *41*,8414-8421.
4. David, M.; Daverana M.-L.; Batut, J.; Dedieu, A.; Domergul, O.; Ghai, J.; Hertig, C.; Boistard, P.; Kahn, D. *Cell*, **1988**, *54*, 671-683.
5. de Philip, P.; Batut, J.; Boistard, P. *J. Bacteriol.* **1990**, *172*, 4255-4262.
6. Gilles-Gonzalez, M.-A.; Ditta, G. S.; Helinski, D. R. *Nature* **1991**, *350*, 170-172.
7. Monson, E. K.; Weinstein, M.; Ditta, G. S.; Helinski, D. R. *Proc. Natl. Acad. Sci. USA* **1992**, *89*, 4280-4294.
8. Lois, A. F.; Ditta, G. S.; Helinski, D. R. *J. Bacteriol.* **1993**, *175*, 1103-1109.
9. Hertig, C.; YaLi, R.; Louarn, A.-M.; Garnerone, A.-M.; David, M.; Batut, J.; Kahn, D.; Boistard, P. *J. Bacteriol* **1989**, *171*, 1736-1738.
10. Virts, E. L.; Stanfield, S. W.; Helinski, D. R.; Ditta, G. S. *Proc. Natl. Acad. Sci. USA* **1988**, *85*, 3062-3065.
11. Anthamatten, D.; Hennecke, H. *Mol. Gen. Genet.* **1991**, *225*, 38-48.
12. Gilles-Gonzalez, M.-A.; Gonzalez, G.; Perutz, M. F. *Biochemistry* **1995**, *34*, 232-236.
13. Delgado-Nixon, V. M., Gonzalez, G.; Gilles-Gonzalez, M.-A. *Biochemistry* **2000**, *39*, 2685-2691.
14. Tomita, T.; Gonzalez, G.; Chang, A. L.; Ikeda-Saito, M.; Gilles-Gonzalez, M.-A. *Biochemistry* **2002**, *41*, 4819-4826.
15. Gong, W.; Hao, B.; Mansy, S. S.; Gonzalez, G.; Gilles-Gonzalez, M.-A.; Chan, M. K. *Proc. Natl. Acad. Sci. USA* **1998**, *95*, 15177-15182.
16. Gong, W.; Hao, B.; Chan, M. K. *Biochemistry* **2000**, *39*, 3955-3962.

17. Miyataka, H.; Mukai, M.; Pary, S.-Y.; Adachi, S.-I.; Tamura, K.; Nakamura, H.; Nakamura, K.; Tsuchiya, T.; Iizuka, T.; Shiro, Y. *J. Mol. Biol.* **2000** *301*, 415-431.

18. Park, H. J.; Suquet, C. M.; Savenkova, M.; Satterlee, J. D.; Kang, C. H. *Acta Crystallog. D* **2002** *D58*, 1504-1506 .

19. Zhulin, I. B.; Taylor, B. L.; Dixon, R. *Trends Biochem. Sci.* **1997**, *22*, 331-333.

20. Gilles-Gonzalez, M.-A.; Gonzalez, G.; Perutz, M.F.; Kiger, L.; Marden, M. C.; Poyart, C. *Biochemistry* **1994**, *33*, 8067-8073.

21. Satterlee, J. D., Suquet, C. M. **2002**, in preparation.

22. Suquet, C. M., Savenkova, M. I.; Satterlee, J. D., **2002**, in preparation.

23. Miyataka, H.; Mucai, M.; Adachi, S.; Nakamura, H.; Tamura, K.; Iizuka, T.; Shiro, Y. J. *J. Biol. Chem.* **1999**, *274*, 23176-23184.

24. Perutz, M. F.; Paoli, M.; Lesk, A. M. *Chem. Biol.* **1999**, *6*, R291-R297

25. Tuckerman, J. R.; Gonzalez, G.; Dioum, E. M.; Gilles-Gonzalez, M.-A. *Biochemistry* **2002**, *41*, 6170-6177.

26. Kahn, D.; Ditta, G. *Molec. Microbiol.* **1991**, *5*, 987-997.

27. Chan, M. K. *Curr. Opin. Chem. Biol.* **2001**, *5*, 216-222.

28. La Mar, G. N.; Satterlee, J. D.; de Ropp, J. S. in *The Porphyrin Handbook* (Kadish, K. M.; Smith, K. M.; Guilard, R.; eds) **2000**, **vol 5**, pp85-298, Academic Press, San Diego, USA.

29. Constantinidis, I.; Satterlee, J. D.; Pandy, R.; Lewis, H.; Smith, K. M. *Biochemistry* **1988** *27*, 3069-3076.

30. La Mar, G. N.; Budd, D. L.; Smith, K. M.; Langry, K. C. *J. Am. Chem. Soc.* **1980**, *102*, 1822-1830.

31. Rodgers, K. R.; Lukat-Rodgers, G. S.; Barron, J. A. Biochemistry 1996, 35, 9539-9548.

32. Bertolucci, C.; Ming, L.-J.; Gonzalez, G.; Gilles-Gonzalez, M. A. *Chem. Biol.* **1996**, *3*, 561-566.

Chapter 14

NMR Investigation of the Iron–Sulfur Cluster Environment in the Hyperthermostable Ferredoxin from *Pyrococcus furiosus*

Luigi Calzolai[1,3], Halvard Haarklau[1], Philip S. Brereton[2], Michael W. W. Adams[2], and Gerd N. La Mar[1,*]

[1]Department of Chemistry, University of California, Davis, CA 95616
[2]Department of Biochemistry and Molecular Biology, Center for Metalloenzyme Studies, University of Georgia, B/22 Life Sciences Building, Green Street, Athens, GA 30602
[3]Current address: Institute for Molecular Biology and Biophysics, ETH-Honggerberger, HPK CH–8093, Zurich, Switzerland

The paramagnetic cluster environment of the ferredoxin, Fd, from the hyperthermophilic archeaon *Pyrococcus furiosus* has been investigated by ^{1}H and ^{15}N NMR to identify the signals from the strongly relaxed residues not identified by conventional ^{1}H/2D 1D/2D NMR, and to characterize their dipolar contacts via appropriately tailored NOESY spectra and the proximity to the cluster via T_1 data. A combination of rapid pulsing and/or short delay/mixing time ^{1}H WEFT spectra, NOESY, TOCSY and [^{1}H-^{15}N]-HSQC maps locate the [15]N-^{1}H fragment of six of the 11 target residues, and partial to complete side chains of the other five. Short mixing time WEFT and NOESY spectra provide key dipolar contacts that confirm a short helix and the role of six residues in two β-sheets. A comparison of T_1 values for the wild type, Asp14-ligand cluster, with the more "conventional" all Cys ligand D14C-Fd mutant reveal that the intrinsic cluster relaxivity is unchanged upon substitutingAsp with Cys cluster ligand and suggest that differences in cluster ligand T_1 between wild-type and mutant results from differences in delocalized spin on the cluster S and/or O atoms.

Introduction

The bacterial-type ferredoxin (Fd) is a small electron transfer protein (1) that possesses cubane, usually cys-ligated, 4Fe $[Fe_4S_4]^{+1,+2}$ and\or 3Fe, $[Fe_3S_4]^{0,+1}$, cluster(s) as redox chromophore(s). The location of the "reducing electron" in the cluster and the structural bases for the discrimination of the micro potentials of pairs of iron is an active area of research to which NMR has made unique contributions (2-7). Each oxidation state for a cubane cluster is either paramagnetic or allows for extensive population of excited paramagnetic states. The contact shifts, and their characteristic temperature dependence, for the cluster ligands allow the identification of the oxidation state of the coordinated iron (2,7,8). However, the paramagnetic relaxation leads to diminution or loss of the dipolar and scalar correlations needed to accurately define the cluster environment in a solution structure determination by ^1H NMR (5,6). In favorable cases, the paramagnetic relaxation can serve as a structural constraint in building a molecular model (9-12). The differential relaxation experienced by the non-equivalent $C_\beta H$ of ligated Cys have also been used to make the essential sterospecific assignments based on the expectation of dipolar relaxation ($\propto R^{-6}$) by the spin density on the iron (13,14). A detailed structural model for a Fd is needed to interpret the structural basis for the differentiation of the micropotentials of the individual iron (4,15). The general folding topology for a single cluster bacterrial Fd is largely conserved (5,11,16-20) and the positions of the cluster and disulfide bond in the Fd are depicted schematically in Figure 1.

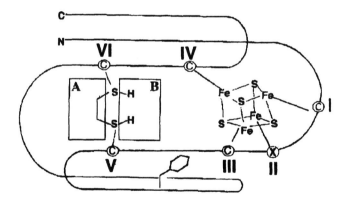

Figure 1. Schematic depiction of a single-cluster, bacterial-type Fd which possess the four cluster ligands CysI, CysIII and CysIV, and a variable ligand II (X). Cys V and VI, can exist as free Cys or participate in a disulfide bridge.

Our interests center on the Fd from the hyperthermophilic archeon *Pyrococcus furiosus, Pf,* which possesses a number of unusual features (21,22), including carboxylate ligation of the cluster by Asp14 (23), extremely slow electron self-exchange (24), a disulfide bridge that is redox active at about the same potential as the cluster (25), as well as extraordinary thermostability (22) (negligible denaturation after 24h at 90°C). Lastly, the cloned gene expressed in *E. Coli* has afforded the point mutants (26) where Asp14 is replaced by Cys or Ser.

The four contact-shifted cluster ligands (23,26,27) (residues 11, 14, 17, and 56) and the residues relatively remote from the cluster (27) (residues 1-10, 19-32, 34-56 and 62-66) have been sequence specifically assigned in both WT and the D14C mutant for the oxidized 4Fe cluster with intact disulfide, designated Fd^A_{ox}. While neither the pattern of the conserved Cys contact shifts (26) nor the chemical shifts and NOESY contacts for the remainder of the protein (27) differed significantly between D14C-Fd and WT Fd^A_{ox}, the Cys $C_\beta H$ in the WT Fd^A_{ox} exhibited more rapid relaxation (26) (in one case, by a factor ~2) than in D14C Fd^A_{ox}. These different T_1s for the conserved cluster Cys suggest that the electronic T_{1e} depends on the nature of the position 14 ligand in each oxidation state, and, indirectly, suggests there would be a distinct advantage in pursuing a solution structure in a particular ligation mutant and/or a particular cluster oxidation state. The amino acid sequence of *Pf* Fd (22) and the identified secondary structural elements (12,27) are illustrated in Figure 2.

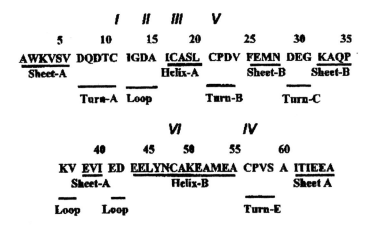

Figure 2. Sequence of Pf *Fd and location of secondary structural elements. The cluster ligands are Cys11(I), Asp14(II), Cys17(III), and Cys56(IV). Cys21(V) and Cys48(VI) participate in a disulfide bond.*
(Reproduced from reference 26. Copyright 1997 American Chemical Society.)

We concern ourselves here with obtaining more information on the cluster environment by locating and assigning the remaining residues (residues 12, 13, 15, 16, 18, 19, 33, 57-61) using appropriately tailored 1D and 2D NMR experiments (27,28) on ^{15}N labeled WT FdA_{ox}, and ascertaining whether the increased Cys $C_\beta H$ relaxation (26)in WT relative to $D14C\text{-}FdA_{ox}$ reflects an altered electronic T_1 by determining the proton T_1s for non-ligated residues in the 4Fe FdA_{ox} cluster vicinity whose relaxation is dominated by paramagnetic influences. A molecular model for Pf 4Fe FdA_{ox} has been generated (12) which has demonstrated that the short helix A and turn B are highly conserved with respect to all other cubane Fd structures.

MATERIAL AND METHODS

Proteins: The 99% randomly labeled *Pyrococcus furiosus*, *Pf*, (DSM 3638) wild type 4Fe FdA_{ox} was prepared as described in detail previously (27). The *Pf* D14C Fd mutant is the same as described previously (26). NMR samples contained 50 mM sodium phosphate buffer, pH 7.6.

NMR spectra: ^1H NMR spectra were recorded at 500 MHz on a GE Omega 500 spectrometer. Chemical shifts are referenced to 2,2-dimethyl-2-sulfonate, DSS, through the residual solvent signal. 1D ^1H spectra were collected under "normal" conditions (repetition rate 1 s^{-1}) and as WEFT (29,30) spectra (repetition rates, 5-10 s^{-1}, relaxation delay τ_r 10-250 ms). T_1 data for resolved protons were collected with a standard inversion recovery pulse sequence in which recycle times were set at ~5 times the T_1 times for the peaks of interest. T_1 data for fast relaxing, non-resolved protons were collected using an Inversion-Recovery-TOCSY (IR-TOCSY) pulse sequence (31) where a (180° pulse τ_r) sequence precede the standard TOCSY (32) pulse sequence. T_1 values from 1D and 2D experiments were calculated from the non-linear square fitting of the intensity of the resolved resonances or of the cross peak respectively.

NOESY (33) and TOCSY (32) spectra were collected with 256 t1 blocks each consisting of 512 scans of 1024 t2 points over a 6.25 kHz bandwidth. Short mixing time NOESY (mixing time 50 ms) and TOCSY (mixing time, τ_m = 15 and 30 ms were collected at 2 s^{-1} using MLEV 17 for the spin-lock. WEFT-NOESY (34) spectra were recorded with a τ_m = 10 ms mixing time at a repetition time of 5 s^{-1} and a relaxation delay of 50 ms. IR-TOCSY spectra (31) utilized a 30 ms spin-lock, a repetition rate of 1 s^{-1} with relaxation delays, τ_r, varying from 10 to 500 ms. The water signal was suppressed with a low-power selective irradiation during the predelay period for all experiments. [^1H-^{15}N]HSQC (35) spectra consisted of 256 t1 blocks of 8-48 transients collected over 2048 t2 points over a 12.5 kHz (^1H) and 5.0 kHz (^{15}N) bandwidth using either a coherence transfer time, Δ = 22.5 ms, at a 1 s^{-1} repetition rate (weakly relaxed NHs) and Δ = 15 ms, at a 3 s^{-1} repetition rate (strongly relaxed NHs).

NMR data were processed on either a Silicon Graphics Indigo workstation or a SUN Sparcstation using the Biosym Felix 2.3 program. A 30°-shifted sine-bell-squared functions were applied in both dimensions for NOESY and TOCSY and data sets were zero-filled to 2048 x 2048 real data points prior to Fourier transformation.

RESULTS

Resonance assignments: In addition to the previously assigned (23,26,27) residues 1-11, 14, 17, 19-32, 34-56 and 62-66, five TOCSY detected, but unassigned spin systems (labeled A-E) were located (27) with chemical shifts, in ppm, A(4.1, 1.3, 0.8), B(5.20, 2.68, 1.11), C(4.79, 2.83, 0.93), D(4.7, 12.2, 1.8) and E(5.07, 1.98, 1.38, 1.08, 0.60). The residues and/or protons of interest are the strongly relaxed, but inconsequentially hyperfine-shifted, and hence unresolved, signals that comprise the cluster loops (Ile12, Gly13, Ala15, Ile16), the beginning of helix A (Ala18), parts of β-sheet B (Ala33) and β-sheet A (Ile61), and turn E (Pro57-Ala60), for a total of 11 unassigned residues, ten of which possess peptide NHs. The protons of these residues are expected to exhibit $T_1s \leq 50$ ms, with many with $T_1s < 10$ ms, and dictate the use of short mixing times and/or rapid pulse repetition conditions to optimize detection and assignments (8,28).

The amide portion of the 500 MHz ^{1}H NMR spectrum (repetition rate 10 s^{-1}, $\tau_r = 50$ ms) is shown in Figure 3A. The WEFT-spectrum (29,30) for this spectral window designed to suppress slowly relaxing protons (Figure 3B) detects moderately to strongly relaxed signals from the previously assigned (27,30) non-labile protons, Cys17 $C_\alpha H$ (α_{17}), the Phe 25 ring (ϕ_{25}), three labile proton peaks from Cys11 (N_{11}), Asp7 (N_7) Cys56 (N_{56}), and seven unassigned peptide NH signals labeled a-g. The "normal" [^{1}H-^{15}N]-HSQC spectrum (35) in Figure 3C exhibits reasonably intense cross peaks for peaks b and c, very weak peaks for peaks f and g, and no detectable cross peaks for peaks a, d and f. The [^{1}H-^{15}N]-HSQC spectrum under more rapid repetition conditions and shorter coherence transfer time strongly enhances the intensity of cross peaks for Cys 11 and peaks e and g, allows detection of a weak cross peak to peaks f and d, but fails to yield a cross peak to peak a, as shown in Figure 3D. The combination of HSQC and WEFT spectra locate the ^{15}N-^{1}H fragments (peaks b-f) of six, and the peptide proton (peak a) for one, of the ten previously unassigned non-Pro residues; no evidence for the other three NH signals is found in the spectral window 0-10 ppm.

Short mixing time (15-30 ms) TOCSY spectra in ^{1}H$_2$O (not shown) connect peak b to spin system E, peak d to spin system C, identify peak c (and spin system C) as originating from a Ser with the characteristic AMX side chain (not shown) that must be Ser59, and the residue origin as Ala for peaks e, f and g. A portion of the rapid repetition rate (3 s^{-1}), short mixing time ($\tau_m = 15$ ms)

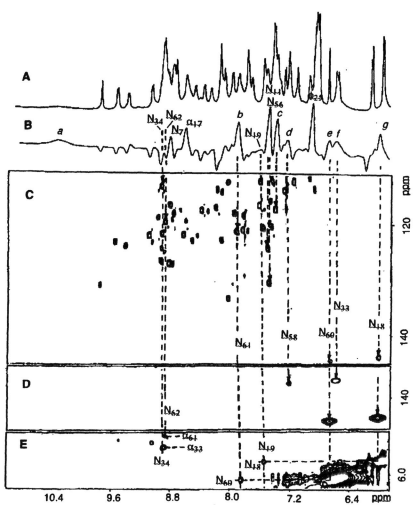

Figure 3. Amide region of the 500 MHz ^1H NMR in 90% ^1H_2O:10% ^2H_2O: **(A)** reference trace (1 s^{-1}); **(B)** WEFT trace (10 s^{-1}), τ_r = 50 ms); **(C)** "normal" $[^1H-^{15}N]$-HSQC spectrum (1 s^{-1}; Δ = 22.5 ms), **(D)** "rapid repetition" $[^1H-^{15}N]$-HSQC spectrum (3 s^{-1}, Δ = 15 ms); and **(E)** WEFT-NOESY spectrum (5 s^{-1}, τ_r = 50 ms, τ_m = 10 ms). Assigned peptide NH (N_i) $C_\alpha H$ (α_i) and aromatic ring (ϕ_i) are labeled, where i = residue number; unassigned peptide NHs are labeled a-g.

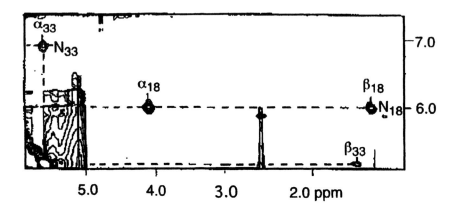

Figure 4. Portion of the 30°C TOCSY spectrum (mixing time 15 ms, repetition rate 3 s^{-1}) of Pf 4Fe FdA^{ox} in ^1H_2O depicting the connectivities for the strongly relaxed Ala18 (labile proton g) and Ala33 (labile proton f).

TOCSY spectra illustrating the connectivities for labile proton f and g is given in Figure 4. The chemical shift for these residues are listed in Table 1.

Short mixing time (50 ms) NOESY spectra (not shown) reveal a cross peak from peak b to the previously assigned (27) $C_\alpha H$ of Thr62, and identifies peak b (spin system E) as Ile61. Similarly, the $C_\alpha H$ spin coupled to proton f (Ala) exhibits a cross peak (Figure 3E) to the previously assigned (27) NH of Gln34 which identifies Ala33. Neither the NHs (peaks c, d or g) nor the coupled non-labile protons for the other detected residues exhibited cross peaks in the 50 ms NOESY map. However, a WEFT-NOESY spectrum under rapid pulsing conditions (Figure 3E) exhibits a cross peak between proton b (Ile61) and proton e (of an Ala), identifying Ala60, and a cross peak between proton g (the NH of an Ala) and the previously assigned (27) Ser19 NH (N_{19}) is observed, thereby identifying Ala18 (Figure 3E).

The highly conserved cluster loop and turn in the structurally characterized single cluster Fds (11,16-20) show that the NHs for the loop residues 12, 13, 15 and 16 are all expected to be ≤ 3.5Å from an Fe, and can be expected to exhibit T_1s ≤ 5 ms. The Val58 NH in the conserved turn E, however, is expected to have $R_{Fe} \sim 5$Å, which translates into $T_1 > 20$ ms. Hence labile proton d, and its spin coupled $C_\alpha HC_\beta HC_\gamma H_3$ fragment, (spin system C), are assigned to Val58.

The 15 ms TOCSY spectrum in ^1H_2O exhibits (not shown) a cross peak between two unassigned, rapidly relaxed protons with chemical shifts indicative of C_αHs, and hence assigns Gly13, as well as the α-β cross peak for a strongly relaxed and unassigned Ala, and hence identifies the remaining Ala15. Left unassigned are spin systems A, B and D for which NHs could not be detected. These three partial spin systems, which exhibit no detectable NOEs, must arise from the only remaining residues, Ile12, Ile16 and Pro57. Saturation of the Asp14 C_βH (not shown) results in steady-state NOEs to two of the protons (23) from spin system B. Inspection of the structure for homologous Fds indicate that these contacts are resonances from Ile12, and that spin system B is so assigned. Spin-system A exhibits one shift indicative of a methyl and hence must be attributed to Ile16, leaving spin system D as arising from the remaining Pro57. The ^{1}H and ^{15}N chemical shift data are summarized in Table 1.

Table 1 ^{1}H NMR spectral parameters for cluster environment residues in
Pf 4Fe Fd$^A_{ox}$

	NH					
Residue	^{15}N shift[a]	N^1H^b Label	N^1H shift[c]	$C_\alpha H$	$C_\beta H$	Others
Ile12	e	d	e	5.20(45)[d]	2.68(53)	1.11(30)
Gly13	e	-	e	4.01(28) 3.59(13)	-	-
Ala15	e	-	e	4.8	2.0	-
Ile16	e	-	e	4.1(75)	1.3	0.8
Ala18	143.9	g	6.01(12)	4.12(95)	1.22(78)	-
Ala33	136.7	f	6.54(15)	5.45(26)	1.50	-
Pro57	-	-	-	4.7	2.8, 1.8	
Val58	117.1	d	7.26(~30)	4.79	2.83(60)	0.93(46)
Ser59	117.3	c	7.38(26)	4.62(30)	4.08,3.8	-
Ala60	144.4	e	6.65	4.7	1.3	-
Ile61	121.2	b	7.91(42)	5.07(48)	1.98	1.38, 1.08 0.60(~15)

a) ^{15}N chemical shift referenced to external NH$_4$Cl at 24.9 ppm at 30°C; b) Peak labels *a-g* used in Figure 3B; c) Chemical shifts in ppm referenced against DSS, in ^{1}H$_2$O, 50 mM phosphate, pH 7.6 at 30°C; d) Estimated T_1s (uncertainty ±20%), in ms, are given in parentheses; e) Note detected because of expected severe ^{1}H relaxation (T_1 < 5 ms).

Proton relaxation: The estimated ^{1}H T_1 values (uncertainties ±20%) determined by 1D and 2D techniques for WT 4Fe Fd$^A_{ox}$ are in parentheses in Table 1 for the presently assigned residues, and in Table 2 for additional residues. Where available, the T_1 values for the same residues in D14C-Fd$^A_{ox}$ are included in Table 2.

Table 2 Proton relaxation behavior in the cluster environment of WT and D14C 4Fe Fd$^A_{ox}$

| | | T_1, ms[a] | | | | T_1, ms[a] | |
Residue	Proton	WT	D14C	Residue	Proton	WT	D14C
Asp7	NH	55	61	Ala33	NH	15	
	C$_\beta$H	110	100		C$_\alpha$H	26	33
Gln8	NH	95	~100	Gln34	NH	52	
	C$_\alpha$H	56	42		C$_\alpha$H	42	
Thr10	NH	130	140	Ala52	C$_\alpha$H	70	
Cys11	NH	50	50		C$_\beta$H$_3$	46	
Ala18	NH	12	13	Met53	NH	140	120
	C$_\alpha$H	95	81		C$_\alpha$H	45	
	C$_\beta$H$_3$	78	70	Ala55	NH	113	
Ser19	NH	86			C$_\alpha$H	110	
Leu20	NH	120	140		C$_{\beta1}$H	68	
Val24	C$_{\gamma2}$H$_3$	180	200	Ser59	NH	30	40
Phe25	C$_{\beta1}$H	77	100		C$_\alpha$H	48	54
	C$_{\beta1}$H	40	33	Ile61	NH	42	30
Met26	NH	175	180				
	C$_\alpha$H	82	95				
Lys27	NH		80				
	C$_\alpha$H	15	15				
	C$_\beta$H	86	97				

a) T_1s in ms (±20% uncertainty), determined by 1D I-R and 2D IR-TOCSY.

Dipolar contacts among relaxed protons: NOESY spectra (not shown) reveal numerous key dipolar contacts between relaxed residues that are clearly detectable when using a 50 ms mixing time and more rapid repetition rates (2-3 s^{-1}) (30), but are either very weak or undetectable when a 250 ms mixing time is employed (27). These contacts include strong cross peaks: α_{18}-N$_{25}$, α_{18}-β^2_{25}, β^1_{18}-N$_{19}$, ε_{25}-α_{49}, ε_{25}-γ^1_{61}, α_{33}-N$_{34}$, β_{52}-N$_{53}$, N$_{60}$-N$_{61}$, α_{61}-N$_{62}$, γ^2_{61}-N$_{62}$; moderate cross peaks: β_{10}-N$_{11}$, α_{17}-β^1_{20}, α_{17}-β_{52}, α_{18}-β^1_{25}, α_{18}-N$_{19}$, α_{18}-N$_{27}$, ε_{25}-γ_{61} and weak cross peaks: α_{17}-β^2_{20}.

DISCUSSION

Assignment protocol: NMR experiments tailored (28) to detect strongly relaxed resonances results in the assignment of at least a part of each of the 66 residues not located by standard 2D NMR techniques (27). Thus, the combination of the long mixing time NOESY/TOCSY/[^{1}H-^{15}N]-HSQC data, the steady-state NOEs (23,26,27,30) for the contact shifted ligands, and the present short mixing time/rapid repetition rate NOESY/TOCSY/[^{1}H-^{15}N]-HSQC data assign at least portions of all 66 residues and 93% of the total number of protons. The combination of the dipolar contacts among those 66 residues and the relaxation times for residues near the density cluster provide the necessary constraints to generate (12) a reasonable molecular model for Pf 4Fe Fd$_A$^{OX}.

The [^{1}H-^{15}N]-HSQC spectra lead to the detection of five residues for which no information was previously available, but do not detect the cross peak for proton a (T_1 ~2 ms). In fact, the HSQC peaks are detected only for labile protons with $T_1 \gtrsim 10$ ms. The NHs of the cluster loop are expected to have T_1 values well below 10 ms. It is also clear that WEFT spectra (see Figure 3B) do not allow even the detection of such strongly relaxed NHs *if they resonate in the diamagnetic envelope*. One C$_\beta$H of both Cys11 and Cys56 resonate in the amide region, as readily observed(23,24) in ^{2}H$_2$O spectra, but, because of their very short T_1s (~3 ms) and large linewidth, they cannot be seen in ^{1}H$_2$O in Figure 2B. However, for protons with $T_1 > 10$ ms, cross peaks useful for both assignment and spatial placement within the protein are achievable if they resonates in a relatively uncrowded spectral window (i.e., NH, C$_\alpha$H); the most difficult to detect and characterize signals are those near the intense methyl envelope.

Secondary structural element near the cluster. Long mixing time 2D experiments (27) provided very limited support for helix A, for which the residues are expected to be primarily within 6Å of the cluster iron(s). The present assignment of Ala18 in short mixing time NOESY for 4Fe Fd$_A$^{OX} and the detection of the expected N$_{18}$-N$_{19}$, α_{18}-N$_{19}$, β_{18}-N$_{19}$, cross peaks, together with the N$_i$-N$_{i+1}$, α_i-N$_{i+1}$ and β_i-N$_{i+1}$ cross peaks based on long mixing time NOESY (27), and the α_i-β_{i+3} detected for the Cys 17 (23), support the presence of the highly conserved helix A in Pf Fd. The conservation of the orientation of helix A and turn B in Pf 4Fe Fd^A$_{OX}$, moreover, is confirmed by observing (23) the Cys 17 C$_\beta$H NOE to both the C$_\beta$Hs and ring of Phe25 (not shown). Assignment of the complete Ala33 (Figure 4), together with the detection of strong α_{33}-N$_{34}$ and α_{27}-N$_{28}$ cross peaks in the $\tau_m = 50$ ms WEFT NOESY spectrum (Figure 3E), which was missing or only weak in long mixing time NOESY spectrum (27), confirm the role of these four residues in β-sheet B. Similarly, the present assignment of the Ile61 backbone and the detection of a strong WEFT-NOESY α_{61}-N$_{62}$ cross peak (Figure 3E) confirm the position of the Ile61/Val6 pair in β-sheet A.

The new assignments, moreover, provide valuable data to define important tertiary contacts among pairs of secondary structural elements near the cluster.

The short mixing time NMR data provide all of the detectable tertiary contacts involving helix A to β-sheet B (Phe25 C_βHs ring), and to helix B (to Ala52, Asn49). Thus, while relatively few new contacts could be observed, they represent important constraints for certain parts of the structure.

Relaxation properties: The relaxation, T_{1i}, of proton i experiencing only dipolar relaxation by the unpaired spin(s) of a paramagnetic cluster is given by (6,10,11,36):

$$T_{1i}^{-1} \sum_j D_j R_{ij}^{-6} \qquad (1)$$

where j are the individual atoms (*i.e.*, Fe, S, O) in the cluster, at a distance R_{ij} from the proton i, that possess unpaired electron spin density (D_j will depend on j). While the unapired spin is centered primarily ($\geq 80\%$) in the four iron, non-negligible spin density is delocalized onto each of the eight non-Fe atoms (15). For protons distant from a cluster atom (*i.e.* >5Å), only the dominant spin density on the iron need be considered, and in the special case of the oxidized 4Fe cluster where the four iron are magnetally equivalent (1,2,7), the relation simplifies to:

$$T_{1i}^{-1} = D \sum_j R_{i-Fej}^{-6} \qquad (2)$$

The cluster, short helix A, and β-turn B (see Figure 2) are among the most strongly conserved structural features among single cubane-cluster Fd (5,11,16-18,20). Hence, backbone protons on helix A can serve as sensitive probles of the constant D. The constant D is proportional to the electron spin relaxation time (6,36). Inspection of the data for WT and D14C-Fd_A^{OX} in Table 2 reveals that the proton T_1s for Ala18 and Ser19 are well within the experimental uncertainties of each other. The very similar constant D for WT and D14C-Fd_A^{OX} is also supported by the T_1 data for numerous other residues in Table 2. Hence we can conclude that the constant D in Eq. (2), and therefore the electron spin relaxation time, are indistinguishable for WT and D14C Fd, so that there is no advantage, from the point of adverse relaxation effects, to pursue in solution NMR structural determination in the D14C-Fd mutant with the more "conventional" all-Cys ligated cluster, relative to the native cluster Asp14-ligated WT Fd_A^{OX}.

It has been observed (23,26), however, that the Cys C_βHs experience shorter T_1s (in the case of Cys11 by factor ~2) in D14C than WT Fd_A^{OX}, in spite of the fact that the Cys C_βH contact shift pattern and magnitude and the same in the two Fd complexes. The most reasonable interpretation of the different Cys relaxation properties is that, because Cys $C_\beta H$ are much closer to the cluster atoms than protons on helix A, the point-dipole approximation breaks down (37), and the Cys protons are relaxed by the spin density on both the iron and the other cluster heteroatoms (bridging S or thiol S or, in the case of WT, Asp14 carboxylate O). Since the WT Fd^A_{ox} possesses the unique Asp14 ligand, it is likely that the spin density among the S and O atoms is rearranged relative to that of the all S-atom cluster in $D14C\text{-}Fd_A^{OX}$.

It has been proposed that the relative T_1s of the two $C_\beta Hs$ of a ligating Cys in a Fd can be used to quantitatively determine the Cys orientation (14). However, the delocalized unpaired spin over both the Fe and S of a cluster make it unlikely that Cys proton T_1s are controlled solely by the point-dipolar relaxation by the cluster iron (37). The different relaxation pattern for the Cys $C_\beta H$ T_1s in WT and D14C-Fd strongly support that Cys $C_\beta H$ relaxation by spin density on the non-metal components of the cluster cannot be ignored. The relative $C_\beta H$ T_1s of Cys ligand can be used to infer the stereospecific assignments (13,14), but are likely not valid for quantitatively inferring the orientation of the Cys relative to the cluster (14).

Acknowledgements

The authors are grateful to Drs. K. Bren and P.-L. Wang for valuable discussions. The research was supported by grants from the National Science Foundation, MCB 96-00759 (GNL) and MCB 99-04624 (MWWA).

References

(1) Cammack, R. *Adv. Inorg. Chem.* **1992**, *38*, 281-322.
(2) Bertini, I.; Ciurli, S.; Luchinat, C. *Struct. Bond.* **1995**, *83*, 2-53.
(3) Cheng, H.; Markley, J. L. *Annu. Rev. Biophys. Biomol. Struct.* **1995**, *24*, 209-237.
(4) Stephens, P. J.; Jollie, D. R.; Warshel, A. *Chem Rev* **1996**, *96*, 2491-2513.
(5) Goodfellow, B. J.; Macedo, A. L. *Repts. NMR Spectros.*; **1999**, *37*, 119-177.
(6) Bertini, I.; Luchinat, C.; Rosato, A. *Adv. Inorg. Chem.* **1999**, *47*, 251-282.
(7) La Mar, G. N. *Meth. Enzyme* **2001**; *334*, 351-389.
(8) Bertini, I.; Luchinat, C. *Coord. Chem. Rev.* **1996**, *150*, 1-296.
(9) Bertini, I.; Luchinat, C.; Rosato, A. *Prog. Biophys. Mol. Biol.* **1996**, *66*, 43-80.
(10) Bertini, I.; Couture, M. M. J.; Donaire, A.; Eltis, L. D.; Felli, I. C.; Luchinat, C.; Piccioli, M.; Rosato, A. *Eur. J. Biochem.* **1996**, *241*, 440-452.

(11) Wang, P. L.; Donaire, A.; Zhou, Z. H.; Adams, M. W. W.; La Mar, G. N. *Biochemistry* **1996**, *35*, 11319-11328.

(12) Sham, S.; Calzolai, L.; Wang, P.-L.; Bren, K.; Haarklau, H.; Brereton, P. S.; Adams, M. W. W.; La Mar, G. N. *Biochemistry* **2002**, *41*, 12498-12508.

(13) Busse, S. C.; La Mar, G. N.; Howard, J. B. *J. Biol. Chem.* **1991**, *266*, 23714-23723.

(14) Davy, S. L.; Osborne, M. J.; Breton, J.; Moore, G. R.; A.J., T.; Bertini, I.; Luchinat, C. *FEBS Lett.* **1995**, *363*, 199-204.

(15) Noodleman, L.; Peng, C. Y.; Mouesca, J.-M.; Case, D. A. *Coord. Chem. Rev.* **1995**, *144*, 199-244.

(16) Kissinger, C. R.; Sieker, L. C.; Adman, E. T.; Jensen, L. H. *J. Mol. Biol.* **1991**, *219*, 693-715.

(17) Séry, A.; Housset, D.; Serre, L.; Bonicel, J.; Hatchikian, C.; Frey, M.; Roth, M. *Biochemistry* **1994**, *33*, 15408-15417.

(18) Macedo-Ribeiro, S.; Darimont, B.; Sterner, R.; Huber, R. *Structure* **1996**, *4*, 1291-1301.

(19) Sticht, H.; Wildegger, G.; Bentrop, D.; Darimont, B.; Sterner, R.; Rösch, P. *Eur. J. Biochem.* **1996**, *237*, 726-735.

(20) Davy, S. L.; Osborne, M. J.; Moore, G. R. *J. Mole. Biol.* **1998**, *277*, 683-706.

(21) Conover, R. C.; Kowal, A. T.; Fu, W.; Park, J.-B.; Aono, S.; Adams, M. W. W.; Johnson, M. K. *J. Biol. Chem.* **1990**, *265*, 8533-8541.

(22) Kim, C.; Brereton, P. S.; Verhagen, F. J. M.; Adams, M. W. W. *Meth. Enzyme* **2001**, *334*, 30-55.

(23) Calzolai, L.; Gorst, C. M.; Zhao, Z.-H.; Teng, Q.; Adams, M. W. W.; La Mar, G. N. *Biochemistry* **1995**, *34*, 11373-11384.

(24) Calzolai, L.; Zhou, Z. H.; Adams, M. W. W.; La Mar, G. N. *J. Am. Chem. Soc.* **1996**, *118*, 2513-2514.

(25) Gorst, C. M.; Zhou, Z. H.; Ma, K. S.; Teng, Q.; Howard, J. B.; Adams, M. W. W.; La Mar, G. N. *Biochemistry* **1995**, *34*, 8788-8795.

(26) Calzolai, L.; Gorst, C. M.; Bren, K. L.; Zhou, Z. H.; Adams, M. W. W.; LaMar, G. N. *J. Am. Chem. Soc.* **1997**, *119*, 9341-9350.

(27) Wang, P.-L.; Calzolai, L.; Bren, K. L.; Teng, Q.; Jenney, F. E., Jr.; Brereton, P. S.; Howard, J. B.; Adams, M. W. W.; La Mar, G. N. *Biochemistry* **1999**, *38*, 8167-8178.

(28) La Mar, G. N.; de Ropp, J. S. *Biol. Magn. Reson.* **1993**, 18.

(29) Gupta, R. K. *J. Magn. Reson.* **1976**, *24*, 461-465.

(30) Gorst, C. M.; Yeh, Y. H.; Teng, Q.; Calzolai, L.; Zhou, Z. H.; Adams, M. W. W.; La Mar, G. N. *Biochemistry* **1995**, *34*, 600-610.

(31) Huber, J. G.; Moulis, J.-M.; Gaillard, J. *Biochemistry* **1996**, *35*, 12705-12711.

(32) Bax, A.; Davis, D. G. *J. Magn. Reson.* **1985**, *65*, 355-360.

(33) Jeener, J.; Meier, B. H.; Bachmann, P.; Ernst, R. R. *J. Chem. Phys.* **1979**, *71*, 4546-4553.

(34) Chen, Z.; de Ropp, J. S.; Hernández, G.; La Mar, G. N. *J. Am. Chem. Soc.* **1994**, *116*, 8772-8783.

(35) Bax, A.; Ikura, M.; Kay, L. E.; torchia, D. E.; Tshudin, R. *J. Magn. Reson.* **1990**, *86*, 304-318.

(36) Banci, L.; Bertini, I.; Luchinat, C. Nuclear and Electron Relaxation; VCH: Weinheim, FRG, **1991**.

(37) Wilkens, S. J.; Xia, B.; Volkman, B. F.; Weinhold, F.; Markley, J. L.; Westler, W. M. *J. Phys. Chem. B* **1998**, *102*, 8300-8305.

Chapter 15

Electronic Isomerism in Oxidized High-Potential Iron–Sulfur Proteins Revisited

Ivano Bertini[1], Francesco Capozzi[2], and Claudio Luchinat[1]

[1]Center of Magnetic Resonance, University of Florence,
Via Luigi Sacconi 6, IT–50019, Florence, Italy
[2]Department of Food Science, University of Bologna, Via Ravennate 1020,
IT–47023, Cesena (FC), Italy

The pair-of-pairs model that successfully explained the Mössbauer and NMR data on oxidized High Potential Iron-sulfur Proteins (HiPIP) in terms of equilibria between two electronic isomers is reviewed. A bridging of the high temperature NMR data with the low temperature EPR data is attempted through extrapolation to low temperature of the species distribution detected at room temperature by NMR. This extrapolation is validated by recent ¹H ENDOR data. With one exception, the more abundant isomer is the one with the mixed-valence iron pair localized either on irons II-III or on irons III-IV. It is suggested that one or two more isomers besides these two, such as I-III or II-IV, may be present in smaller amount.

Dedication

This work is dedicated to our friend Brian M. Hoffmann on the occasion of the celebration of his 60th birthday. In a way this contribution has been prompted by this occasion, and would not have been conceived otherwise. We have worked on the electronic structure and electronic isomerism of HiPIP for several years in the recent past, and we believe we have succeeded, using NMR, to uniquely determine the partial localization of the iron oxidation states within the protein frame. However, as it often happens, after drafting the grand picture, our attention (as everybody else's attention!) has been drawn by other exciting developments in biology, and neither we nor anybody else cared of the details. Now, the availability of as yet unpublished EPR and ENDOR data make us think that maybe it is time to try and add some detail to the NMR picture. And we take this opportunity to do so because we feel that the approach that we have used for HiPIPs, which was only successful because we combined information from NMR with information from Mössbauer, EPR and ENDOR, is particularly suitable to honor Brian's career.

Introduction

High potential iron-sulfur proteins (HiPIP) are a class of small bacterial proteins containing a single iron-sulfur cluster (1). In their oxidized state they formally contain three ferric and one ferrous ion (2). These ions are strongly antiferromagnetically coupled, yielding a total spin 1/2, with EPR spectra having all g components larger than 2 (3). Through several Mössbauer investigations (4-9) it was shown that the electronic spin system can be described by two antiferromagnetically coupled high spin pairs, a mixed valence and a ferric pair, with the former having a subspin larger than the latter by 1/2. Moreover, the analysis of the isomer shifts and the magnetic hyperfine tensors from the Mössbauer fits of several HiPIPs shows that the extra electron (referred to the all ferric iron cluster) is completely delocalized over the mixed valence pair, yielding two essentially equivalent $Fe^{2.5+}$ (as a consequence of double exchange between one pure Fe^{2+} and one pure Fe^{3+}) and two Fe^{3+} iron ions (4-9).

The organization in subspin pairs rather than, for example, spin triplets, seems to be the lowest magnetic energy organization in four iron clusters (10-13). Possible subspin values for these pairs could be 9/2,4 (4), 7/2,3 (14) or a mixture of 9/2,4 and 7/2,4 with the former prevailing (15, 16). Whatever the exact description, which in any case is based on high symmetry - and therefore approximate - analytical solutions of the spin-Hamiltonian, in a magnetic field the mixed valence pair will tend to align its subspin along the field, so that the ferric pair will align opposite to the field (Figure 1).

This was the key to interpret the Mössbauer data, which showed the two iron pairs to have seemingly opposite hyperfine couplings with the ^{57}Fe magnetic nuclei. In reality, the hyperfine coupling has always the same sign, while it is the polarization of the electron spin in the ferric pair to be opposite to the field (4).

This same feature has been later exploited to define the localization of the two pairs in the protein frame using NMR (17-19). Indeed, the ^1H NMR spectra show distinct hyperfine-shifted signals (20-22) arising from the β-CH_2 protons of all four cluster-coordinated cysteines (17-19). These signals could be assigned sequence-specifically in a number of different HiPIPs (17-19, 23-35). Interestingly, while the signals from two of the four cysteines were shifted

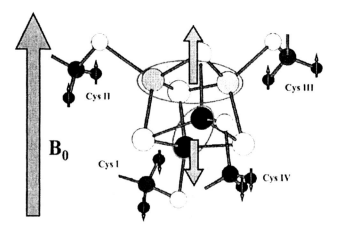

Figure 1. Iron pairs within the major "electronic isomer"in HiPIP II from E. halophila. Mixed valence iron ions are represented as light grey spheres, while pure ferric ions are showed as large black spheres.

downfield (as expected from direct spin delocalization over the β-CH₂ protons due to hyperconjugation with sulfur p orbitals involved in π-bonding to the metal (36, 37), those from another cysteine were shifted upfield, and those from the last cysteine were either shifted upfield or, if downfield, extrapolated to upfield shifts at low temperature (Figure 2).

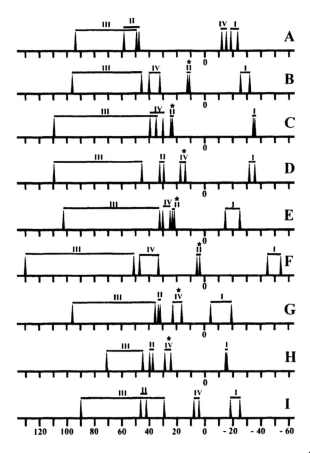

Figure 2. Schematic representation of hyperfine shifted signals in ¹H-NMR spectra of HiPIPs from: A) E. halophila (iso-II), B) R. gelatinosus, C) C. vinosum, D) R. globiformis, E) E. vacuolata (iso-II), F) R. tenue, G) E. halophila (iso-I), H) C. vinosum (C77S mutant), I) C. vinosum (C63S mutant). Signal pairs showing anti-Curie temperature dependence, when present, are indicated with asterisks.

Given the electronic structure outlined above, the downfield signals must arise from the mixed valence pair, which has a normal magnetic behavior, and the upfield (or extrapolating upfield) to the ferric pair, which displays an inverted electron spin polarization (Figure 1). With the sequence specific assignment at hand, it clearly emerged that the signals from the third cysteine in the sequence (Cys III) were always downfield (*i.e.* the attached iron was always part of the mixed valence pair) and those from the first cysteine (Cys I) were always upfield (*i.e.* the attached iron was always part of the ferric pair) (38, 39). With the exception of the HiPIP II from *E. halophila* (19) where cysteine II clearly behaved as cysteine III and cysteine IV behaved as cysteine I, so that irons II-III and I-IV could be clearly assigned to the mixed valence and ferric pairs, respectively, in all other HiPIPs cysteine II and IV showed an intermediate behavior.

The two-isomers model

The simplest model that could explain the NMR data is that of an equilibrium situation between *two* isomers where the mixed valence pair would switch from irons II-III to irons III-IV (and conversely the ferric pair would switch between irons I-IV and I-II) (Figure 3) (27), and where the equilibrium position depends on the particular HiPIP involved (30). Of course, the II-III and III-IV would only be the two most representative of the six possible mixed valence pairs, and the presence of smaller amounts of other pairwise distributions of the electron subspins could not be ruled out.

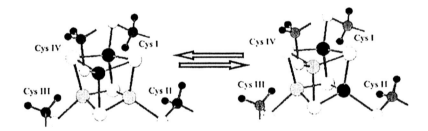

Figure 3. Equilibrium in HiPIP from C. vinosum according to the two "electronic isomers" model.

These HiPIP isomers differing only by the localization of the electronic subspin pairs were later termed "electronic isomers" (39). As the localization of the subspin pairs in the protein frame implies at least a partial displacement of charges (*i.e.* the mixed valence pair contains a nominally ferrous ion, and thus must carry a slightly higher electron density than the ferric pair) (40), their localization may provide information on the individual (or microscopic) reduction potentials of the four ferric iron ions.

If this is true, it would appear that in all HiPIPs the ferric iron ion coordinated by cysteine III is the most reducible, that coordinated by cysteine I is the least reducible, and those coordinated by cysteines II and IV have intermediate microscopic reduction potential, whose order varies from one HiPIP to another. Free energy calculations on two HiPIPs, HiPIP from *C. vinosum* and HiPIP II from *E. halophila*, that were found to differ in the localization of the subspin pairs, provided microscopic reduction potentials in remarkable agreement with the experimental findings (41): the II-III pair was calculated to be the most reducible in HiPIP II from *E. halophila* (experimentally found to be the mixed-valence pair), whereas both the II-III pair and the III-IV pair were found equally reducible in the HiPIP from *C. vinosum*. Indeed, in the latter a substantial amount of both mixed-valence pairs was experimentally found. Interestingly, the values of the reduction potentials calculated for the other pairs suggested that, besides the dominant pair in HiPIP I from *E. halophila* and the two dominant pairs in the HiPIP from *C. vinosum*, at least two more mixed valence pairs could be present in smaller amounts, although in different orders. The least reducible ferric pairs were in both cases the I-II and the I-IV pairs (41).

The above calculations and their agreement with experimental findings have an important implication which may have not been fully appreciated. In principle, there are two possible determinants for the electronic structure of oxidized HiPIPs: one is magnetic and the other is electrostatic. In the first case, the localization of the pairs depends on the values of the magnetic parameters, such as the superexchange J values within the cluster and the double exchange B values within the mixed valence pairs. In this case, the charge displacement and therefore the microscopic reduction potentials are a consequence of the magnetic structure. In the second case, electrostatic/chemical bonding forces determine the charge displacement and the preference for the mixed valence pair to settle on one (or more) of the six possible iron pairs. In this case, the magnetic structure with its organization in subspin pairs simply adapts to the electrostatic driving force. Since free energy calculations take into account electrostatic and chemical bonding forces, but not magnetic interactions, their success in predicting the localization of the mixed valence pairs in two diverse HiPIPs strongly suggests that the magnetic structure is a consequence rather than the origin of the microscopic reduction potentials. We are thus in an ideal situation of having detailed reporters of the magnetic structure (the hyperfine shifts of the cysteine β-CH_2 protons) which somehow act as amplifiers of the small charge displacements among the iron ions in the cluster.

Low temperature spectroscopies

While high resolution NMR is bound to solution state measurements, i.e. around and above freezing, Mössbauer, EPR and ENDOR spectroscopies on HiPIPs can only provide information at low temperature, i.e. liquid nitrogen or below. It is the link between these two temperature ranges that is still partially lacking. In particular, it is not clear whether the two (or possibly more) electronic isomers detected at room temperature still exist at low temperature and, if so, whether they can be detected by low temperature spectroscopies. As already discussed, Mössbauer measurements at low temperature and in the presence of a magnetic field provided the first experimental evidence of the pair-of-pairs model. However, no evidence for the presence of more than one pair-of-pairs was ever obtained by Mössbauer (4-9, 40, 42, 43). This could of course be due to the lack of multiple electronic isomers at low temperature, but also to the limited resolution of the technique. Indeed, it is likely that electronic isomers that differ in the orientation of the pair of pairs would yield indistinguishable Mössbauer patterns.

The same reasoning holds for ^{57}Fe ENDOR. Q-Band ^{57}Fe ENDOR, indeed, reveals inequivalency among iron ions in FeS proteins and provides the orientation, relative to the g-tensor, for the hyperfine tensors of iron sites (44-46). As a consequence, ENDOR provides information to place iron pairs within the cubane clusters of oxidized HiPIPs and reduced ferredoxins, but, again, although in principle four inequivalent iron ions are present, only two signals can be discerned – one attributable to the mixed valence pair and the other to the ferric pair – on the ground of the different A values. Therefore, if two or more electronic isomers are present, the ENDOR parameters of the mixed valence pair and ferric pair in each of them may be indistinguishable within the resolution of the technique (47-50).

The situation is different for the EPR spectra. It has long been observed that the EPR spectra of all HiPIPs, with the notable exception of HiPIP II from *E. halophila*, are mixtures of at least two species (8, 51-55). A recent reexamination of the EPR properties of a series of HiPIPs, based on spectral simulations including g-strain, suggest that (again with the exception of HiPIP II from *E. halophila*), a minimum of three species is often required (56). It would be tempting to speculate that the low temperature EPR species correspond to the electronic isomers detected by NMR. That this is a realistic possibility is also suggested by ^{1}H ENDOR studies where $[Fe_4S_4]^{3+}$ centers are obtained in low yield by γ-irradiation of a $[Fe_4S_4]^{2+}$-tetrathiolate model compound (48, 51, 57-59). This random oxidation of cluster units within a crystal causes the appearance of EPR spectra constituted by more than one component. Their selective saturation yields ^1H ENDOR spectra that could be assigned to species where the mixed valence and ferric pairs have different localization within the crystal, revealed by the assignment of the different ^1H ENDOR lines to different thiolate donors. These experiments could also show that the direction of the lowest field EPR feature (termed g_1) is always along an axis which is perpendicular to both the vector joining the two mixed valence irons and the one joining the two ferric irons (48, 51, 57-59).

If indeed different electronic isomers are able to give rise to different EPR spectra in model compounds, one could try to see if any relationship exists between electronic isomers detected by NMR and EPR isomers by finding a way of classifying the latter. Examination of the EPR spectra (56) suggests that the most diagnostic feature is the absorption-like low-field one, i.e. the g_1 value. One type of EPR spectra has g_1 in the range 2.13-2.15; another has g_1 in the range 2.11-2.13; a third one has g_1 in the range 2.06-2.08. Possibly, a fourth species with g_1 in the range 2.09-2.11 may be present. In all HiPIPs investigated there is always a species which is largely dominant, characterized by either the highest (2.13-2.15) or the second highest (2.11-2.13) g_1 species.

Bridging room temperature NMR with low temperature EPR

Following this line of reasoning, we take as a starting point the single EPR species with $g_1 = 2.145$ observed for HiPIP II from *E. halophila*. For this protein, NMR shows absolute predominance of the isomer with the mixed valence pair on irons II-III. The first goal is to demonstrate that the mixed valence pair is localized on irons II-III also at the EPR temperature. A breakthrough came from multiple irradiation (or crystal-like) ^1H ENDOR experiments on frozen HiPIP samples (60). By irradiating throughout the EPR spectrum, and thus by selecting different protein orientations, it was possible to follow the ^1H ENDOR signals and assign them on the basis of the known protein structure and orientation within the dipolar field of the cluster. Subtraction of the dipolar contribution to the ^1H ENDOR shift left the contact contribution, which came out of positive sign for two cysteine proton pairs (mixed valence pair) and negative for the other two (ferric pair). After proper scaling for the Boltzmann population of the excited states at room temperature, these ^1H contact shifts were in very good agreement with the room temperature NMR experiment. Therefore, ^1H ENDOR on *E. halophila* HiPIP II constitutes the first safe link between high temperature NMR and low temperature EPR spectra. Based on this link alone one could speculate that the electronic isomers having the mixed valence pair on irons II-III would correspond to the EPR species with $g_1 = 2.13$-2.15. If so, this species would be dominant also in *E. halophila* HiPIP I. Conversely, the other major species observed in EPR, that with $g_1 = 2.11$-2.13, which is dominant in

HiPIPs I and II from *E. vacuolata* and in the HiPIPs from *C. vinosum, R. gelatinosus* and *R. tenuis*, could correspond to the III-IV isomer (HiPIP from *R. globiformis* (g_1 = 2.128) would be a borderline case). Recent unpublished ^{1}H ENDOR results on *E. halophila I, C. vinosum* and *R. tenuis* seem to fully confirm this assignment (61).

In the pair-of-pairs NMR model we were already able to indicate which of the two electronic isomers, the II-III or the III-IV mixed valence pair, was the dominant one at room temperature (30, 62). It is also possible to extrapolate to low temperature based on temperature dependent NMR data in the relatively narrow range between freezing and high temperature denaturation. Qualitatively, the extrapolation is based on a comparison between the temperature dependence of the contact shifts of the four cysteine β-CH_2 pairs. For a single isomer, the temperature dependence is essentially originated from the different population of excited S' states above the ground S' = 1/2 state. On top of this temperature dependence, however, any change in ferric vs. mixed valence character of each iron ion would result in an additional contribution which should be easily discernible. A clear manifestation of this change is the anti-Curie behavior of one of the cysteine β-CH_2 pairs with intermediate behavior (Cys III or Cys IV), often paralleled by a more marked Curie behavior for the other pair (Cys IV or Cys III). So, following this additional contribution, it is easy to predict that the dominant species at low temperature will have the mixed valence pair located on irons II-III in HiPIPs I and II from *E. halophila* and in the HiPIP from *R. globiformis*, while the mixed valence pair will be dominantly located on irons III-IV in the HiPIP II from *E. vacuolata* and in the HiPIPs from *R. gelatinosus, C. vinosum* and *R. tenuis*. This prediction is in remarkable agreement with the classification of the EPR spectra based on the g_1 values. According to the NMR prediction, the borderline EPR spectrum of *R. globiformis* would be assigned to the II-III pair.

Cys→Ser mutations and minor species

While it is relatively straightforward to relate the major isomer in the EPR spectra of each HiPIP to one of the two major isomers (II-III and III-IV) detected by NMR, it would be much more difficult to identify the minor EPR isomers because there is no clear indication of the presence of a third, or of a third and fourth, isomer in the NMR spectra. However, some considerations can be made. For example, although in principle all six possible isomers may be present, isomers with the mixed valence pair on irons I-II and I-IV would be much disfavored based on the free energy calculations discussed previously (41). Any of the other two possible isomers present at room temperature besides the II-III and the II-IV would imply that either iron III is not 100% involved in a mixed valence pair (as it would happen for a II-IV isomer) or iron I is not 100% involved in a ferric pair (as it would happen for a I-III isomer). As a consequence, the shifts and temperature dependences displayed by the β-CH$_2$ pairs of the cysteines bound to irons I and III could not represent limiting behaviors. In particular, the presence of II-IV and/or I-III mixed valence isomers would cause the shifts of the β-CH$_2$ pairs of the cysteine bound to iron III to be less downfield and/or the shifts of the β-CH$_2$ pairs of the cysteine bound to iron I to be less upfield. Examination of the ^{1}H NMR spectra in the light of these considerations suggests that one could take the HiPIP with the largest extremal shift values for cysteines I and III as that having the minimal (or zero) involvement of additional pairs. Such HiPIP is that from *R. tenuis* (63). Thus, based on the assumption that the extremal shifts of Cys III and Cys I in *R. tenuis* HiPIP represent 100% and 0% involvement in a mixed valence pair, one can calculate the percentage of involvement in mixed valence pairs for each iron ion in each other HiPIP, again as a function of temperature. The results are reported in Table I, in the assumption that at most four of the six possible pairs can be present.

In all cases the results of Table I are in qualitative agreement with the simpler pair-of-pairs model, in the sense that all the predictions for the dominant isomer at low temperature are maintained. Such isomer is already the most abundant at room temperature for the two HiPIPs from *E. halophila* and for the HiPIPs from *R. gelatinosus*, *C. vinosum*, and *R. tenuis*, while it just becomes the most abundant around room temperature for the HiPIP II from *E. vacuolata* and the HiPIP from *R. globiformis*. It is also interesting to notice that a second minor isomer seems to increase in percentage with decreasing temperature in the case of the HiPIP from *R. gelatinosus* (isomer II-III), and of the HiPIP I from *E. halophila* (isomer III-IV).

Table I provides a particularly interesting insight in the behavior of two Cys→Ser mutants of the HiPIP from *C. vinosum*, namely the C77S (serine coordinating iron IV) (32, 40) and C63S (serine coordinating iron III) (64). These mutations were conceived with the aim of decreasing the tendency of the coordinated iron to be involved in a mixed valence pair, due to the stabilization of the ferric form brought about by the serinate donor. In the case of the C77S mutant, the mutation decreases the tendency of iron IV to belong to a mixed valence pair. Therefore, with respect to the WT HiPIP, the percentage of III-IV isomer strongly decreases, as already noted (32). However, it looks like this decrease is accompanied by a redistribution over the other three pairs, resulting in a low temperature extrapolation where the dominant species is the II-III but the second species is still the III-IV.

Even more interesting is the case of the C63S mutant, where the serinate now coordinates the iron that in the WT has the strongest tendency to belong to a mixed valence pair. This mutant is the only case where there are three cysteine proton pairs in the downfield region all three with a Curie type temperature dependence (64). The distribution of the electronic isomers is the most even, with no isomer really dominant, and with isomer II-IV as the most probable candidate for the low temperature dominant species. This prediction has yet to be verified by EPR.

Table I. Electronic isomer distribution in various HiPIPs around room temperature. Extrapolated dominant (bold) and secondary (underlined) species at low temperature are shown.

II-III	I-III	III-IV	II-IV	HiPIP	Temp.
0.542	0.204	0.088	0.165	Hal II	308
0.570	0.192	0.087	0.150		278
<u>0.258</u>	0.160	**0.405**	0.177	R Gel	313
<u>0.264</u>	0.134	**0.452**	0.151		284
0.347	0.093	**0.390**	0.170	C Vin	328
0.348	0.091	**0.407**	0.165		298
0.437	0.103	0.339	0.121	R Glob	310
0.448	0.102	0.335	0.114		286
0.284	0.212	**0.284**	0.220	Vac II	321
0.280	0.191	**0.335**	0.194		280
0.364	0.000	**0.636**	0.000	R Tenue	313
0.347	0.000	**0.653**	0.000		279
0.423	0.195	<u>0.145</u>	0.237	Hal I	304
0.425	0.188	<u>0.153</u>	0.234		292
0.290	0.253	<u>0.208</u>	0.249	C77S	304
0.315	0.241	<u>0.214</u>	0.230		288
0.245	<u>0.272</u>	0.166	**0.316**	C63S	298
0.228	<u>0.283</u>	0.158	**0.330**		278

A final comment is due to the possible determinants of the different electronic isomer distribution in the various HiPIPs. As suggested by free energy calculations (41) electrostatic interactions may play an important role. As the latter are of enthalpic origin, their temperature dependence should be modest, and the variation of the equilibrium distibution with temperature should obey Boltzmann law. In other words, of the various electronic isomers detected at room temperature the one lying lowest in energy should be dominantly present at low temperature. Indeed, this is observed, in the sense that the EPR spectra, while being often heterogeneous, always show that one species is in much larger amount the others (56). Only in the case of HiPIP II from *E. vacuolata* NMR shows evidence of a possible inversion of abundance between two species above room temperature, so that the dominant one at low temperature does not coincide with the most abundant at high temperature. This can only be due to a temperature dependence of the free energy of the system, possibly through a modest entropic contribution of unknown nature.

Acknowledgments

We would like to acknowledge the proficuous interactions that our lab has had over the years with M. Belinski, L.D. Eltis, J. J. Girerd, W.R. Hagen, J. Hüttermann, F. Parak, M. Teixeira, that allowed us to develop and refine the present electronic isomers model.

References

1. Beinert, H.; Holm, R. H.; Munck, E. *Science* **1997**, 277, 653-659.
2. Carter, C. W. J.; Kraut, J.; Freer, S. T.; Alden, R. A.; Sieker, L. C.; Adman, E. T.; Jensen, L. H. *Proc.Natl.Acad.Sci.USA* **1972**, 69, 3526-3529.
3. Guigliarelli, B.; Bertrand, P. In Advances in Inorganic Chemistry; Academic Press Inc.: San Diego, CA. 1999; Vol. 47, 421-497.
4. Middleton, P.; Dickson, D. P.; Johnson, C. E.; Rush, J. D. *Eur.J.Biochem.* **1980**, 104, 289-296.
5. Dickson, D. P.; Cammack, R. *Biochem.J.* **1974**, 143, 763-765.
6. Dickson, D. P.; Johnson, C. E.; Cammack, R.; Evans, M. C.; Hall, D. O.; Rao, K. K. *Biochem.J.* **1974** , 139, 105-108.
7. Bertini, I.; Campos, A. P.; Luchinat, C.; Teixeira, M. *J.Inorg.Biochem.* **1993**, 52, 227-234.
8. Dunham, W. R.; Hagen, W. R.; Fee, J. A.; Sands, R. H.; Dunbar, J. B.; Humblet, C. *Biochim.Biophys.Acta* **1991**, 1079, 253-262.
9. Papaefthymiou, V.; Millar, M. M.; Münck, E. *Inorg.Chem.* **1986**, 25, 3010-3014.
10. Bominaar, E. L.; Borshch, S. A.; Girerd, J.-J. *J.Am.Chem.Soc.* **1994**, 116, 5362-5372.
11. Noodleman, L. *Inorg.Chem.* **1988**, 27, 3677-3679.
12. Noodleman, L. *Inorg.Chem.* **1991**, 30, 246-256.

13. Blondin, G.; Girerd, J.-J. *Chem.Rev.* **1990**, 90, 1359-1376.
14. Mouesca, J.-M.; Noodleman, L.; Case, D. A.; Lamotte, B. *Inorg.Chem.* **1995**, 34, 4347-4359.
15. Belinskii, M. I.; Bertini, I.; Galas, O.; Luchinat, C. *Z.Naturforsch.* **1995**, 50a, 75-80.
16. Belinskii, M. I.; Bertini, I.; Galas, O.; Luchinat, C. *Inorg.Chim.Acta* **1996**, 243, 91-99.
17. Bertini, I.; Briganti, F.; Luchinat, C.; Scozzafava, A.; Sola, M. *J.Am.Chem.Soc.* **1991**, 113, 1237-1245.
18. Banci, L.; Bertini, I.; Briganti, F.; Luchinat, C.; Scozzafava, A.; Vicens Oliver, M. *Inorg.Chem.* **1991**, 30, 4517-4524.
19. Banci, L.; Bertini, I.; Briganti, F.; Scozzafava, A.; Vicens Oliver, M.; Luchinat, C. *Inorg.Chim.Acta* **1991**, 180, 171-175.
20. Phillips, W.D. In *NMR of Paramagnetic Molecules, Principles and Applications*; La Mar, G.N.; Horrocks, W.D., Jr.; Holm, R.H., Eds.; Academic Press: New York, NY. 1973.
21. Phillips, W.D.; Poe, M. In *Iron-sulfur Proteins*; Lovenberg, W., Ed.; Academic Press: New York, NY. 1977.
22. Nettesheim, D. G.; Meyer, T. E.; Feinberg, B. A.; Otvos, J. D. *J.Biol.Chem.* **1983**, 258, 8235-8239.
23. Bertini, I.; Capozzi, F.; Luchinat, C.; Piccioli, M.; Vicens Oliver, M. *Inorg.Chim.Acta* **1992**, 198-200, 483-491.
24. Banci, L.; Bertini, I.; Capozzi, F.; Carloni, P.; Ciurli, S.; Luchinat, C.; Piccioli, M. *J.Am.Chem.Soc.* **1993**, 115, 3431-3440.
25. Banci, L.; Bertini, I.; Carloni, P.; Luchinat, C.; Orioli, P. L. *J.Am.Chem.Soc.* **1992**, 114, 10683-10689.
26. Bertini, I.; Capozzi, F.; Luchinat, C.; Piccioli, M. *Eur.J.Biochem.* **1993**, 212, 69-78.
27. Banci, L.; Bertini, I.; Ciurli, S.; Ferretti, S.; Luchinat, C.; Piccioli, M. *Biochemistry* **1993**, 32, 9387-9397.
28. Bertini, I.; Ciurli, S.; Dikiy, A.; Luchinat, C. *J.Am.Chem.Soc.* **1993**, 115, 12020-12028.
29. Bertini, I.; Felli, I. C.; Kastrau, D. H. W.; Luchinat, C.; Piccioli, M.; Viezzoli, M. S. *Eur.J.Biochem.* **1994**, 225, 703-714.
30. Bertini, I.; Capozzi, F.; Eltis, L. D.; Felli, I. C.; Luchinat, C.; Piccioli, M. *Inorg.Chem.* **1995**, 34, 2516-2523.
31. Bertini, I.; Dikiy, A.; Kastrau, D. H. W.; Luchinat, C.; Sompornpisut, P. *Biochemistry* **1995**, 34, 9851-9858.
32. Babini, E.; Bertini, I.; Borsari, M.; Capozzi, F.; Dikiy, A.; Eltis, L. D.; Luchinat, C. *J.Am.Chem.Soc.* **1996**, 118, 75-80.
33. Bentrop, D.; Bertini, I.; Capozzi, F.; Dikiy, A.; Eltis, L. D.; Luchinat, C. *Biochemistry* **1996**, 35, 5928-5936.

34. Bertini, I.; Capozzi, F.; Ciurli, S.; Luchinat, C.; Messori, L.; Piccioli, M. *J.Am.Chem.Soc.* **1992**, 114, 3332-3340.

35. Luchinat, C.; Ciurli, S.; Capozzi, F. In *Perspective in Coordination Chemistry*; Williams, A.F.; Floriani, C.; Merbach, A.E., Eds.; Verlag Helvetica Chimica Acta: Basel (CH). 1992; pp 245-269.

36. Bertini, I.; Luchinat, C.; Parigi, G. *Solution NMR of Paramagnetic Molecules*; Elsevier:Amsterdam (NL). 2001.

37. Bertini, I.; Capozzi, F.; Luchinat, C.; Piccioli, M.; Vila, A. J. *J.Am.Chem.Soc.* **1994**, 116, 651-660.

38. Bertini, I.; Ciurli, S.; Luchinat, C. *Struct.Bonding* **1995**, 83, 1-54.

39. Bertini, I. and Luchinat, C. (1996) Electronic *isomerism* in oxidized Fe₄S₄ high-potential iron-sulfur proteins. In *Transition Metal Sulfur Chemistry*. ACS Symposium Series. pp. 57 - 73.

40. Dilg, A. W.; Capozzi, F.; Mentler, M.; Iakovleva, O.; Luchinat, C.; Bertini, I.; Parak, F. G. *J.Biol.Inorg.Chem.* **2001**, 6, 232-246.

41. Banci, L.; Bertini, I.; Gori Savellini, G.; Luchinat, C. *Inorg.Chem.* **1996**, 35, 4248-4253.

42. Dilg, A. W.; Grantner, K.; Iakovleva, O.; Parak, F. G.; Babini, E.; Bertini, I.; Capozzi, F.; Luchinat, C.; Meyer-Klaucke, W. *J.Biol.Inorg.Chem.* **2002**, 7, 691-703.

43. Dilg, A. W.; Mincione, G.; Achterhold, K.; Iakovleva, O.; Mentler, M.; Luchinat, C.; Bertini, I.; Parak, F. G. *J.Biol.Inorg.Chem.* **1999**, 4, 727-741.

44. Werst, M. M.; Kennedy, M. C.; Houseman, A. L. P.; Beinert, H.; Hoffman, B. M. *Biochemistry* **1990**, 29, 10533-10540.

45. Telser, J.; Smith, E. T.; Adams, M. W. W.; Conover, R. C.; Johnson, M. K.; Hoffman, B. M. *J.Am.Chem.Soc.* **1995**, 117, 5133-5140.

46. Telser, J.; Huang, H. S.; Lee, H. I.; Adams, M. W. W.; Hoffman, B. M. *J.Am.Chem.Soc.* **1998**, 120, 861-870.

47. Lindskog, S.; Coleman, J. E. *Proc.Natl.Acad.Sci.USA* **1973**, 70, 2505-2508.

48. Mouesca, J. M.; Rius, G. J.; Lamotte, B. *J.Am.Chem.Soc.* **1993**, 115, 4714-4731.

49. Anderson, R. E.; Anger, G.; Petersson, L.; Ehrenberg, A.; Cammack, R.; Hall, D. O.; Mullinger, R.; Rao, K. K. *Biochim.Biophys.Acta* **1975**, 376, 63-71.

50. Kappl, R.; Hüttermann, J. In *Advanced EPR*; Elsevier: Amsterdam (NL). 1989; Vol. pp 501-540.

51. Gloux, J.; Gloux, P.; Lamotte, B.; Mouesca, J. M.; Rius, G. J. *J.Am.Chem.Soc.* **1994**, 116, 1953-1961.

52. Antanaitis, B. C.; Moss, T. H. *Biochim.Biophys.Acta* **1975**, 405, 262-279.

53. Le Pape, L.; Lamotte, B.; Mouesca, J.-M.; Rius, G. J. *J.Am.Chem.Soc.* **1997**, 119, 9757-9770.
54. Moulis, J.-M.; Lutz, M.; Gaillard, J.; Noodleman, L. *Biochemistry* **1988**, 27, 8712-8719.
55. Heering, H. A.; Bulsink, Y. B.; Hagen, W. R.; Meyer, T. E. *Eur.J.Biochem.* **1995**, 232, 811-817.
56. Priem, A. H.; Klaasen, A. A. K.; Reijerse, E. J.; Meyer, T. E.; Luchinat, C.; Capozzi, F.; Hagen, W. R., in preparation.
57. Mouesca, J. M.; Lamotte, B.; Rius, G. J. *Inorg.Biochem.* **1991**, 43, 251-251.
58. Rius, G. J.; Lamotte, B. *J.Am.Chem.Soc.* **1989**, 111, 2464-2469.
59. Gloux, J.; Gloux, P.; Hendriks, H.; Rius, G. J. *J.Am.Chem.Soc.* **1987**, 109, 3230-3224.
60. Kappl, R.; Ciurli, S.; Luchinat, C.; Hüttermann, J. *J.Am.Chem.Soc.* **1999**, 121, 1925-1935.
61. Hüttermann, J., personal communication.
62. Capozzi, F.; Ciurli, S.; Luchinat, C. *Struct.Bonding* **1998**, 90, 127-160.
63. Krishnamoorthi, R.; Cusanovich, M. A.; Meyer, T. E.; Przysiecki, C. T. *Eur.J.Biochem.* **1989**, 181, 81-85.
64. Babini, E.; Bertini, I.; Borsari, M.; Capozzi, F.; Luchinat, C.; Turano, M., in preparation.

Chapter 16

Paramagnetic NMR of Electron Transfer Copper Proteins

Claudio O. Fernández[1] and Alejandro J. Vila[2,*]

[1]LANAIS-RMN, CONICET, University of Buenos Aires, Junín 956, 1113 Buenos Aires, Argentina
[2]Instituto de Biología Molecular y Celular de Rosario (IBR), University of Rosario, Suipacha 531, 2000 Rosario, Argentina

Paramagnetic ^{1}H NMR studies of electron transfer copper sites in proteins have been possible due to recent progress in the field. This has enabled the analysis of the electron spin delocalization onto the different ligands in blue copper and Cu_A sites. The weakly interacting axial ligands play an important role in tuning the electronic structure of both sites.

Copper in living systems exists in two oxidation states: Cu(I) and Cu(II). This redox couple is adaptable to the biological redox potential range. Intra- and intermolecular electron transfer by copper centers is performed by the mononuclear blue (type 1) copper and binuclear Cu_A sites (*1*). Type 1 centers are present in the so-called blue copper proteins (BCP's) found in bacteria and plants, as well as in multicopper enzymes, present in different organisms (*2;3*). The binuclear purple Cu_A center is the electron entry port of terminal oxidases, and it has not been found in single-domain soluble proteins (*4*).

Cu(II) is a d^9, open shell transition metal ion, with one unpaired electron located in the $d_{x^2-y^2}$ orbital. Instead, Cu(I) possesses a d^{10} closed shell, being diamagnetic. Blue copper sites use the Cu(II)/Cu(I) redox pair. In the binuclear

Cu$_A$ center, the two physiologically relevant redox states are the fully reduced (Cu(I),Cu(I)), and the oxidized, mixed-valence Cu(I),Cu(II). Therefore, the oxidized states of both blue and Cu$_A$ sites are S=1/2. In these forms, the coupling between the unpaired electron and the nearby nuclear spins in the protein affects both the chemical shifts and the relaxation rates of these nuclei. Paramagnetic Cu(II) centers in proteins were considered for a long time as not amenable for NMR studies, since the electron relaxation rates of Cu(II) gave rise to extreme line broadening. In 1996, the first paramagnetic ^{1}H NMR spectra of a blue copper center and of a Cu$_A$ site in the oxidized forms were reported by Canters (Leiden) (5) and Bertini and Luchinat (Florence) groups (6), respectively. More recently, the full assignment of the paramagnetic spectrum of an oxidized blue copper site was pursued by the Florence lab (7), and the first solution structure of an oxidized blue copper protein was reported (8). These efforts gave breadth to several NMR studies, that we will review in this Chapter.

The Electron-Nucleus Coupling

The electron-nucleus hyperfine coupling occurs via two mechanisms: contact (through chemical bonds) and dipolar (through space). Both mechanisms affect the chemical shifts and nuclear relaxation rates. The observed chemical shifts in paramagnetic systems are the result of three contributions:

$$\delta_{obs} = \delta_{dia} + \delta_{con} + \delta_{pc} \tag{1}$$

where δ_{dia} is the chemical shift of the observed nucleus in an analogous diamagnetic system, δ_{con} is the Fermi contact shift due to the unpaired electron density on the nucleus, and δ_{pc} is the pseudocontact shift (see below).

The Fermi contact contribution is given by (9;10):

$$\delta_{con} = \frac{A}{\hbar} \frac{g_e \mu_B S(S+1)}{3\gamma_I kT} \tag{2}$$

where A is the hyperfine coupling constant and γ_I is the nuclear gyromagnetic ratio. This contribution is operative through chemical bonds, and reflects electron delocalization through the molecular orbitals. Contact shifts directly provide an estimate of the the spin density delocalization and, indirectly, information on the metal-ligand orbital overlap. In blue copper proteins, contact shift values of protons from the copper ligands range from ~ 900-300 ppm for cysteine signals to less than 100 ppm for histidine resonances (7;11;12).

The electron-nucleus dipolar interaction gives rise to a pseudocontact contribution (δ_{pc}) to the shifts (*10*). This term can be evaluated if the protein structure and the magnetic susceptibility anisotropy tensor are known. Due to the low magnetic anisotropy of Cu(II) centers, δ_{pc} can be calculated by neglecting the rhombic contribution (*5;7*), according to:

$$\delta_{pc} = \frac{\mu_o}{4\pi} \frac{\mu_B^2 S(S+1)}{9kTr^3} (3\cos^2\theta - 1)(g_{\parallel}^2 - g_{\perp}^2) \times 10^6 \tag{3}$$

where r is the proton-copper distance, θ is the angle between the metal-proton vector and the magnetic z axis, and the g tensor parameters are defined by $g_{\parallel} = g_{zz}$ and $g_{\perp} = (g_{xx} + g_{yy})/2$. In Cu(II) proteins, δ_{pc} values on metal ligand protons are expected to be smaller than 3.2 ppm (*5;7;11;12*). This value is negligible for the large shifts observed for β-Cys protons (we will consider $\delta_{con} = \delta_{obs}$).

The possibility of detecting ^{1}H NMR signals in Cu(II) proteins depends on the nuclear relaxation rates, that depend on the electron relaxation rates (τ_s^{-1}). Copper(II) sites in proteins possess distinctive electron relaxation times which confer idiosyncratic spectral properties to each type of metal center (*13*). Catalytic mononuclear Cu(II) centers display slow electron relaxation rates ($\tau_s^{-1} \le 10^9$ s^{-1}), thus giving rise to very broad NMR lines which escape direct detection. In blue copper sites, the unpaired electron relaxes faster ($\tau_s^{-1} \ge 10^9$ s^{-1}), due to the existence of low-lying electronic excited states arising from the trigonal nature of the metal center (*14*). Paramagnetic NMR spectra of blue copper proteins can then be recorded, even if the resonances are still broad, and the assignment is not straightforward (*5;7*). The oxidized form of Cu$_A$ is a mixed-valence S=1/2 system, that displays even faster electron relaxation rates ($\ge 10^{11}$s^{-1}) (*15*), thus giving rise to sharp lines, which can be assigned through NOE and 2D experiments (*6*).

Mononuclear Blue Copper Sites

Blue copper sites in the oxidized form are characterized by unusual spectroscopic features such as an intense charge transfer band in the absorption spectrum at 600 nm, and a small hyperfine coupling constant in the EPR spectrum (*1;16*). All blue sites present a single copper ion coordinated to a Cys and two His residues, located approximately in the same plane (Figure 1) (*2*). Despite these conserved features, the electronic structure of blue sites is finely tuned by small perturbations that give rise to different spectroscopic fingerprints (*16;17*). Blue copper sites have been classified into "classic" and perturbed sites by Solomon and coworkers, based on their spectroscopic properties (*16;18*).

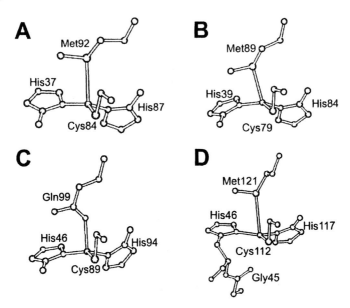

Figure 1. Schematic drawing of the blue copper sites in: (A) plastocyanin, (B) pseudoazurin, (C) stellacyanin and (D) azurin.

Classic blue copper sites (present in plastocyanin, amicyanin and azurin) exhibit an intense absorption feature at ~600 nm, a weak band at ~450 nm, and nearly axial EPR spectra. These centers are characterized by a weak bonding interaction with an axial methionine ligand (Figure 1A) (*19;20*). In azurin, a backbone oxygen from a Gly residue is located on the opposite side of the equatorial plane, at ~ 3 Å from the copper ion (Figure 1D). As a consequence, the copper ion is located almost coplanar to the $N_{\delta1}$(His)-$N_{\delta1}$(His)-S_γ(Cys) plane.

Perturbed centers can be distinguished by an increased absorption ratio of the 450 nm and 600 nm bands and markedly rhombic EPR spectra (*17*). These perturbations occur when stronger copper-axial ligand interactions are present. Perturbed centers can arise from a tetragonal distortion with an axial Met ligand in a different orientation, as in pseudoazurin (Figure 1B), the cucumber basic protein and rusticyanin,(*21-23*) or from a tetrahedral distortion, when a Gln ligand replaces the Met, as in stellacyanin (Figure 1C) (*24;25*).

^{1}H NMR Spectra and Signal Assignment Strategy

Assignment of paramagnetic resonances in blue Cu(II) sites cannot be performed by NOE or 2D techniques, due to the large line widths. Instead, they may be assigned by saturation transfer experiments in a sample containing a mixture of the oxidized and reduced forms of the protein. Irradiation of signals of the Cu(II) species gives rise to a saturation transfer to the corresponding resonances of the same nuclei in the diamagnetic Cu(I) state (*5;7;26*). These

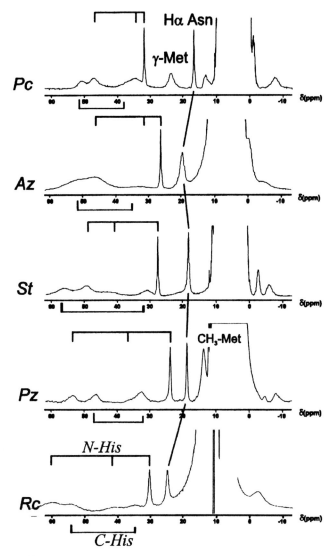

Figure 2. ¹H NMR spectra of Cu(II) plastocyanin (Pc), azurin (Az), stellacyanin (St), pseudoazurin (Pz) and rusticyanin (Rc) (7;11;12). The groups of signals labeled with N and C correspond to the N-and C-terminal His ligands, respectively. Resonances from the Met and Asn residues are indicated.

resonances can be assigned by standard NMR protocols, and the assignments can be transferred to the paramagnetic species. This strategy has led to the assignment of resonances corresponding to the His and the axial ligands in different blue copper sites (Figure 2) (5;8;11;12;26;27).

The β-CH$_2$ Cys signals bound to Cu(II) cannot be directly detected due to their large line widths. However, they can be assigned through "blind" saturation transfer, a strategy devised by Bertini, Luchinat and coworkers.(7) In this experiment, selective saturation is applied over a large spectral range (± 2000 ppm) by shifting the position of the decoupler, even if no signals are observed. If a signal broadened beyond detectable limits is present in the irradiated region, a saturation transfer will be observed. This effect will be maximum when the decoupler exactly matches the frequency of the broad (unobserved) signal, and will fade out when irradiation is off-resonance. The Cys signals can be reconstructed by plotting the saturation transfer response *vs.* the irradiation frequency. This strategy has been recently applied to record, assign and analyze the paramagnetic ^{1}H NMR spectra of Cu(II) plastocyanin, azurin (two classic blue sites), stellacyanin (a tetrahedrally distorted site), rusticyanin and pseudoazurin (two tetragonally distorted sites). These studies have allowed us and others to map the electron delocalization on ^{1}H nuclei from the different metal ligands in blue copper proteins (8;11;12).

Metal-ligand interactions

The Cu-His Interaction

The chemical shifts for each type of proton span rather narrow shift ranges for both His resonances (Figure 2). This suggests that: (1) the unpaired spin density in both histidine ligands is fairly constant, either in classic or distorted sites; and (2) a similar electron delocalization pattern is operative in the imidazole groups (12). These conclusions are feasible provided the orientation of the copper orbitals in the xy plane with respect to the histidine ligands is similar in all blue sites. This concurs with the proposal from Van Gastel et al., who pointed out that the magnetic z axis in different BCP's is nearly perpendicular to the plane defined by the copper ion and the coordinated N$_{\delta 1}$(His)(28).

The Cu-Cys Interaction

The β-CH$_2$ Cys protons display large hyperfine coupling constants, consistent with a high Cu(II)-Cys covalence (Figure 3). In addition, these signals display drastically different chemical shifts among the different proteins (from 240 to 850 ppm, that correspond to hyperfine coupling constants from 8 to 28 MHz). These variations reveal changes in the Cu(II)-Cys covalence of the different sites.

The observed β-CH$_2$ Cys shifts depend on the Cu-S-C$_\beta$-H$_\beta$ dihedral angles. Instead, the average chemical shift of these nuclei ($\delta_{1/2}$) does not reflect a strong dependence on the Cys conformation.(12;29) The β-CH$_2$ Cys $\delta_{1/2}$ values follow this trend: Cu(II)-azurin (825 ppm) > Cu(II)-plastocyanin (570 ppm) > Cu(II)-pseudoazurin (450 ppm) > Cu(II)-stellacyanin (415 ppm) > Cu(II)-rusticyanin (270 ppm) (Figure 3).

Pseudoazurin and rusticyanin exhibit almost identical UV-vis and EPR spectra,(30;31) but quite different Cu(II)-Cys covalencies. These differences in the NMR spectrum parallel results from X-ray data (the Cu-S$_\gamma$(Cys) distance in rusticyanin is 0.13 Å longer than in pseudoazurin).(22;23) A plot of $\delta_{1/2}$ vs. r_{CuS} for the studied proteins shows a linear trend, except for azurin (Figure 4A). In blue copper sites, the unpaired electron is mostly confined to the copper $d_{x^2-y^2}$ orbital. Then, at a fixed r_{CuS}, the copper-sulfur overlap is expected to depend on the position of the sulfur atom respect to the xy plane. If $\delta_{1/2}$ is plotted towards a corrected r_{CuS} parameter, in which the copper-sulfur distance is weighed by the cos of the angle between the Cu-S bond and the xy plane, a linear trend is found for all blue sites (Figure 4B). The fact that the β-CH$_2$ Cys $\delta_{1/2}$ values can be accounted for by geometric features of the metal site that determine the copper-sulfur overlap further validates the use of this parameter as a reliable estimate of the electron spin density on the S$_\gamma$(Cys) atom.

The Cu-Met Interaction

The axial ligand is weakly bound in blue copper proteins, and thus axial ligand resonances display small or null electron spin density. In azurin, no hyperfine-shifted resonances corresponding to the axial Met or Gly ligands could be detected. This correlates with the long Cu-S$_\delta$(Met) bond in azurin (3.0 Å). Instead, hyperfine contributions to the Met resonances have been identified in plastocyanin, amicyanin, pseudoazurin and rusticyanin. In Cu(II)-plastocyanin (8) and Cu(II)-amicyanin,(5) the γ-CH$_2$ Met resonances show net electron spin density. Instead, hyperfine shifts in the ε-CH$_3$ Met moieties have been identified in Cu(II)-pseudoazurin and rusticyanin.(12) The different electron delocalization patterns can be attributed to the fact that the Met side chain adopts a *trans* conformation in classic blue sites (plastocyanin and amicyanin), and a *gauche* conformation in tetragonally distorted sites (pseudoazurin and rusticyanin). In all cases, the larger hyperfine shifts correlate with shorter Cu-S$_{Met}$ bonds.

Electron delocalization through hydrogen bonds

In most blue copper proteins, the amide NH proton of an Asn residue forms a hydrogen bond with the S$_\gamma$(Cys) atom (2). A net electron spin density has been found on the Asn NH nucleus in four different blue copper proteins, that reveals

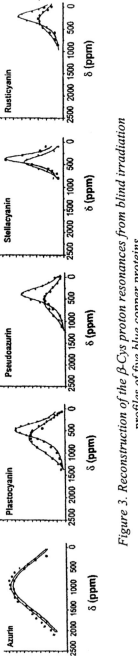

Figure 3. Reconstruction of the β-Cys proton resonances from blind irradiation profiles of five blue copper proteins.

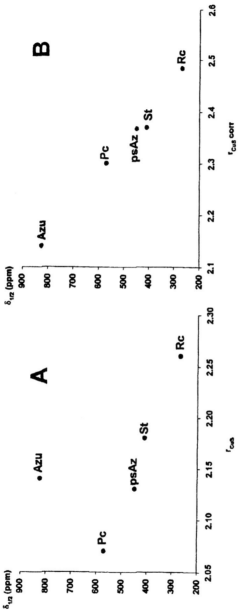

Figure 4. Plot of the β-Cys $\delta_{1/2}$ parameter vs. (A) the experimental copper-sulfur distance, and (B) the copper-sulfur distance corrected for the Cys deviation from the equatorial ligand plane.

that electron delocalization is operative through this hydrogen bond. The contact shift of the Asn NH signal shows a linear correlation with the electron spin density on the $S_\gamma(Cys)$ (quantified by the β-CH_2 Cys $\delta_{1/2}$ value) in all studied blue sites, except in rusticyanin (Figure 5) (*12*). Rusticyanin lacks the conserved Asn, that is replaced by a Ser. This results in a higher unpaired spin density in the Ser NH proton compared to that expected according to the trend shown in Figure 5. This suggests that the replacement of an Asn by a Ser gives rise to a perturbed hydrogen bond interaction. It has been suggested that this perturbation could contribute to the unusually high redox potential (680 mV) of rusticyanin (*12*).

Metal-ligand interplay in blue copper sites

A comparative analysis of the different copper sites reveals a subtle interplay between the different copper-ligand interactions. In azurin, the Cu(II) ion lies in the His_2Cys plane, and the axial ligands (Met121 and Gly45) are located at ~ 3Å from the metal ion at both sides of the equatorial ligand plane (Figure 1B) (*20*). This coplanarity gives rise to the largest Cu(II)-Cys covalence among blue sites, with no electron spin density in the axial ligands. In

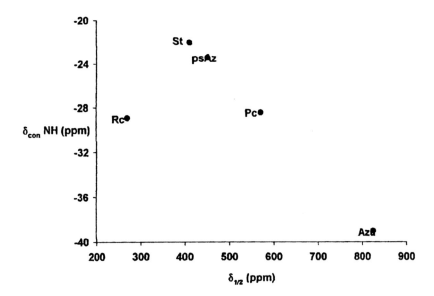

Figure 5. Plot of the NH amide contact shift vs. the $\delta_{1/2}$ β-Cys parameter.

plastocyanin and amicyanin, with one axial ligand (a Met), the copper moves closer to the Met ligand (*19;32*), resulting in a net electron spin density in the Met side chain and reducing the Cu(II)-Cys overlap. In stellacyanin, the axial bond between copper and the Gln oxygen imposes a tetrahedral distortion to the site, which further weakens the Cu(II)-Cys bond (*24;25*). These results reveal that in classic and tetrahedrally distorted blue sites, a stronger interaction with the axial ligand results in a weakened electron delocalization in the Cu(II)-Cys bond (*11*).

The electron spin density in the Cys ligands in tetragonally distorted blue sites (pseudoazurin and rusticyanin) is decreased compared to classic sites (azurin and plastocyanin), but this is not related to a stronger interaction with the axial ligand. Instead, the Cu-Cys and the Cu-Met interaction are independently tuned by the protein folding around the metal site in tetragonally distorted blue copper sites (*12*).

Electronic structure of Cu$_A$ sites

Cu$_A$ is a binuclear center, where the copper ions are bridged by the two sulfur atoms of Cys ligands, forming an almost planar Cu$_2$S$_2$ rhombic structure with a metal-to-metal distance ranging between 2.3-2.6 Å (*33-36*). One of the copper ions is coordinated also to N$_{\delta 1}$(His) and a S$_\delta$(Met) donor atoms, whereas the other copper ion is coordinated by a N$_{\delta 1}$(His) and a backbone carbonyl (Figure 6). These geometrical features are conserved in all structurally characterized Cu$_A$ centers (*33-36*). The weak Met and carbonyl ligands may help to maintain the site architecture and to regulate its properties (*37*).

Cu$_A$ centers exist in two redox states: [Cu(II),Cu(I)] and [Cu(I),Cu(I)]. The oxidized species is a fully delocalized type III mixed-valence pair (formally two Cu +1.5 ions), as early revealed by EPR spectroscopy (*38*). These systems display sharper NMR lines than BCP's due to a shorter electron relaxation time

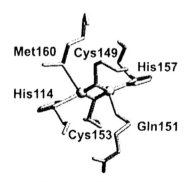

Figure 6. Schematic drawing of T.thermophilus Cu$_A$ site) (36).

of the paramagnetic center ($\approx 10^{-11}$ s), giving rise to well-resolved paramagnetic ^{1}H NMR spectra (Figure 7) (6). NMR studies are available for the native Cu_A centers from the soluble fragments of the *T. thermophilus* (6;39), *P. denitrificans* (40), *P. versutus* (41), and *B.subtilis* oxidases (42), *P. stutzeri* N_2O reductase (43) as well as for engineered Cu_A sites in amicyanin (44)and *E. coli* quinol oxidase (45).

The proton signals of each His ligand, which are found up to 40 ppm (Figure 7), can be assigned by 1D NOE or NOESY experiments (6). They can be assigned to specific residues based on the fact that only one of them is solvent exposed, as it also occurs in BCP's. The two His residues show different amounts of delocalized spin density in each system. These differences might reflect different tiltings of the imidazoles with respect to the Cu_2S_2 rhombus, likely being modulated by the Cu-axial ligand distances (37).

The four β-CH_2 Cys resonances are shifted well downfield, spanning a broad range of chemical shifts from 50 to 450 ppm (Figure 7). Three are very broad, and usually are found between 200 and 450 ppm whereas a fourth, sharper one, falls in the range 50-110 ppm. In some cases, line broadening is so drastic that the broader resonances cannot be detected. This problem can be overcome by selective deuteration of the β-CH_2 Cys protons, and collection of the ^{2}H NMR spectrum (40). The assignment of the β-CH_2 Cys signals has been performed from the analysis of the Cu-S_γ-C_β-H_β dihedral angles and on the basis of the shifts temperature dependence. The Cys resonances exhibit strong deviations from Curie behavior, that can be explained by assuming partial population of low-energy excited states (41;46). Analysis of the Cys signal pattern has confirmed that the ground level in Cu_A sites is a σ_u^* orbital, and the excited state is a π_u level. The σ_u^*/π_u energy gap can be estimated by analysis of the temperature dependence of the Cys chemical shifts (41).

No signals from any of the axial ligands are observed significantly shifted outside of the diamagnetic envelope. This is indicative of a very weak interaction of both Met and Glu ligands with the copper ions and a small or negligible amount of spin density delocalized onto them.

The Met160Gln mutant in the Cu_A soluble fragment from *Thermus thermophilus* provided the first stable, mixed valence axial ligand mutant in a soluble Cu_A fragment (39;47). A ^{1}H NMR study of this mutant reveals no net electron spin density in the axial ligands, and that the Cu-Cys covalence is unaltered (39). These results indicate that no interplay between axial ligands and the Cu-Cys bonds is present in Cu_A sites, in contrast with blue copper proteins. Instead, other changes were observed. The σ_u^*/π_u energy gap is increased with respect to the WT protein. This picture is consistent with changes in the electron relaxation time in the mutant. Therefore, the energy of the low-lying excited state can be tuned by the interaction of the copper atoms with the axial ligands (39).

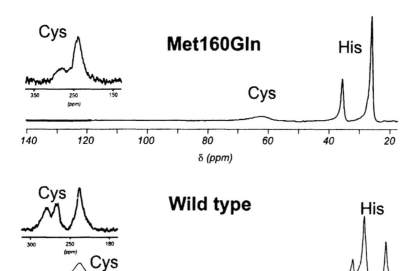

*Figure 7. Downfield region of the ^1H NMR spectra of wild type and
Met160Gln mutant of the Cu_A-Thermus fragment (6, 39)*

The two His ligands are less alike in Met160Gln than in the WT protein due to the selective perturbation of His114, whereas His157 resonances remain unaltered. His 114 is bound to the same copper atom as the engineered Gln160 residue (Figure 6). The engineered Gln ligand in the mutant affects the copper-His interaction, either by tilting His114 relative to its orientation in the WT protein, or by strengthening the Cu-His bond. These results provide experimental evidence that each of the two halves of the Cu_A core can be modulated independently. Axial ligand perturbations of the Cu_1 or Cu_2 coordination environment thus may provide unique regulatory possibilities of the Cu_A reactivity in the full oxidase.

Comparison of Cu_A features with blue copper sites sheds some light on the role of axial ligands in these related systems. Stronger copper-axial ligand interactions in blue copper sites (such as a Met→Gln mutation) induce changes in the relative intensities of the charge transfer bands (*16;17*), give rise to rhombic EPR spectra and affect the Cu(II)-Cys covalence (*16*). In contrast, the charge transfer band pattern in Met160Gln-Cu_A is only slightly altered with respect to that of the WT protein (*47*), the EPR spectrum is axial, and the Cu-Cys covalence is not affected (*39*). It can be concluded that axial ligands in Cu_A sites can exert perturbations in the Cu_2S_2 rhombus that tune the energy of the π_u excited state, and determine the orientation of the His ligands. Both mechanisms could play a functional role in terminal oxidases by regulating intra- and intermolecular electron transfer.

Acknowledgments

Alejandro J.Vila thanks the Fundación Antorchas and the Fogarty International Center- NIH (R03-TW000985-01) for supporting studies on copper proteins.

References

1. Vila, A. J. and Fernandez, C. O. (2001) in *Handbook of Metalloproteins*; Bertini, I., Sigel, A., and Sigel, H., Eds.; Marcel Dekker, New York.
2. Adman, E. T. (1991) *Adv. Protein Chem. 42*, 144-197.
3. Messerschmidt, A. (1998) *Struct. Bonding 90*, 37-68.
4. Beinert, H. (1997) *Eur. J. Biochem. 245*, 521-532.
5. Kalverda, A. P., Salgado, J., Dennison, C., and Canters, G. W. (1996) *Biochemistry 35*, 3085-3092.

6. Bertini, I., Bren, K. L., Clemente, A., Fee, J. A., Gray, H. B., Luchinat, C., Malmström, B. G., Richards, J. H., Sanders, D., and Slutter, C. E. (1996) *J. Am. Chem. Soc. 118*, 11658-11659.

7. Bertini, I., Ciurli, S., Dikiy, A., Gasanov, R., Luchinat, C., Martini, G., and Safarov, N. (1999) *J. Am. Chem. Soc. 121*, 2037-2046.

8. Bertini, I., Ciurli, S., Dikiy, A., Fernandez, C. O., Luchinat, C., Safarov, N., Shumilin, S., and Vila, A. J. (2001) *J. Am. Chem. Soc. 123*, 2405-2413.

9. Bertini, I. and Luchinat, C. (1986) *NMR of Paramagnetic Molecules in Biological Systems;* Benjamin/Cummings, Menlo Park, CA.

10. Kurland, R. J. and McGarvey, B. R. (1970) *J. Magn. Reson. 2*, 286-301.

11. Bertini, I., Fernandez, C. O., Karlsson, B. G., Leckner, J., Luchinat, C., Malmström, B. G., Nersissian, A. M., Pierattelli, R., Shipp, E., Valentine, J. S., and Vila, A. J. (2000) *J. Am. Chem. Soc. 122*, 3701-3707.

12. Donaire, A., Jimenez, B., Fernandez, C. O., Pierattelli, R., Kohzuma, T., Moratal, J. M., Hall, J. F., Kohzuma, T., Hasnain, S. S., and Vila, A. J. (2002) *J. Am. Chem. Soc. 124*, 13698-13708.

13. Banci, L., Pierattelli, R., and Vila, A. J. (2001) *Adv. Protein Chem. 60*, 397-449 (2002).

14. Kroes, S. J., Salgado, J., Parigi, G., Luchinat, C., and Canters, G. W. (1996) *J. Biol. Inorg. Chem 2*, 559-568.

15. Pfenninger, S., Antholine, W. E., Barr, M. E., Hyde, J. S., Kroneck, P. M., and Zumft, W. G. (1995) *Biophys. J. 69*, 2761-2769.

16. Randall, D. W., Gamelin, D. R., LaCroix, L. B., and Solomon, E. I. (2000) *J. Biol. Inorg. Chem. 5*, 16-29.

17. Lu, Y., LaCroix, L. B., Lowery, M. D., Solomon, E. I., Bender, C. J., Peisach, J., Roe, J. A., Gralla, E. B., and Valentine, J. S. (1993) *J. Am. Chem. Soc. 115*, 5907-5918.

18. La Croix, L. B., Randall, D. W., Nersissian, A. M., Hoitink, C. W., Canters, G. W., Valentine, J. S., and Solomon, E. I. (1998) *J. Am. Chem. Soc. 120*, 9621-9631.

19. Guss, J. M. and Freeman, H. C. (1983) *J. Mol. Biol. 169*, 521-563.

20. Nar, H., Messerschmidt, A., Huber, R., van de Kamp, M., and Canters, G. W. (1991) *J. Mol. Biol. 221*, 765-772.

21. Guss, J. M., Merritt, E. A., Phizackerley, R. P., and Freeman, H. C. (1996) *J. Mol. Biol. 262*, 686-705.

22. Inoue, T., Nishio, N., Suzuki, S., Kataoka, K., Kohzuma, T., and Kai, Y. (1999) *J. Biol. Chem. 274*, 17845-17852.

23. Walter, R. L., Ealick, S. E., Friedman, A. M., Blake, R. C., Proctor, P., and Shoham, M. (1996) *J. Mol. Biol. 263*, 730-751.

24. Hart, P. J., Nersissian, A. M., Herrman, R. G., Nalbandyan, R. M., Valentine, J. S., and Eisenberg, D. (1996) *Protein Sci. 5*, 2175-2183.

25. Vila, A. J. and Fernandez, C. O. (1996) *J. Am. Chem. Soc. 118*, 7291-7298.

26. Salgado, J., Kalverda, A. P., and Canters, G. W. (1997) *J. Biomol. NMR 9*, 299-305.

27. Salgado, J., Kroes, S. J., Berg, A., Moratal, J. M., and Canters, G. W. (1998) *J. Biol. Chem. 273*, 177-185.

28. van Gastel, M., Canters, G. W., Krupka, H., Messerschmidt, A., de Waal, E. C., Warmerdam, G. C., and Groenen, E. J. (2000) *J. Am. Chem Soc. 122*, 2322-2328.

29. Fernandez, C. O., Sannazzaro, A. I., and Vila, A. J. (1997) *Biochemistry 36*, 10566-10570.

30. Kohzuma, T., Dennison, C., McFarlane, W., Nakashima, S., Kitagawa, T., Inoue, T., Kai, Y., Nishio, N., Shidara, S., Suzuki, S., and Sykes, A. G. (1995) *J. Biol. Chem. 270*, 25733-25738.

31. Hall, J. F., Kanbi, L. D., Harvey, I., Murphy, L. M., and Hasnain, S. S. (1998) *Biochemistry 37*, 11451-11458.

32. Romero, A., Nar, H., Huber, R., Messerschmidt, A., Kalverda, A. P., Canters, G. W., Durley, R., and Mathews, F. S. (1994) *J. Mol. Biol. 236*, 1196-1211.

33. Tsukihara, T., Aoyama, H., Yamashita, E., Tomizaki, T., Yamaguchi, H., Shinzawa-Itoh, K., Nakashima, R., Yaono, R., and Yoshikawa, S. (1996) *Science 272*, 1136-1144.

34. Iwata, S., Ostermeier, C., Ludwig, B., and Michel, H. (1995) *Nature 376*, 660-667.

35. Soulimane, T., Buse, G., Bourenkov, G. P., Bartunik, H. D., Huber, R., and Than, M. E. (2000) *EMBO J. 19*, 1766-1776.

36. Williams, P. A., Blackburn, N. J., Sanders, D., Bellamy, H., Stura, E. A., Fee, J. A., and McRee, D. E. (1999) *Nature Struct. Biol. 6*, 509-516.

37. Robinson, H., Ang, M. C., Gao, Y. G., Hay, M. T., Lu, Y., and Wang, A. H. (1999) *Biochemistry 38*, 5677-5683.

38. Kroneck, P. M. H., Antholine, W. E., Riester, J., and Zumft, W. G. (1988) *FEBS Lett. 242*, 70-74.

39. Fernández, C. O., Cricco, J. A., Slutter, C. E., Richards, J. H., Gray, H. B., and Vila, A. J. (2001) *J. Am. Chem. Soc. 123*, 11678-11685.

40. Luchinat, C., Soriano, A., Djinovic, K., Saraste, M., Malmström, B. G., and Bertini, I. (1997) *J. Am. Chem. Soc. 119*, 11023-11027.

41. Salgado, J., Warmerdam, G. C., Bubacco, L., and Canters, G. W. (1998) *Biochemistry 37*, 7378-7389.

42. Dennison, C., Berg, A., De Vries, S., and Canters, G. W. (1996) *FEBS Lett. 394*, 340-344.

43. Holz, R. C., Alvarez, M. L., Zumft, W. G., and Dooley, D. M. (1999) *Biochemistry 38*, 11164-11171.

44. Dennison, C., Berg, A., and Canters, G. W. (1997) *Biochemistry 36*, 3262-3269.

45. Kolczak, U., Salgado, J., Siegal, G., Saraste, M. , and Canters, G. W. (1999) *Biospectroscopy. 5*, S19-S32.

46. Shokhirev, N. V. and Walker, F. A. (1995) *J. Phys. Chem. 99*, 17795-17804.

47. Slutter, C. E., Gromov, I., Richards, J. H., Pecht, I., and Goldfarb, D. (1999) *J. Am. Chem. Soc. 121*, 5077-5078.

Chapter 17

Applications of Paramagnetic NMR Spectroscopy for Monitoring Transition Metal Complex Stoichiometry and Speciation

Debbie C. Crans[1], Luqin Yang[1], Ernestas Gaidamauskas[1], Raza Khan[1], Wenzheng Jin[1], and Ursula Simonis[2]

[1]Department of Chemistry, Colorado State University, Fort Collins, CO 80523–1872
[2]Department of Chemistry and Biochemistry, San Francisco State University, 1600 Holloway Avenue, San Francisco, CA 94132–4163

Although it is well established that paramagnetic NMR spectroscopy is a powerful tool to derive structural information, the methodology is still not yet universally applied to paramagnetic small molecule complexes. In this paper paramagnetic ^1H NMR spectroscopy is investigated as a convenient method for the experimental inorganic chemist to elucidate solution structures and speciation of small molecule metal complexes derived from 2,6-pyridinedicarboxylic acid as ligand. Spectra of complexes with O_h geometry, in which the spin states of the metal ion range from d^3 (Cr^{3+}), d^5 (Fe^{3+}), d^6 (Fe^{2+}), d^7 (Co^{2+}) to d^8 (Ni^{2+}), were recorded and analyzed. For all complexes the ^1H NMR spectra give well-resolved, easy detectable lines, which depending on the spin state and electron relaxation time of the metal ion and the pH of the solution can be fairly broad. Regardless, the spectra allow complexes of 1:1 and 1:2 stoichiometries to be distinguished in spite of the metal nucleus short nuclear correlation and relaxation times, and the magnitude of the hyperfine shift spread. The pH stability profile and the ability of the complexes to undergo ligand exchange reactions were also investigated for each of the complexes. This work demonstrates that paramagnetic ^1H NMR spectroscopy is very useful for characterizing small molecule complexes and their solution chemistry without requiring a detailed analysis of the hyperfine shifts and relaxivities.

Introduction.

NMR spectroscopy remains one of the major tools for examining solution structures of inorganic complexes and for determining their purity and identity. Although NMR spectroscopy is most commonly used to characterize diamagnetic inorganic compounds (*1*), it is well established that NMR spectroscopy provides valuable information on complex structure and spin and oxidation states of the metal ion. Most paramagnetic NMR applications focus on large biomolecules (*2-18*), however, paramagnetic NMR spectroscopy can also be a powerful tool in small molecule chemistry, providing valuable information on solution species (*19-28*). In this work we describe the application of paramagnetic ^{1}H NMR spectroscopy to characterize the structures and speciation of a series of related small molecule metal complexes. The ^{1}H NMR spectra presented here were recorded to ultimately determine the effects of first row transition metal complexes on streptozotocin (STZ)-induced diabetic rats (*29-33*). To this end, the structure and stability of a series of 2,6-pyridinedicarboxylate (dipicolinate) complexes were investigated (Fig. 1) (*29-33*). The complexes contain five first row transition metal ions, namely Cr^{3+} (d^3), Fe^{3+} (d^5), Fe^{2+} (d^6), Co^{2+} (d^7) and Ni^{2+} (d^8), in which each of the metal ions has at a minimum two unpaired electrons (Fig. 2). These complexes are, thus, paramagnetic. In spite of the fast nuclear relaxation times and short correlation times of the complexes, it was possible to characterize unambiguously all complexes of either 1:1 or 1:2 stoichiometry and to investigate their pH stability, thereby demonstrating the general applicability of paramagnetic NMR spectroscopy to the characterization of such complexes. It will be shown that all complexes with comparable electron relaxation times of the metal ion (Cr^{3+} and Fe^{3+} vs. Co^{2+}, Fe^{2+} and Ni^{2+}) exhibit similar paramagnetic shift trends in spite of their different electron relaxation times, Fermi contact, dipolar coupling and relaxation mechanisms. These shifts can be attributed to either the formation of a mono- or a bis-dipicolinate complex.

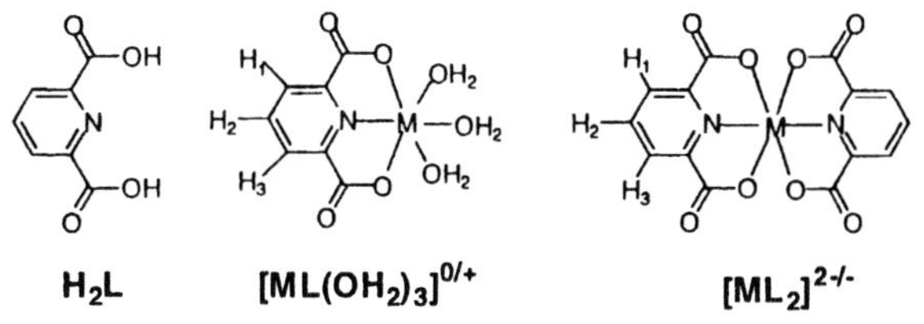

H_2L $[ML(OH_2)_3]^{0/+}$ $[ML_2]^{2-/-}$

Fig. 1. The structures of ligand, the 1:1 and 1:2 dipicolinate complexes and the proton numbering schemes are shown. Both the 1:1 and 1:2 complex are presumed to coordinate dipic^{2-} in a tridentate manner based on experimental evidence for this series of complexes described elsewhere in detail (31). Other possible structural alternatives exist but are not discussed further here.

The interactions of the spin of the nucleus under investigation and the spins of the unpaired electrons of the metal ion greatly affect the relaxation times of the complexes (*14-17, 34, 35*). The nuclear relaxation times of transition metal ions are very short, and, thus, spectra with a large chemical shift range are obtained (Fig. 3). Depending on the relaxation times of the nucleus under investigation, the resonances are typically much broader than those of diamagnetic complexes (Fig. 3). The broadness of the signals leads to a decrease in peak height, which can affect the signal-to-noise ratio. To avoid a decrease in signal-to-noise ratio, it is essential, that the spectral parameters be adjusted according to the short nuclear relaxivities. Selection of proper acquisition parameters does not always guarantee that all signals can be detected. For some complexes the relaxation times are so short that their resonance lines are rendered undetectable or are hardly discernable from the baseline due to their broadness. Depending on the electronic spin state of the complex, which imparts a certain geometry onto the complex, a wide range of resonance shifts can be observed, which depending on the degree of unpaired electron spin delocalization into ligand orbitals can span a range of hundreds of ppm (*14-17, 36, 37*).

The much greater differences that are observed for the chemical shift range of paramagnetic compounds than of the corresponding diamagnetic complexes, and their origins have been discussed in detail by several authors (*14, 16, 17*). Furthermore, the optimization of spectral parameters to obtain spectra with good signal-to-noise ration and differences between the analysis of NMR spectra of diamagnetic and paramagnetic compounds have been described in detail elsewhere (*14-17*). The objective of this paper is to describe the practical applicability of ^{1}H NMR spectroscopy to paramagnetic complexes to encourage the novice investigator to use this technique to derive structural information without having to analyze hyperfine shifts and patterns. We will show that the resonance assignment and complex quantization is straightforward for the complexes under investigation. The signals can be assigned unambiguously by a titration study in which spectra are recorded at different metal ion:ligand ratios. However, additional structural information can be obtained such as whether the complexes are coordinated by one or two dipicolinate ligands. Furthermore, the determination of the chemical shift ranges of the complexes allows characterization of the spin states. In addition, qualitative and quantitative data can be obtained by peak integration, and if used cautiously, it is an effective tool in characterizing the stability of the complex.

Dipicolinate complexes were selected to illustrate that ^{1}H NMR spectra can be recorded for most of the first row transition metal ions (Cr^{3+} (d^3), Fe^{3+} (d^5), Fe^{2+} (d^6), Co^{2+} (d^7) and Ni^{2+} (d^8)). Previously, paramagnetic NMR studies were reported for characterizing the reaction chemistry of complexes with the metal ions Fe(II) (*2, 19, 22, 26, 27, 38-41*), Fe(III) (*9, 11, 21, 25, 32, 39, 40, 42-47*), Co(II) (*4, 20, 24, 31, 41, 48-51*), Ni(II) (*6, 12, 23, 28, 51-56*) and Cr(III) (*25, 30 57-61*) including the analysis of electron relaxation times, reaction mechanisms, and electron-nuclear correlation times. Dipicolinic acid contains two acidic protons (pK$_a$ values of 2.0 and 4.5 (*62*)), and deprotonation in the presence of

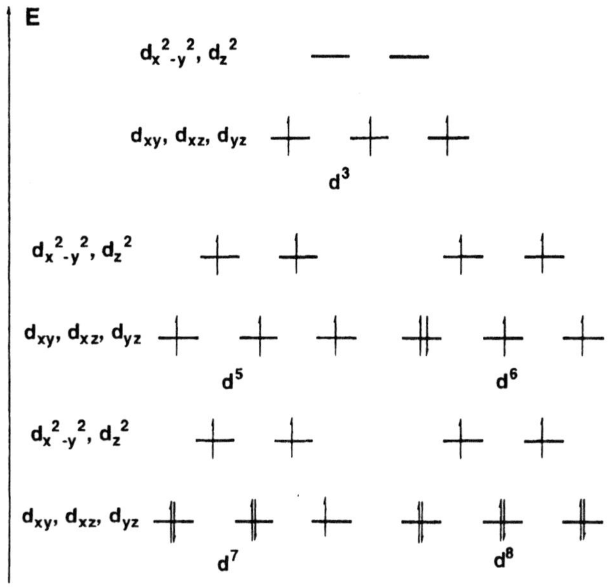

Fig. 2. The d^3, d^5, d^6, d^7 and d^8 high spin states for the complexes with O_h geometry under investigation.

metal ions leads to complex formation with Cr(III), Fe(III), Fe(II), Co(II) and Ni(II). As will be discussed below in more detail, the ^{1}H NMR spectra shown in Fig. 3 reflect the different spin states of the metal ion, their nuclear relaxation processes, and the differences in stabilities of the dipicolinate complexes. It will also be shown that distinction between the 1:1 and 1:2 complexes and their quantitation is possible (Fig. 3). The spectroscopic studies presented in this work serve to document the general usefulness of paramagnetic ^{1}H NMR spectroscopy to the experimental synthetic inorganic chemist.

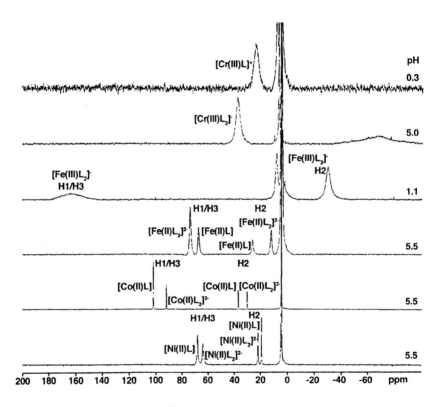

Fig. 3. The paramagnetic ^{1}H NMR spectra are shown of transition metal dipicolinate complexes. The spectra shown are the 1:1 and the 1:2 Cr(III) complex (30), the 1:2 Fe(III) complex (32), the 1:2 and 1:1 Fe(II) complexes (32), the 1:1 and 1:2 Co(II) complexes (31) and the 1:1 and the 1:2 Ni(II) complexes (33).

Materials and Methods.

Preparation of Complexes. The complexes [Co(dipic)₂Co(H₂O)₅]•2H₂O (*31*), K₂[Co(dipic)₂] (*31*), Co(dipic) (*31*), [Co(H₂dipic)(dipic)] (*31*), Na₂[Fe(dipic)₂]•2H₂O (*63*), Na[Fe(dipic)₂]•2H₂O (*63*), [Fe(dipic)(OH₂)(μ-OH)]₂ (*64*) and K[Cr(dipic)₂] (*30*) where 2,6-pyridinedicarboxylate (dipicolinate) is abbreviated as dipic, were prepared as described previously. The K₂[Ni(dipic)₂]•7H₂O and [Ni(dipic)]₂•2H₂O complexes were prepared from 2,6-pyridinedicarboxylic acid (H₂dipic) and NiCl₂ as described elsewhere in detail (*33*).

Solution Preparation. Unless otherwise noted, the samples for NMR analysis were prepared by dissolving crystalline complexes and free ligand in deuterium oxide. The composition of the solutions varied and depended on whether the complex was prepared in the presence and absence of free ligand or solutions of metal ion and free ligand depending on complex stability profile. When necessary the pH was adjusted with DCl or NaOD solutions. The pH values reported are those measured and not adjusted for the presence of D₂O. In some cases (such as solutions containing Fe(II) at pH ≥ 9.1 or Fe(III) at pH ≥ 5.0) insoluble species (presumably including metal hydroxides) were removed, and supernatant solutions were used to acquire the ¹H NMR spectra.

NMR Spectroscopy. ¹H NMR spectra were recorded on Varian INOVA–300 and 400 spectrometers operating at 300.118 and 400.107 MHz, respectively, and temperatures of 300 ± 1K. The ¹H NMR spectra were recorded with the standard pulse sequence (INOVA/s2pul) using a spectral width of 200-250 kHz, a data size of 2048-24000 points, a relaxation delay of 50-100 ms to yield acquisition times between 0.5-1.2 s for Co(II) (*31*), 0.12 s for the Ni(II) (*33*), 0.1 s for the Fe(II) (*32*), 0.05 s for the Fe(III) (*32*), and 0.9 s for the Cr(III) (*30*) compounds. After Fourier transformation and apodization with a 2 to 20 Hz exponential line broadening factor, depending on the specific complex, the data were phase corrected to yield the NMR spectra. Chemical shifts are referenced relative to the residual HOD peak of D₂O at 4.80 ppm or external DSS (sodium salt of 3-methyl(silyl)propane sulfonic acid).

The 2D ¹H chemical exchange (EXSY) NMR spectra were recorded on the Varian INOVA-300 NMR spectrometer operating at 300.138 MHz and at 293 ± 1 K using the standard pulse sequence (INOVA/noesy) (*31*). The 2D maps were acquired in D₂O using 128 complex pairs in t₁ (States-TPPI) 32 signal averaging transients each over a spectral bandwidth ranging from 31000 Hz to 39000 Hz with 4096 complex pairs in t₂. Recycle delays ranged from 0.01-0.3 s. Mixing times of 0.015-0.8 s were used and optimized for each of the complexes to obtain maximum cross peak intensity. The data were processed with Gaussian apodization fit to the linear predicted data in t₁ and Gaussian weighted apodization in t₂. After zerofilling and Fourier transformation, the final 2D spectral matrix consisted of 4 K × 4 K complex pairs.

Results and Discussion.

Complexes of Cr(III), Fe(II), Fe(III), Co(II), and Ni(II) with dipicolinic acid were prepared as described previously (*30, 31, 33, 63, 64*). Both 1:1 and 1:2 complexes are formed with each of the metal ion assuming that dipic^{2-} coordinated to the metal ion in a tridentate manner as has been shown previously for the Co-complexes (*31*). Although other possible modes of binding have been described (*31*), the NMR spectra presented below support formation of complexes with the stoichiometries 1:1 and 1:2 as is shown in Fig. 1. Below we will describe in detail each complex, but at this point some general statements will be made for all the transition complexes investigated. The ^{1}H NMR spectra of the dipicolinate complexes are shown in Fig. 3; the free ligand in the absence of any paramagnetic metal ion gives rise to a pH dependent spectrum with three signals resonating from 7 to 9 ppm (*29*) and this spectrum is, thus, not shown. However, the resonances of free ligand are observed in the spectrum of [Fe(III)(dipic)$_2$]$^-$ as one broad signal (at ~8 ppm) and one close to the HOD signal (~5 ppm). Although the electronic spin states of the metal ion in these complexes vary dramatically, several similarities are apparent. Regardless of the reported solid state structure of the respective complexes, the observed aqueous solution species in the absence of evidence to the contrary is anticipated to coordinate to the metal ion in a tridentate manner to form dipicolinate complexes of both 1:1 and 1:2 stoichiometries (Fig. 1). The two sets of resonances in the ^{1}H NMR spectra are attributable to either the 1:1 or the 1:2 complex (Fig. 3). For each complex we have carried out studies, in which the metal to ligand ratio has been monitored by examining the concentration dependences and/or Job plots, respectively. The complex obtained in solution at a high metal to ligand ratio has been attributed to a 1:1 complex. The second complex formed upon decreasing the metal to ligand ratio and in the presence of a 2-fold ligand (and larger) excess is assigned to the 1:2 complex.

Considering the chemical shifts for all 10 complexes, it can be concluded that in the case of the Cr(III), Fe(II) and Fe(III) dipic complexes the chemical shifts for the resonances of the 1:2 complex are further downfield than for 1:1 complex. This pattern is reversed for the Ni(II) and Co(II) complexes. For these complexes the signals of the 1:1 complex are downfield from those of the 1:2 complex. Whereas in diamagnetic complexes the chemical shifts reflect the electron density provided to the ligand by the metal ion, a similar interpretation cannot be used for these paramagnetic complexes, since upfield and downfield shifts are related to the contact interactions, i.e., the transfer of unpaired electron spin density into ligand orbitals, the dipolar interaction, and/or the mechanisms of unpaired electron spin polarization. Regardless of the effects contributing to the chemical shifts of paramagnetic complexes, a study in which the metal ion:ligand ratio is varied will provide information on the stoichiometry of the complex. In addition, a correlation of signal linewidth and chemical shift with distance between the proton and metal center will unambiguously assign many of the signals.

Independent of complex stoichiometries all dipicolinate complexes span a much greater chemical shift range (Table I) than the corresponding diamagnetic complex [VO$_2$dipic]$^-$ (29). All complexes have two signals for the three dipic-protons and fail to show the pH dependence of the free ligand reflecting the sequential deprotonation of H$_2$dipic. Protons H1 and H3 in the complexes are chemically equivalent and give rise to the most downfield shifted signals in each of the complexes (Fig. 3). This resonance is expected to be the most downfield shifted, since protons H1 and H3 are closest to the paramagnetic metal ion center, and, thus, are influenced the most by the presence of the unpaired electrons. The signal of H2 is shifted less and gives rise to the other peak observed for the complexes of either 1:1 or 1:2 stoichiometry, which is consistent with the greater distance to the paramagnetic center. Examining the spectra in Fig. 3, the magnitude of the paramagnetic shifts and the broadness of the resonance signals correlate with the number of unpaired electrons and the electron relaxation time of the metal ion. However, it should be pointed out the distance to the paramagnetic center and the number of unpaired electrons are only two factors that influence the paramagnetic shifts. In general, paramagnetic shifts are governed by contact and dipolar interactions. In terms of the contact shift, the hyperfine coupling controls the chemical shift, whereas for the dipolar shift, the distance and the angle of the distance vector and the magnetic susceptibility tensor contribute to the chemical shift.

The Cr(III) complexes contain three unpaired electrons, and, consequently, the chemical shifts of the two resonances span a large chemical shift range of 100 ppm with the resonance spread ranging from − 70 to + 40 ppm, (Table I). The high spin Fe(III)-dipicolinate complexes contain five unpaired electrons and their signals spread by 190 ppm from -30 to 160 ppm (Table I). The large resonance spread reveals electron spin delocalization from metal ion orbitals into ligand orbitals although a fraction of the electron spins remains localized on the metal ion.

Table I. ^{1}H Chemical Shifts of Paramagnetic Metal Dipicolinate Complexes

M^{z+}	[ML]		[ML$_2$]	
	H1/H3	H2	H1/H3	H2
Cr^{3+}	-50	24	-70	35
Fe^{3+}	NA^a	-16	160	-30
Fe^{2+}	68	27	74	13
Co^{2+}	101	36	93	31
Ni^{2+}	68	20	64	22

^{a}- not distinct from the baseline.

The corresponding high spin Fe^{2+} complex contains four unpaired electrons. The NMR lines are much sharper, and the chemical shift range (10 to 80 ppm, Table I) is smaller than that observed for the signals of the Fe(III) complex. The smaller chemical shift range and the sharper lines correlate well with the reduced electron spin density of Fe(II). The high spin Co(II)-dipicolinate complexes with three unpaired electrons have a resonance spread of 30 - 110 ppm, which is similar to the one of the complex of the Cr(III) ion that has also three unpaired electrons. However, the Co(II) complex, has much sharper lines than the Cr(III) complexes, which is expected, since the metal centered dipolar relaxation is smaller for the Co(II) complex. In accordance with its two unpaired electrons and its smaller electron relaxation time, the resonances are the least shifted for the Ni(II) complex and span a range from 20 to 70 ppm. The line widths are the narrowest observed in this series of complexes making the NMR spectra of the Co(II) and Ni(II) complexes the easiest to observe. These observations are in agreement with results of Bertini and coworkers who reported that the metal ions most suitable for recording NMR spectra with relatively narrow lines are high-spin tetrahedral Ni(II), penta- and hexacoordinate high-spin Co(II), and high-spin Fe(II) (14). Thus, it can be concluded that the electron-nuclear interactions are such that the two signals of the dipicolinate complexes can be easily observed, and that the two sets of signals observed in the spectra for the 1:1 and 1:2 complexes are sufficiently different to be unambiguously distinguished.

Examination of the spectra in Fig. 3, indicates that Cr(III) and Fe(III) complexes yield spectra that are distinct from Fe(II), Co(II) and Ni(II) complexes. The spectra in question contain upfield, shifted resonances for H2, which reflects differences in the delocalization mechanisms of the spins of the unpaired electrons (14-17). At first, the discussion below focuses on the Co-complexes. These complexes have been studied in detail (31). Following this discussion, the Ni-complexes (33), the Fe-complexes (63, 64), and finally the Cr-complexes will be described (30).

The d^7 Co-dipicolinate complexes. Dissolving 9.0 mM [Co(H₂dipic)(dipic)] and varying the pH from 1.8 to 12.8 result in the spectra shown in Fig. 4 (31). In the pH range of 2.5 to 12.8 one major complex is observed. Since little free ligand is present, the observed shifts at 30.6 (H2) and 92.6 (H1/H3) ppm can be attributed to the $[Co(dipic)_2]^{2-}$ complex. In addition to paramagnetic line width broadening, peak broadness can also result from exchange broadening (see Fig. 4) (1, 65). As described in detail elsewhere, this class of complexes was subjected to detailed characterization both in the solid state and in solution. In the solid state X-ray crystallography and IR spectroscopy both showed the tridentate coordination of $dipic^{2-}$, $Hdipic^-$ and H_2dipic in most complexes (31). Below pH 2, the 1:1 complex is observed giving rise to signals at 36 (H2) and 101 (H1/H3) ppm.

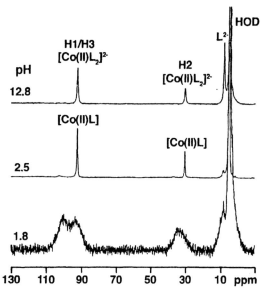

Fig. 4. ^1H NMR spectra of solutions of 9.0 mM [Co(H₂dipic)(dipic)] at pH
12.8, 2.5 and 1.8. Reproduced from reference 31.
(Reproduced from reference 31. Copyright 2002 American Chemical Society.)

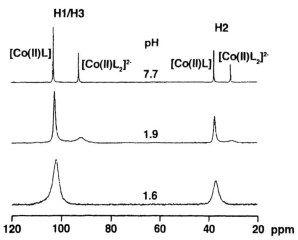

Fig. 5. ^1H NMR spectra of solutions of 28 mM [Co(dipic)] at pH values of (from
the top) 7.7, 1.9 and 1.6. Reproduced from reference 31.
(Reproduced from reference 31. Copyright 2002 American Chemical Society.)

In Fig. 5 the spectra are shown of 28 mM [Co(dipic)] at acidic pH. At low pH the 1:1 complex remains the most stable one, however, as the pH is increased, the concentration of the 1:2 complex increased suggesting that at higher pH, the 1:2 complex is the most stable one.

Spectra of four different Co(II)-dipic complexes $K_2[Co(dipic)_2]$, $[Co(dipic)_2Co(H_2O)_5]\cdot 2H_2O$, Co(dipic), and $[Co(H_2dipic)(dipic)]$ are shown in Fig. 6 (31). The spectrum obtained by dissolution of $K_2[Co(dipic)_2]$ illustrates the stability of this 1:2 complex. At neutral pH (the spectrum is shown at pH 7.3 in Fig. 6) only this complex, which has sharp lines, is observed. Dissolution of solid $[Co(dipic)_2Co(H_2O)_5]\cdot 2H_2O$ at pH 6.4 results in the formation of both the 1:2 and the 1:1 complex. The sharp NMR lines suggest that these complexes are not interconverting, at least not at pH 6.4. Dissolution of the 1:1 material ([Co(dipic)]) resulted in a solution containing the 1:1 complex at pH 1.6. However, the resulting broad lines indicate that this complex is undergoing ligand exchange at low pH. The latter observation is supported by dissolution of $[Co(H_2dipic)(dipic)]$ at low pH. At pH 1.8 this complex resulted in a solution containing two broad lines with chemical shifts assigned to both the 1:1 and 1:2 complexes. These spectra clearly show the presence of two complexes, and the pH of the solution defines the stability of each of the complexes. These studies demonstrate as previously has been reported that ^1H NMR spectra of Co(II) complexes are very informative (4, 20, 24, 41, 48-51).

The dynamics of the complexes were investigated using variable temperature and 2D EXchange SpectroscopY (EXSY) spectroscopy as has been

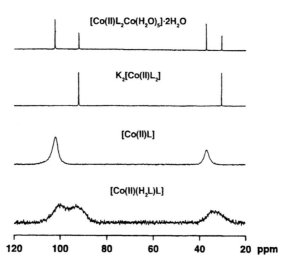

Fig. 6. *^1H NMR spectra of solutions of 14.0 mM $[Co(dipic)_2Co(H_2O)_5]\cdot 2H_2O$ (pH 6.4), 10 mM $K_2[Co(dipic)_2]$ (pH 7.3), 28 mM [Co(dipic)] (pH 1.6) and 9.0 mM $[Co(H_2dipic)(dipic)]$ (1.8). Reproduced from reference 31.*
(Reproduced from reference 31. Copyright 2002 American Chemical Society.)

described previously (*1, 31, 67-69*). Variable temperature studies showed that at both low pH and high pH, the signals shifted and the line widths increased consistent with the system approaching coalescence (data not shown). In solutions of $[Co(dipic)_2]^{2-}$ and $dipic^{2-}$ the EXSY experiment is specifically addressing the question if free ligand will exchange with ligand coordinated to the Co(II) as shown in eq 1. The ligand exchanging with complex both at low and high pH was confirmed by the cross peaks between ligand and complex protons observed in the EXSY maps (Fig. 7). A representative ^1H EXSY spectrum is shown of a solution containing 28 mM $[Co(dipic)_2]^{2-}$ and 27 mM H_2dipic at pH 3.3. The observation of cross peaks between the proton signal for the 1:2 complex and free ligand documents exchange between these species. The fundamental reaction between Co^{2+} and free ligand is not favorable in terms of entropy. The reaction between the 1:2 complex and free ligand is shown in eq 1 and this reaction is likely to take place in solutions with excess ligand. By recording the ^1H EXSY spectra at varying pH the ligand exchange was shown to be pH-dependent as described in detail elsewhere (*31*).

$$[Co(dipic)_2]^{2-} + \mathbf{dipic}^{2-} \rightleftharpoons [Co(dipic)(\mathbf{dipic})]^{2-} + dipic^{2-} \qquad (1)$$

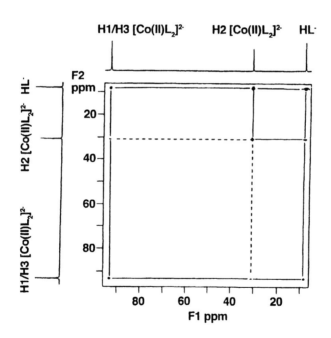

Fig. 7. ^1H EXSY maps of a solution containing 28 mM $[Co(dipic)_2]^{2-}$ and 27 mM H_2dipic at pH 3.3. Reproduced from reference 31.
(Reproduced from reference 31. Copyright 2002 American Chemical Society.)

By examining the ligand exchange in solutions with higher Co^{2+}:$dipic^{2-}$ ratios, evidence for a different ligand exchange reaction was observed. The ^1H EXSY spectrum in Fig. 8 of 8.8 mM [Co(dipic)] and 5.1 mM [Co(dipic)$_2$]$^{2-}$ at pH 6.4 revealed cross-signals between [Co(dipic)$_2$]$^{2-}$ and [Co(dipic)]. This spectrum illustrates that these complexes interconvert with each other as is shown in eq 2. Since little ligand is observed under these conditions, ligand exchange between complexes is not observed. The EXSY maps (Fig. 7 and 8) demonstrate that the 1:2 complex is kinetically labile since ligand exchange does take place between the 1:2 complex and ligand or the 1:1 complex.

$$[Co(dipic)_2]^{2-} + [Co(\textbf{dipic})] \rightleftarrows [Co(dipic)(\textbf{dipic})]^{2-} + [Co(dipic)] \qquad (2)$$

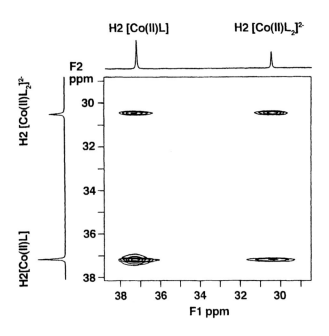

Fig. 8. ^1H EXSY map of a solution containing 8.8 mM [Co(II)dipic] and 5.1 mM [Co(II)(dipic)$_2$]$^{2-}$ at pH 6.4. Reproduced from reference 31.
(Reproduced from reference 31. Copyright 2002 American Chemical Society.)

The d^8 Ni-dipicolinate complexes. The spectrum is shown of a solution containing 22 mM K$_2$[Ni(dipic)$_2$] (33) and 34 mM H$_2$dipic from pH 10.4 to 1.8 (Fig. 9). In the presence of excess ligand in this entire pH range only the 1:2 complex is observed with signals at 22.1 and 63.7 ppm. However, when dissolving [Ni(dipic)]$_2$ (which contains a 1:1 ratio of ligand and Ni(II)) at pH 5.5 (Fig. 3), both the 1:1 complex with signals at 19.5 and 67.9 ppm and the 1:2 complex are observed. Two species are observed upon preparation of 14 mM

K$_2$[Ni(dipic)$_2$] at pH 1.2; both the 1:2 complex and free ligand form (eq.3) (no 1:1 complex). These studies show that the 1:2 complex is very stable over a wide pH range analogous to [Co(dipic)$_2$]$^{2-}$.

The ^{1}H NMR spectra of a Ni-complex solution at acidic pH were measured at temperatures ranging from 298 K to about 345 K. As was observed in case of the Co(II) ion, the temperature dependent studies confirm that [Ni(dipic)]$_2$ is in exchange with free ligand, since the ligand and complex signals approach coalescence with increasing temperatures. This observation was confirmed by the formation of cross peaks in the ^{1}H EXSY map (Fig. 10).

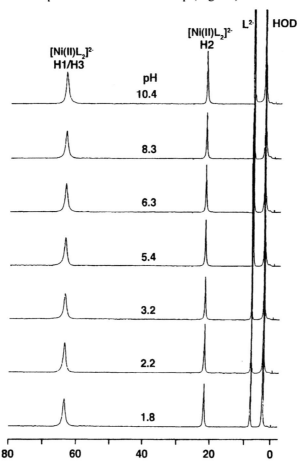

Fig. 9. ^{1}H NMR spectra of solutions containing 22 mM K$_2$[Ni(dipic)$_2$] and 34 mM H$_2$dipic at varying pH values (10.4, 8.3, 6.3, 5.4, 3.2, 2.2 and 1.8).

318

The cross peaks observed between the ligand and the protons in complex directly demonstrate exchange between free ligand and $[Ni(dipic)_2]^{2-}$ at acidic pH. No evidence for ligand exchange by EXSY spectroscopy was observed at neutral pH (*33*). These studies demonstrate as previous examples that ^1H NMR spectra of Ni(II) complexes are very informative (*6, 12, 23, 28, 51-56*).

$$[Ni(dipic)_2]^{2-} + \mathbf{dipic}^{2-} \rightleftharpoons [Ni(dipic)(\mathbf{dipic})]^{2-} + dipic^{2-} \qquad (3)$$

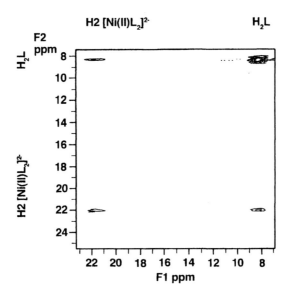

Fig. 10. *^1H EXSY spectrum of a solution of 14 mM $K_2[Ni(dipic)_2]$ at pH 1.2*

The d^5 and d^6 Fe-dipicolinate complexes. A number of Fe-dipicolinate complexes were previously prepared and characterized in the solid state (*63, 64*). Several of these complexes were characterized in the solution state and their solution properties investigated using potentiometry and UV-Vis spectroscopy (*66, 70, 71*). In Fig. 11 a solution of 20 mM Fe^{2+} containing varying amounts of dipic-ligand is shown at pH 4.2. The 1:1 complex was the major complex when ligand was present at 7 mM, and the resonances for protons H1, H3 and H2 were observed at 68 and 27 ppm, respectively. As the concentration of ligand was increased the signals attributed to the 1:2 complex at 74 and 12.5 ppm increased. In a pH study, the spectra of a solution containing 20 mM $[Fe(II)(dipic)_2]^{2-}$ were recorded at a pH ranging from 0 and to 12.8 (Fig. 12). The 1:2 complex remains completely intact until about pH 8, when free ligand is observed in the spectra (Fig. 12). Increasing amounts of ligand in the spectra as the pH is increased show that the 1:2 complex is hydrolytically unstable at alkaline pH. In contrast, the complex is very stable in the acidic pH range and remains in solution down to a pH of about 2.5. Below pH 2 the signals of the complex are significantly broadened indicative of exchange and deprotonation processes, all of which result in reduced complex stability.

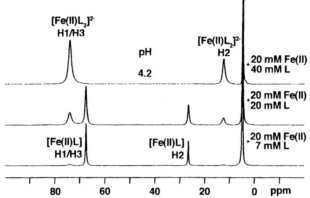

Fig. 11. ¹H NMR spectra of 20 mM Fe²⁺ solutions containing (from the top) 40, 20 and 7 mM dipic at pH 4.2.

In Fig. 12 the peak (–30 ppm) with an asterisk over it grows in at pH 3.5 (and below). This signal is assigned to the 1:2 [Fe(III)(dipic)$_2$]⁻ complex with the metal ion in its +3 oxidation state and indicates that, although the Fe(II) complex may be somewhat stable to hydrolysis, it is oxidized at low pH. Since higher pH commonly results in a greater propensity of metal complexes to oxidize, the opposite trend observed here may reflect that the small fraction of Fe(II)-complex which hydrolyzes is rapidly oxidized. The resonances in the ¹H NMR spectra of the Fe(III)-system are more difficult to detect due to the five unpaired electrons of the metal ion when compared to the four unpaired electrons of the metal ion in the Fe(II)-system. The ¹H NMR spectra of 20 mM Fe(III) in the presence of 15, 30 and 40 mM dipicolinate ligand at pH 1.95 resulted in spectra with the signals spanning a range from –30 to 160 ppm depending on the nature of the complex (Fig.13). Two signals were observed at ppm values below 0. In these spectra the proton signals are shifted both upfield and downfield reflecting differences in spin polarization in the delocalization of the unpaired Fe³⁺ spins (*14-17*). The signal of protons H1, H3 centered at 160 ppm is very broad and is difficult to distinguish from the baseline, and is, therefore, not useful for monitoring the existence of the complexes. However, the signal at –16 ppm assigned to the proton H2 accurately report on the nature of the complex existing in solution. As the ligand concentration increases up to 30 mM, a second signal at –30 ppm emerges. This signal is assigned to the H2 proton of the 1:2 complex, which is the major species in solution with metal ion to ligand ratios of 1:2. Spectra were also recorded of solutions containing Fe(III) complex at various pH values (Fig. 14).

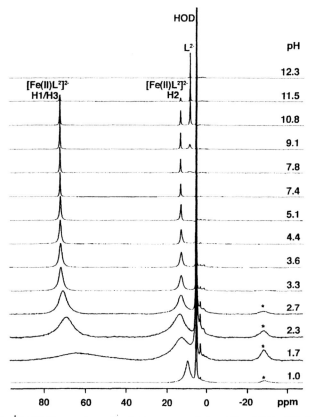

Fig. 12. ^1H NMR spectra of solutions containing 20 mM Na$_2$[Fe(dipic)$_2$] recorded at varying pH values (12.3, 11.5, 10.8, 9.1, 7.8, 7.4, 5.1, 4.4, 3.6, 3.3, 2.7, 2.3, 1.7 and 1.0). The * indicates the presence of the Fe(III) dipicolinate complex.

These spectra show that the 1:2 complex is stable at acidic pH but not in a solution of basic pH. Above pH 5.0 a brown precipitate formed; supernatant solution spectra show that the Fe(III) was removed by precipitation because the free ligand to complex ratio increases with increasing pH (Fig.14). When comparing the data of the Fe(II)-dipic and Fe(III)-dipic complexes, it can be shown that [Fe(II)(dipic)$_2$]$^{2-}$ is stable in neutral and slightly alkaline pH regions, whereas [Fe(III)(dipic)$_2$]$^-$ is stable at acidic pH.

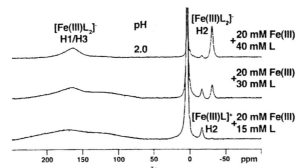

Fig. 13. ^1H NMR spectra of 20 mM Fe^{3+} solutions containing (from the top) 40, 30 and 15 mM dipic at pH 2.0±0.2.

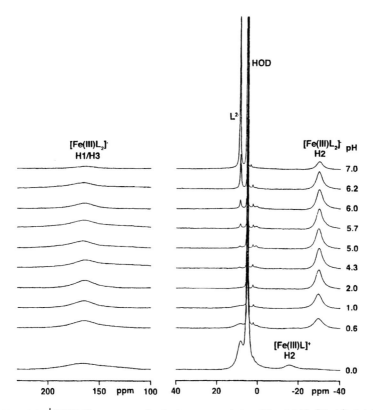

Fig. 14. ^1H NMR spectra of solutions containing 20 mM $Na[Fe(dipic)_2]$ recorded at varying pH values (7.0, 6.2, 6.0, 5.7, 5.0, 4.3, 2.0, 1.0, 0.6 and 0.0).

The d³ Cr-dipicolinate complexes. Dissolution of K[Cr(dipic)₂] results in a spectrum with signals at −70 ppm and 35 ppm at a pH of about 5 (Fig. 15). As the pH of this solution is decreased to pH 0.9 and 0.4, two new signals are observed at −50 and 24 ppm (Fig. 15). Since free ligand also is formed, the new signals are assigned to the 1:1 species. The chemical shifts of the complex were monitored from pH 1 to 10, and no change in complex identity was observed. In this pH range the 1:2 complex was stable in solution, and no change was detected in the chemical shifts of the two protons, suggesting that neither deprotonation takes place in the pH range from 2-10, nor is ligand exchange observed. However, these studies showed that only at pH below 1 did the 1:1 complex become a stable species in solution. A variable temperature study was performed to confirm whether ligand exchange was occurring. Although the signals were shifting as if they approached coalescence with the uncomplexed ligand signal, examination of the line width showed that the continuous decrease was not consistent with a ligand-exchange process.

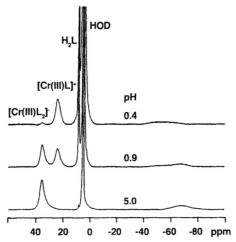

Fig. 15. ¹H NMR spectra of solutions containing 45.3 mM K[Cr(dipic)₂] at pH 0.4, 0.9, and 5.0.

Summary. In this manuscript we have shown that paramagnetic ¹H NMR spectroscopy is an effective tool in investigating complex stoichiometry and speciation in Cr(III), Fe(II), Fe(III), Co(II) and Ni(II) dipicolinate complexes with O_h geometries. We demonstrated that spectra can be obtained for a wide range of metal ion electronic spin states, and although only minimal information is available regarding relaxation mechanisms, the species existing in solution and their quantization can be derived. Despite the significant differences in line widths and extensive line broadening that occurs for some of the complexes, this method can be characterized as generally applicable to this type of structural and stability analysis.

Acknowledgement.

DCC thanks the American Diabetes Association and NIH (The Institute for General Medicine at The National Institutes of Health) for funding. We thank Dr. Christopher D. Rithner for technical assistance.

Referenes:

1. Crans, D. C.; Shin, P. K.; Armstrong, K. B. In *ACS Symp. Ser.* **1995**; *246*, pp 303-328.
2. Ming, L. J.; Lynch, J. B.; Holz, R. C.; Que, Jr. L. *Inorg. Chem.* **1994**, *33* , 83-87.
3. Satterlee, J. D.; Erman, J. E. *Biochemistry* **1991**, *30*, 4398-4405.
4. Epperson, J. D.; Ming, L.-J. *Biochemistry* **2000**, *39*, 4037-4045.
5. Savenkova, M. I.; Satterlee, J. D.; Erman, J. E.; Siems, W. F.; Helms, G. L. *Biochemistry* **2001**, *40*, 12123-12131.
6. Al-Mjeni, F.; Ju, T.; Pochapsky, T. C.; Maroney, M. J. *Biochemistry* **2002**, *41*, 6761-6769.
7. Jain, N. U.; Pochapsky, T. C. *J. Am. Chem. Soc.* **1998**, *120*, 12984-12985.
8. de Ropp, J. S. S., S.; Asokan, A.; Newmyer, S.; Ortiz de Montellano, P. R.; La Mar, G. N. *J. Am. Chem. Soc.* **2002**, *124*, 11029-11037.
9. Hu, B. H., J. B.; Tran, A.-T. T.; Kolczak, U.; Pandey, R. K.; Rezzano, I. N.; Smith, K. M.; La Mar, G. N. *J. Am. Chem. Soc.* **2001**, *123*, 10063-10070.
10. La Mar, G. N.; Kolczak, U.; Tran, A.-T. T.; Chien, E. Y. T. *J. Am. Chem. Soc.* **2001**, *123*, 4266-4274.
11. Asokan, A. d. R., J. S.; Newmyer, S. L.; Ortiz de Montellano, P. R.; La Mar, G. *J. Am. Chem. Soc.* **2001**, *123*, 4243-4254.
12. Kostic, M.; Pochapsky, S. S.; Pochapsky, T. C. *J. Am. Chem. Soc.* **2002**, *124*, 9054-9055.
13. Yamamoto, Y.; Terui, N.; Tachiiri, N.; Minakawa, K.; Matsuo, H.; Kameda, T.; Hasegawa, J.; Sambongi, Y.; Uchiyama, S.; Kobayashi, Y.; Igarashi, Y. *J. Am. Chem. Soc.* **2002**, *124*, 11574-11575.
14. Bertini, I.; Turano, P.; Vila, A. J. *Chem. Rev.* **1993**, *93*, 2833- 2932.
15. Bertini, L.; Luchinat, C. NMR of Paramagnetic Substances In *Coordination Chemistry Reviews*, Lever A. B. P., Ed.; Vol. 150; Elsevier: Amsterdam, 1996.
16. Ming, L.-J.; Nuclear Magnetic Resonance of Paramagnetic Metal Centers in Proteins and Synthetic Complexes. In *Physical Methods in Bioinorganic*

Chemistry: Spectroscopy and Magnetism, Que L. Jr., Ed.; University Science Books: Sausalito, CA, 2000; pp 375-464.

17. Bertini, I.; Luchinat, C.; Parigi, G. In *Current Methods in Inorganic Chemistry*; Solution NMR of Paramagnetic Molecules: Application to Metallobiomolecules and models, Vol. 2; Elsevier : Amsterdam 2001.

18. Bertini, I.; Luchinat, C. *Curr. Opin. Chem. Biol.* **1999**, *3*, 145-151.

19. Hagadorn, J. R.; Que, L. Jr.; Tolman, W. B. *Inorg. Chem.* **2000**, *39*, 6086-6090.

20. Blake, A. J.; Gilby, L. M.; Parsons, S.; Rawson, J. M.; Reed, D.; Solan, G. A.; Winpenny, R. E. P. *J. Chem. Soc., Dalton Trans.* **1996**, 3575-3581.

21. Heistand R. H. II; Lauffer R. B.; Fikrig E.; Que L. Jr. *J. Am. Chem. Soc.* **1982**, *104*, 2789-2796.

22. Chiuo, Y.-M.; Que, L. Jr. *J. Am. Chem. Soc* **1995**, *117*, 3999-4013.

23. Belle, C.; Bougault, C.; Averbuch, M.-T.; Durif, A.; Pierre, J.-L.; Latour, J.-M.; Le Pape, L. *J. Am. Chem. Soc.* **2001**, *123*, 8053-8066.

24. Cotton, A. F.; Murillo, C. A.; Wang, X. *Inorg. Chem.* **1999**, *38*, 6294-6297.

25. La Mar, G. N.; Walker, F. A. *J. Am. Chem. Soc* **1973**, *92*, 6950-6956.

26. Goff, H.; La Mar, G. N. *J. Am. Chem. Soc.* **1977**, *99*, 6599-6606.

27. Panda, A.; Stender, M.; Wright, R. J.; Olmstead, M. M.; Klavins, P.; Power, P. P. *Inorg. Chem.* **2002**, *41*, 3909-3916.

28. Chivers, T.; Krahn, M.; Schatte, G. *Inorg. Chem.* **2002**, *41*, 4348-4354.

29. Crans, D. C.; Yang, L.; Jakush, T.; Kiss, T. *Inorg. Chem* **2000**, *39*, 4409-4416.

30. Crans, D. C.; Yang, L.; Alfano, J.; Austin, L.-T.; Wenzheng, J.; Elliot, C. M.; Gaidamauskas, E.; Godzala, M. E. III.; Hutson, A. D.; Kostyniak, P. J.; Willsky, G. R. *Proc. Nat. Acad. Sci.,* submitted.

31. Yang, L.; Crans, D. C.; Miller, S. M.; la Cour, A.; Anderson, O. P.; Kaszynski, P. M.; Godzala, M. E.; Austin, L. D.; Willsky, G. R. *Inorg. Chem.,* **2002**, 41, 4859-4871; see also references therein.

32. Crans D. C.; Gaidamauskas E.; Khan A. R.; Chi L.-H.; Gaidamauskiene E.; Willsky G. W.; to be submitted.

33. Crans D. C. Yang L.; Jin W.; manuscript in preparation.

34. Satterlee, J. D. *Concepts Magn. Reson.* **1990**, *2*, 119-129.

35. Sharp, R. R.; Lohr, L.; Miller, J. *Prog. Nucl. Magn. Reson. Spectrosc.* **2001**, *38*, 115-158.

36. Satterlee, J. D. *Concepts Magn. Reson.* **1990**, *2*, 69-79.

37. Sharp, R. R.; *Nucl. Magn. Reson.* **2001**, *30*, 477-526.

38. Lehmann T. E.; Ming L.-J.; Rosen M. E.; Que L. Jr. Biochemistry, **1997**, 36, 2807-2816.

39. Ming, L.-J.; Lauffer, R. B.; Que, L. Jr. *Inorg. Chem.* **1990**, *29*, 3060-3064.

40. Sowrey, F. E.; MacDonald, J. C.; Cannon, R. D. *J. Chem. Soc., Faraday Trans.* **1998**, *94*, 1571-1574.

41. Diebold, A.; Elbonadili, A.; Hagen, K. S. *Inorg. Chem.* **2000**, *39*, 3915-3923.
42. La Mar, G. N.; Satterlee, J. D.; De Ropp, J. S. *The Porphyrin Handbook K. M. Kadish, K. M. Smith, R. Guilard Eds.* **2000**, *5*, 185-286.
43. Walker, F. A.; Simonis, U. *Encyclopedia of Inorganic Chemistry* **1994**, *R. B. King, Ed., John Wiley & Sons*, 1-58.
44. Walker, F. A. *The Porphyrin Handbook* **2000**, *5*, 81-175.
45. La Mar, G. N.; Eaton, G. R.; Holm, R. H.; Walker, F. A. *J. Am. Chem. Soc.* **1973**, *95*, 63-75.
46. Hage, R.; Gunnewegh, E. A.; Niel, J.; Tjan, F. S. B.; Weyhermuller, T.; Wieghardt, K. *Inorg. Chim. Acta* **1998**, *268*, 43-48.

47. Kojima, T.; Leising, R. A.; Yan, S.; Que, L. Jr. *J. Am. Chem. Soc.* **1993**, *115*, 11328-11335.
48. Duran, N.; Clegg, W.; Cucurull-Sanchez, L.; Coxal, R. A.; Jimenez, H. R.; Moratal, M.-J.; Lloret, F.; Gonzalez-Duarte, P. *Inorg. Chem.* **2000**, *39*, 4821-4832.
49. Pulici, M.; Caneva, E.; Crippa, S. *J. Chem. Research (S)* **1997**, *5*, 160-161.
50. Trofimenko, S.; Rheingold, A. L.; Sands, L. M. L. *Inorg. Chem.* **2002**, *41*, 1889-1896.
51. Ming, L.-J.; Epperson, J. D. *J. Inorg. Biochem.* **2002**, *91*, 46-58.
52. Robert J. M.; Evilia R. F.; Inorg. Chem. **1987**, 26, 2857-2861.
53. Rosenfield, S. G.; Berends, H. P.; Lucio, G.; Stephan, D. W.; Mascharak, P. K. *Inorg. Chem.* **1987**, *26*, 2792-2797.
54. Frydendahl, H.; Toftlund, H.; Becher, J.; Dutton, J. C.; Murray, K. S.; Taylor, L. F.; Anderson, O. P; Tiekink, E. R. T. *Inorg. Chem.* **1995**, *34*, 4467-4476.
55. Arion, V.; Wieghardt, K.; Weyhermueller, T.; Bill, E.; Leovac, V.; Rufinska, A.; *Inorg. Chem.* **1997**, *36*, 661-669.
56. Kasuga, N. C.; Sekino, K.; Koumo, C.; Shimada, N.; Ishikawa, M.; Nomiya, K. *J. Inorg. Biochem.* **2001**, *84*, 55-65.
57. Kingry, K. F.; Royer, A. C.; Vincent, J. B. *J. Inorg. Biochem.* **1998**, *72*, 79-88.
58. Broadhurst, C. L.; Schmidt, W. F.; Reeves, J. B. III.; Polansky, M. M.; Gautschi, K.; Anderson, R. A. *J. Inorg. Biochem.* **1997**, 66, 120-130.
59. Royer, A. C.; Rogers, R. D.; Arrington, D. L.; Street, S. C.; Vincent, J. B. *Polyhedron* **2002**, *21*, 155-165.
60. Kohler, F. H.; Metz, B.; Strauss, W. *Inorg. Chem.* **1995**, *34*, 4402-4413.
61. Blom, R.; Swang, O. *Eur. J. Inorg. Chem.* **2002**, 411-415.
62. Tichane, R. M.; Bennett, W. E. *J. Am. Chem. Soc.* **1957**, *79*, 1293-1296.
63. Laine, P.; Gourdon, A.; Launay, J.-P. *Inorg. Chem.* **1995**, *34*, 5129-5137.
64. Thich, J. A.; Ou, C. C.; Powers, D.; Vasiliou, B.; Mastropaolo, D.; Potenza, J. A.; Schugar, H. J. *J. Am. Chem. Soc.* **1976**, *98*, 1425-1433.

65. Epperson, J.D.; Ming, L.-J.; Baker, G.R.; Newkome, G.R. *J. Am. Chem. Soc.* **2001**, *123*, 8583-8592.
66. Anderegg G. *Helv. Chim. Acta* **1960**, *43*, 1530-1545.
67. Ming, L.-J.; Jang, H. G.; Que, L. Jr. *Inorg. Chem.* **1992**, *31*, 359-364.
68. Ming, L.-J.; Wei; X. *Inorg. Chem.* **1994**, *33*, 4617-4618.
69. Wei, X.D.; Ming, L.-J. *Inorg. Chem.* **1998**, *37*, 2255-2262.
70. Hartkamp H. *Z. Anorg. Chem.* **1962**, *190*, 66-76.
71. Morimoto I.; Tanaka S. *Anal. Chem.* **1963**, *35*, 141-144.

Magnetic Circular Dichroism and Related Techniques

Chapter 18

Variable-Temperature Variable-Field Magnetic Circular Dichroism Combined with Electron Paramagnetic Resonance: Polarizations of Electronic Transitions in Solution

Edward I. Solomon, Mindy I. Davis, Frank Neese, and Monita Y. M. Pau

Department of Chemistry, Stanford University, 333 Campus Drive, Stanford, CA 94305

The behaviour of saturation magnetization in VTVH MCD is determined by the polarization of an electronic transition relative to the magnetic ground state **D** tensor and/or g matrix. The latter can be derived from molecular orbital calculations including spin-orbit coupling. Thus the combination of VTVH MCD, EPR and molecular orbital theory allow the determination of polarizations of electronic transitions in a randomly oriented solution, providing new experimental insight into electronic structure and its contribution to reactivity.

Introduction

Many active sites in bioinorganic chemistry are highly colored as they exhibit intense low energy ligand-to-metal charge transfer (LMCT) transitions in absorption which reflect highly covalent ligand-metal bonds. These are often paramagnetic and in many cases (particularly Kramers ions) also exhibit rich electron paramagnetic resonance (EPR) spectra. Variable-temperature variable-field (VTVH) magnetic circular dichroism (MCD) provides a method to combine these to obtain new insight into electronic structure that can be important to function.

For example, Fe^{III} active sites play key roles in catalysis in the substrate activating enzymes, intradiol dioxygenases and lipoxygenases which are high spin, and in the peroxide level intermediates present in bleomycin (BLM) (activated BLM, a low-spin Fe^{III}-OOH complex) and possibly in the non-heme iron enzymes (which would be high-spin) (1). In contrast to Fe^{II} centers, ferric sites are EPR active and exhibit intense LMCT transitions in their absorption spectra. Variable temperature (and possibly variable frequency) EPR provide spin hamiltonian parameters (g values for $S = \frac{1}{2}$, and zero field splitting (ZFS) parameters (D, E) for $S = \frac{5}{2}$) which, however, are not generally interpreted in terms of geometric and electronic structure. For the LMCT transitions, absorption, CD (circular dichroism), MCD and resonance Raman excitation profiles (the latter analyzed with time domain Heller theory (2, 3)) allow resolution of transitions (based on differences in selection rules) and general assignments to specific ligands (based on excited state distortions) (4). However, there is more information in the temperature and field dependence (VTVH) of the MCD data, particularly in the non-linear region, approaching the saturation limit (i.e., high H and low T). This allows correlation of excited state with ground state properties and the determination of the polarization of electronic transitions in a randomly oriented frozen solution (rather than in a single crystal). However, this is a complex problem because the spin sub-levels of the ground state can cross and mix for specific orientations of the magnetic field which in turn is dependent on the polarization of the electronic transition from the ground to the excited state.

Methodology

We have derived expressions (5) which allow the analysis of VTVH MCD saturation magnetization curves. Equation 1 is for an $S = \frac{1}{2}$ system, while eq 2 is the general expression, where the ZFS of the i sub-levels of the ground state is included.

$$\frac{\Delta \varepsilon_{av}}{E} = -const \int_0^\pi \int_0^{2\pi} \tanh\left(\frac{\gamma\beta H}{2kT}\right)\frac{\sin\theta}{\gamma}\left(l_x^2 g_x M_{yz}^{eff} + l_y^2 g_y M_{xz}^{eff} + l_z^2 g_z M_{xy}^{eff}\right) d\theta d\phi \tag{1}$$

where $\gamma = \sqrt{(G_x^2 + G_y^2 + G_z^2)}$ and $G_p = l_p g_p$ with $p = x, y, z$

$$\frac{\Delta \varepsilon_{av}}{E} = \frac{const}{S} \int_0^\pi \int_0^{2\pi} \sum_i N_i \left(l_x \langle S_x \rangle_i M_{yz}^{eff} + l_y \langle S_y \rangle_i M_{xz}^{eff} + l_z \langle S_z \rangle_i M_{xy}^{eff}\right) \sin\theta d\theta d\phi \tag{2}$$

For both equations, θ and ϕ are the angles between the incident light and the molecular z axis and the xy plane, respectively, l_x, l_y and l_z describe the orientation of the magnetic field relative to the molecular coordinate system, and the M_{ij}^{eff} are the products of the polarizations of the electronic transitions. Two perpendicular transition moments i and j are required for MCD intensity. In a low symmetry protein site a transition is uni-directional, so $M_{ij}^{eff} \neq 0$ is accomplished by spin-orbit mixing with an excited state that has transition moments with perpendicularly polarized components. In eq 1 for the $S = \frac{1}{2}$ case, the g_p values are input from experimental EPR data. Equation 2 is dependent on N_i the Boltzmann population, and $\langle S_p \rangle_i$, the spin expectation value in the p direction, for the spin sub-levels i of the ground state. These are obtained directly from the energies and wave-functions of the spin hamiltonian, eq 3, solved with values of D, E and g_p obtained from EPR data (6).

$$H = D(S_z^2 - \frac{1}{3}S^2) + E(S_x^2 - S_y^2) + \beta(g_x H_x S_x + g_y H_y S_y + g_z H_z S_z) \tag{3}$$

Thus the combination of VTVH MCD and EPR data allow the determination of the relative polarization products, M_{ij}^{eff}, for a transition to a given excited state, which through eq 4 (with cyclic permutations of indices to obtain % y and z) give the polarizations of electronic transitions for a randomly oriented protein solution (5).

$$\%x = 100 \times \frac{\left(M_{xy}^{eff} M_{xz}^{eff}\right)^2}{\left(M_{xy}^{eff} M_{xz}^{eff}\right)^2 + \left(M_{xy}^{eff} M_{yz}^{eff}\right)^2 + \left(M_{xz}^{eff} M_{yz}^{eff}\right)^2} \tag{4}$$

Experimentally, VTVH MCD data require spectroscopy on a strain free optical quality glass. The study of intermediates, an application important to our research, often requires these to be trapped using rapid freeze quench (RFQ) methods. Thus we have also developed a protocol (Figure 1) to obtain good quality VTVH MCD data on a RFQ intermediate. This involves trapping the intermediate in liquid nitrogen and then making a 50% glycerol glass sample at -30° where the intermediates we are studying are stable. This is then transferred to an MCD cell also maintained at -30° for spectroscopy (7).

The polarizations of electronic transitions obtained above are in the g matrix or **D** tensor coordinate system. The g matrix or **D** tensor must then be correlated to the molecular structure to obtain electronic structure information. We have been generally interested in the information content of spin hamiltonian parameters in terms of electronic structure and have shown that the standard ligand field model used to interpret these in terms of d orbital energy splitting must be modified to include differential orbital covalency (8). Each d orbital has a different covalent interaction with its ligand environment which modifies the spin-orbit interactions between the ground and excited states that give the ZFS and g value deviations from 2.00.

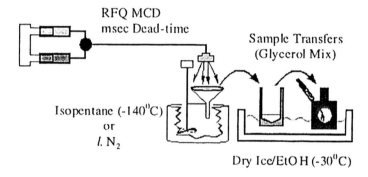

Figure 1. Protocol for trapping rapid freeze quench (RFQ) intermediates.

The absolute orientation of the g matrix and the **D** tensor can be measured with single crystal EPR spectroscopy. However, this depends on the availability of single crystals. Alternatively, the orientation of the g matrix and the **D** tensor in the molecule can be calculated by quantum chemical methods. In general, the elements of the **D** tensor are defined as (9):

$$D_{pq}^{(0;\pm1)} \sim -\sum_{B} \Delta_{B}^{-1} \langle AS_{A}M_{S,A} | H_{SOC}^{p} | BS_{B}M_{S,B} \rangle \times \langle BS_{B}M_{S,B} | H_{SOC}^{q} | AS_{A}M_{S,A} \rangle \quad (5)$$

with $|AS_{A}M_{S,A}\rangle$ being the ground state function of total spin S_{A} and corresponding $M_{S,A}$ value; the $|BS_{B}M_{S,B}\rangle$ are the excited states which appear in the sum over B; the indizes $(p,q) =$ x, y, z refer to the cartesian components of the **D** tensor; H^p_{SOC} is the spin-orbit coupling operator; the superscripts $(0; \pm1)$ represent the allowed $\Delta S = S_A - S_B$ following the selection rule of H_{SOC} and $\Delta_B = E_B - E_A$.[1] The elements of the Δg matrix which define the shifts of the g-values relative to the free electron value (2.00) are obtained from the equation (9):

$$\Delta g_{pq} \sim -\sum_B \Delta_B^{-1} \left[\begin{matrix} \langle AS_A M_{S,A} |L^p| BS_B M_{S,B} \rangle \langle BS_B M_{S,B} |H^q_{SOC}| AS_A M_{S,A} \rangle \\ + \langle AS_A M_{S,A} |H^p_{SOC}| BS_B M_{S,B} \rangle \langle BS_B M_{S,B} |L^q| AS_A M_{S,A} \rangle \end{matrix} \right] \quad (6)$$

with L^p being the orbital angular momentum operator.[1] The elements of the **D** tensor in eq 5 arise from products of spin-orbit coupling matrix elements whereas the g shifts in eq 6 emerge from products of spin-orbit coupling and Zeeman matrix elements. This is the reason why only excited states of the same total spin as the ground state ($S_B = S_A$) contribute to the Δg matrix whereas the **D** tensor has additional contributions from states with $S_B = S_A \pm 1$.

These molecular orbital expressions for g and **D** have been implemented in an INDO/S-CI semi-empirical MO program to calculate the D_{ij} tensor and g_{ij} matrix for a low symmetry active site and by diagonalization determine the orientation of these principle magnetic directions in the molecular frame (9). This allows the evaluation of possible structural models for an active site (e.g., intermediates), determines specific bonding contributions reflected by the spin Hamiltonian parameters, and, in particular, fixes the polarization of electronic transitions from VTVH MCD to specific acive site structural features. This has, as an example, proved critical in determining the different phenolate-Fe^{III} bonding interactions in the intradiol dioxygenases as described below.

Application: Intradiol dioxygenases – Nature of the Tyr $\rightarrow$ Fe^{III} Bonds and Their Contributions to Reactivity (11)

The crystal structures of the intradiol dioxygenases, including protocatechuate 3,4-dioxygenase (3,4-PCD) show a trigonal bipyramidal Fe^{III} site with axial Tyr and His, and equatorial Tyr, His and OH$^-$ ligands, Figure 2 left (12, 13). Protocatechuate (PCA, 3,4-dihydroxybenzoate) binds as the dianion, replacing the axial Tyr and equatorial OH$^-$ forming a square pyramidal structure (14). The catecholate is asymmetrically bound with a long Fe-O bond trans to the

---

[1] By applying the Wigner-Eckhard theorem to the definition of H_{SOC}, $M_{S,A} = S_A$ and $M_{S,B} = S_B$ are the only components needed in the definition of $D^{(0;\pm1)}_{pq}$ and Δg_{pq} (9, 10).

equatorial Tyr. Resting 3,4-PCD has a deep burgundy red color reflecting low energy intense phenolate-to-Fe^{III} charge transfer transitions (15). As we have emphasized, low energy intense charge transfer transitions reflect highly covalent ligand-metal bonds which can activate the metal site for reactivity (16). The absorption spectrum shows a broad band at ~ 22,000 cm^{-1} (ε ~ 3,000 M^{-1}cm^{-1}) and a second feature at ~ 30,000 cm^{-1} (ε ~ 8,000 M^{-1}cm^{-1}). The combination of absorption, CD, and MCD data (each with a different selection rule) demonstrates that at least seven transitions, Figure 3, are required to fit the Tyr $\rightarrow$ Fe^{III} charge transfer region. The VTVH MCD data for these transitions given in Figure 4 reveal different saturation behaviors reflecting different polarizations for the electronic transitions. Fitting these saturation magnetization curves using eq 2 shows that bands 1, 3 and 6 are y polarized and that bands 2, 4 and 7 are dominantly z polarized (Table I).

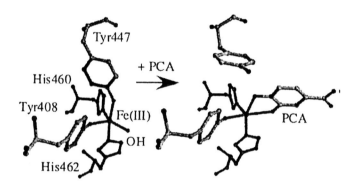

Figure 2. Schematic representation of the binding of substrate protocatechuate (PCA) to protocatechuate 3,4-dioxygenase (3,4-PCD). (Structures were generated using crystallographic coordinates from PDB files 2PCD (from reference 13) and 3PCA (from reference 14))

As developed above these polarizations are relative to the **D** tensor orientation which was mapped onto the trigonal bipyramidal resting site in Figure 2 left using INDO/S-CI calculations. These calculations allow for the second-order spin-orbit coupling of the $^6A_1$ ground state with the complete manifold of quartet and sextet excited states including differential orbital covalency. The calculations give $E/D = 0.32$ and $D = -1.3$ cm^{-1} which are in fortuitously good agreement with experiment ($E/D = 0.33$ and $D = 1.2$ cm^{-1}). Note that in the rhombic limit, while the sign of D does not affect the EPR or magnetic susceptibility data, it does have a strong effect on the MCD analysis as it determines the specific orientations of the effective g values of the first and third Kramers doublets of the ground state. These calculations show that the z-direction of the **D** tensor is along the axial Tyr-Fe^{III} direction and the y axis is approximately along the equatorial Tyr-Fe^{III} direction. Thus we can assign the charge transfer transitions associated with each Tyr-Fe^{III} bond. From the above studies we find that there are at least three Tyr $\rightarrow$ Fe^{III} charge transfer transitions associated with each Tyr-Fe bond, and that the two sets of CT transitions hence the axial and equatorial Tyr-Fe bonds are inequivalent.

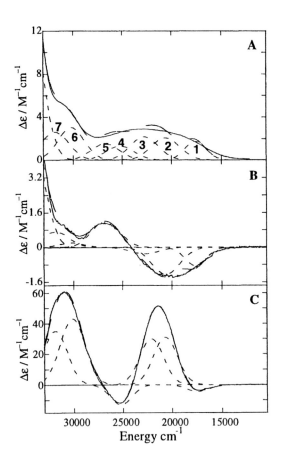

Figure 3. Abs at 4 °C (A), CD at 4 °C (B), and MCD spectra at 5 K and 7 T (C) of 3,4-PCD. Gaussian resolution (- - -) is shown along with spectra (——). (Reprinted from reference 11. Copyright 2002 American Chemical Society.)

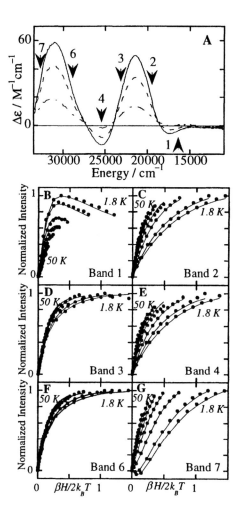

Figure 4. Variable-temperature variable-field MCD data for 3,4-PCD. (A) Temperature dependence of the MCD spectrum at 7 T for 5 K (——), 15 K (- - -) and 50 K (– • –). The arrows indicate energies where VTVH MCD data were taken. VTVH MCD data (•) collected between 0 T and 7 T and between 1.6 K and 50 K and the fit (——) at (B) 32575 cm^{-1} Band 7. (C) 28820 cm^{-1} Band 6. (D) 25250 cm^{-1} Band 4. (E) 23150 cm^{-1} Band 3. (F) 19300 cm^{-1} Band 2. (G) 16390 cm^{-1} Band 1. (Reprinted from reference 11. Copyright 2002 American Chemical Society.)

Table I. Gaussian Resolution and Spectroscopic Parameters for Transitions Observed in MCD data collected at 7T and 5K for 3,4-PCD

band	MCD energy (cm^{-1})	$\Delta\varepsilon$ $(M^{-1}cm^{-1})$	MCD $\nu_{1/2}$ (cm^{-1})	Polarization
1	17 600	-2.33	3000	y
2	20 700	12.6	3400	z
3	22 100	9.20	3400	y
4	25 200	-1.21	3000	z
6	30 150	15.0	3600	y
7	31 900	11.5	3400	z

SOURCE: Adapted from reference 11. Copyright 2002 American Chemical Society.

Each Tyr (axial (ax) and equatorial (eq)) has two occupied valence orbitals (π_{op} (out-of-plane), π_{ip} (in-plane), Figure 5) available for donor bonding interactions with the five ½ occupied Fe d orbitals that are split in energy by the trigonal bipyramidal ligand field. This donor bonding strongly depends on the Fe-O-C angle θ (and to some extent on the Fe-O-CC dihedral angle). For $\theta = 180°$, both Tyr π orbitals in Figure 5 would be involved in π donor bonding with the Fe d orbitals; as θ decreases one (or both depending on dihedral angle) Tyr π level becomes pseudo-σ interacting with the dσ orbitals on the Fe, the CT spectrum (which reflects donor/acceptor orbital overlap) acquires more bands, and the Tyr-Fe^{III} bond strength increases. For the equatorial Tyr, $\theta = 133°$ while for the axial Tyr, $\theta = 148°$. This large difference in structure affects the strengths of the Tyr-Fe^{III} bonds where from Table II the axial Tyr has less donor interaction with the Fe^{III} and a weaker bond. From geometry optimization using DFT, the axial Tyr-Fe^{III} bond angle decreases and becomes equivalent to the equatorial Tyr in its donor interaction with the Fe^{III}. This indicates that the large θ of the axial Tyr-Fe bond is imposed on the active site by the protein and labilizes the axial Tyr for substitution by catecholate substrate. Alternatively, the equatorial Tyr-Fe^{III} bond involves a strong donor interaction which is *trans* to the second oxygen of the catecholate and can contribute to the latter's weak bonding interaction with the Fe^{III} which is thought to activate the substrate for attack by O_2. Thus the active site in 3,4-PCD has evolved to use the same residue, Tyr, in two essential but different roles by controlling the coordination geometry of the Tyr-Fe^{III} bonds.

Substrate coordination, Figure 2 right, produces new very low energy charge transfer transitions associated with the catecholate-Fe^{III} bonds (19, 20). Studies are now underway to use the above methodology to probe the nature of this bond in the intradiol dioxygenases and evaluate the mechanism of substrate activation by this non-heme ferric active site.

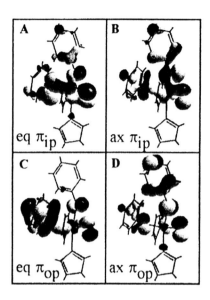

Figure 5. Molecular orbitals representing the π_{op} (out-of-plane) and π_{ip} (in-plane) interactions of both the axial (ax) and equatorial (eq) Tyr. (Reprinted from reference 11. Copyright 2002 American Chemical Society.)

Table II. Bond Order and Charges for DFT and INDO/S-CI Calculations

	Charge Mulliken (ADF[a])	*Charge Mulliken (INDO)*	*Charge Hirschfeld (ADF[a])*	*Fe-Frag Overlap[b] (INDO)*
tyr (ax)	-0.8	-0.7	-0.5	0.4
tyr (eq)	-0.5	-0.6	-0.4	0.5

[a] Amsterdam density functional package (17, 18).

[b] Fragment-Fragment overlap, i.e. Fe with tyr (ax) and Fe with tyr (eq).

SOURCE: Reprinted from reference 11. Copyright 2002 American Chemical Society.

338

Acknowledgement

This research was supported by NIH grant GM40392 (E.I.S.). F.N. thanks the Deutsche Forschungsgemeinshaft for a post-doctoral fellowship. M.I.D. thanks the Evelyn Laing McBain Fund for a doctoral fellowship. The studies in PCD were done in collaboration with professors John D. Lipscomb and Allen M. Orville.

References

1. Solomon, E. I.; Brunold, T. C.; Davis, M. I.; Kemsley, J. N.; Lee, S. K.; Lehnert, N.; Neese, F.; Skulan, A. J.; Yang, Y.-S.; Zhou, J. *Chem. Rev.* **2000**, *100*, 235-349.
2. Lee, S. Y.; Heller, E. J. *J. Chem. Phys.* **1979**, *71*, 4777-4788.
3. Tannor, D. J.; Heller, E. J. *J. Chem. Phys.* **1982**, *77*, 202-218.
4. Solomon, E. I.; Hanson, M. A. In *Inorganic Electronic Structure and Spectroscopy Vol. 2*; Solomon, E. I., Lever, A. B. P., Eds.; John Wiley: New York, 1999, pp 1-129.
5. Neese, F.; Solomon, E. I. *Inorg. Chem.* **1999**, *38*, 1847-1865.
6. Abragam, A.; Bleaney, B. *Electron Paramagnetic Resonance of Transition Ions*; Dover: New York, 1986.
7. Lee, S.-K.; George, S. D.; Antholine, W. E.; Hedman, B.; Hodgson, K. O.; Solomon, E. I. *J. Am. Chem. Soc.* **2002**, *124*, 6180-6193.
8. Deaton, J. C.; Gebhard, M. S.; Solomon, E. I. *Inorg. Chem.* **1989**, *28*, 877-889.
9. Neese, F.; Solomon, E. I. *Inorg. Chem.* **1998**, *37*, 6568-6582.
10. McWeeny, R. *Methods of Molecular Quantum Mechanics*; Academic Press: London, 1992.
11. Davis, M. I.; Orville, A. M.; Neese, F.; Zaleski, J. M.; Lipscomb, J. D.; Solomon, E. I. *J. Am. Chem. Soc.* **2002**, *124*, 602-614.
12. Ohlendorf, D. H.; Weber, P. C.; Lipscomb, J. D. *J. Mol. Bio.* **1987**, *195*, 225-227.
13. Ohlendorf, D. H.; Orville, A. M.; Lipscomb, J. D. *J. Mol. Biol.* **1994**, *244*, 586-608.
14. Orville, A. M.; Lipscomb, J. D.; Ohlendorf, D. H. *Biochemistry* **1997**, *36*, 10052-10066.
15. Fujisawa, H.; Hayaishi, O. *J. Biol. Chem.* **1968**, *243*, 2673-2681.
16. Solomon, E. I.; Lowery, M. D. *Science* **1993**, *259*, 1575-1581.

17. Baerends, E. J.; Ellis, D. E.; Ros, P. *Chem. Phys.* **1973**, *2*, 41-51.
18. te Velde, G.; Baerends, E. J. *J. Comp. Phys.* **1992**, *99*, 84-98.
19. Bull, C.; Ballou, D. P.; Otsuka, S. *J. Biol. Chem.* **1981**, *256*, 12681-12686.
20. Elgren, T. E.; Orville, A. M.; Kelly, K. A.; Lipscomb, J. D.; Ohlendorf, D. H.; Que, L. J. *Biochemistry* **1997**, *36*, 11504-11513.

Chapter 19

Magnetic Circular Dichroism Spectroscopy of Pyranopterin Molybdenum Enzymes

Martin L. Kirk

The Department of Chemistry, The University of New Mexico,
MSC03 2060, 1 University of New Mexico, Albuquerque, NM 87131–0001

MCD spectroscopy is a powerful probe of metalloenzyme active site electronic structure. The technique is usually used in concert with other methods such as electronic absorption, resonance Raman, and EPR spectroscopies in order to develop a comprehensive electronic structure description. Presented here is a brief introduction to MCD spectroscopy, followed by specific examples relating to each of the three primary pyranopterin Mo enzyme families; sulfite oxidase, DMSO reductase, and xanthine oxidase. As the technique is particularly sensitive to paramagnetic chromophores, d^1 Mo^V sites in molybdenum enzymes that possess additional endogenous chromophores can be probed site selectively.

Introduction

This chapter is an extended version of a tutorial on magnetic circular dichroism (MCD) spectroscopy presented at a symposium on magnetic resonance spectroscopies at the 2002 National ACS Meeting in Orlando. Here we will provide an introduction to MCD spectroscopy and discuss results from our laboratories that highlight the role MCD has played in increasing our understanding of pyranopterin molybdenum enzyme active site electronic structure. The pterin-containing molybdenum enzymes form a general class of metalloproteins that function to catalyze various two-electron redox reactions that are coupled to a formal oxygen atom transfer *(1,2)*. This large class of metalloenzymes has recently been grouped according to the specific types of reactions they catalyze, namely proper oxygen atom transfer to or from substrate, and the oxidative hydroxylation of aldehyde and heterocyclic substrates *(1)*. The former enzymes include sulfite oxidase (SO) and DMSO reductase (DMSOR), while enzymes of the xanthine oxidase (XO) family constitute the molybdenum hydroxylases.

Structural Details of Sulfite Oxidase and Xanthine Oxidase

A combination of X-ray crystallograpic, EXAFS, and spectroscopic studies have resulted in a set of consensus structures for the oxidized and reduced forms of each enzyme family. X-ray crystallography indicates that the catalytically competent oxidixed form of XO *(3-5)* adopts a distorted square pyramidal geometry with an unusual apical sulfido ligand. The equatorial ligation consists of a terminal oxo, $H_2O/HO^{\cdot}$, and two sulfur donors from the ene-1,2-dithiolate side chain of the pyranopterin cofactor. Upon reduction, the terminal sulfido is converted to a sulfhydryl moiety. The crystal structure of chicken liver SO indicates a five-coordinate, approximately square pyramidal, active site *(6)*. The molybdenum is coordinated by two different types of sulfur donors, one type arises from the ene-1,2-dithiolate side chain of the pyranopterin, and the second derives from the thiolate side chain of a catalytically essential cysteine residue. Upon reduction, it is believed that an equatorial oxo is lost and replaced by solvent water. In this Chapter, we will focus on the MCD spectroscopy of chicken SO and bovine milk XO. Key insights into the electronic structure of their respective active sites will be presented and discussed.

MCD Spectroscopy

MCD is a powerful probe of metalloenzyme active site electronic structure *(7)*. In this section, we present a brief overview of the theory of magnetic circular dichroism spectroscopy. More complete descriptions of MCD theory and experimental protocols have appeared elsewhere *(8-10)*. Qualitatively, MCD spectroscopy is the combination of circular dichroism (CD) with the application of a longitudinal magnetic field (H_{app} ∥ propagation of CP light). Consider a molecule possessing ground state (*A*) and excited state (*J*) both possessing angular momentum (J) with J = 1/2. The magnetic field splits both the *A* and *J* J -values into M_j = ±1/2 components. The magnitude of this splitting is governed by the Zeeman operator, and for J = 1/2 the splitting is $g_i\beta H$. The selection rules for these transitions are simple, with ΔM_j = +1 for absorption of left circularly polarized radiation (LCP) and ΔM_j = -1 for absorption of right circularly polarized (RCP) light. Here, $M_{\pm} = \mp \dfrac{1}{\sqrt{2}}(M_x \pm M_y)$ where M_x and M_y are *linear* polarizations. It should be noted that the magnetic field (Zeeman) splitting is on the order of 10 cm^{-1}, while the absorption bandwidth is considerably larger, on the order of 1000 cm^{-1}.

Quantitatively, expressions have been developed for describing the broad band MCD dispersion, $\Delta A/E$, within the rigid shift approximation in the limit where the magnetization varies linearly with the applied magnetic field (linear limit) *(10)*.

$$\frac{\Delta\varepsilon}{E} = \gamma\beta H\left[\mathcal{A}_1\left(-\frac{\partial f(E)}{\partial E}\right) + \left(\mathcal{B}_0 + \frac{C_0}{kT}\right)f(E)\right]$$

Here $\Delta\varepsilon$ is the differential molar absorptivity for LCP and RCP light, γ is a collection of constant terms, β is the Bohr magneton, *H* is the applied magnetic field, $f(E)$ is a bandshape function, and $\partial f(E)/\partial E$ is the derivative of this bandshape function. The $\mathcal{A}$-, $\mathcal{B}$- and C-terms describe the nature of the MCD curves, i.e. the MCD dispersion, while $\mathcal{A}_1$-, $\mathcal{B}_0$- and C_0 are the effective magnitudes of these effects and related to specific matrix elements defined below. $\mathcal{A}$-terms are observed to possess a derivative bandshape with intensity that is magnetic field dependent and temperature independent. Similarly, $\mathcal{B}$-terms possess an absorptive bandshape that is magnetic field dependent and temperature independent. Absorptive bandshapes are also observed for C-terms, however they possess a 1/T temperature dependence in the linear limit. Notably, C-terms are usually much more intense than $\mathcal{A}$- or $\mathcal{B}$-terms, and this is often true even at 300K. Furthermore, decreasing the temperature from 300K to 4K results in an increase in C-term intensity of approximately 2 orders of magnitude.

Notably, a given MCD band can have contributions from $\mathcal{A}$-, $\mathcal{B}$- and C-terms.

Equations have also been derived for the case when the light propagation vector and magnetic field directions are both along the molecular z-axis (single crystal) however, this more often than not an unrealized situation. Therefore, we need expressions for randomly oriented molecules, and these are given below (10).

$$\overline{\mathcal{A}}_1 = \frac{i}{3|d_g|}\sum\left(\langle J|\vec{\mu}|J\rangle - \langle A|\vec{\mu}|A\rangle\right)\bullet\left(\langle A|\vec{m}|J\rangle\times\langle J|\vec{m}|A\rangle\right)$$

$$\overline{\mathcal{B}}_0 = -\frac{2}{3|d_g|}\,\mathrm{Im}\sum\left[\begin{array}{l}\displaystyle\sum_{K(K\neq J)}\frac{\langle J|\vec{\mu}|K\rangle}{\Delta E_{KJ}}\bullet\left(\langle A|\vec{m}|J\rangle\times\langle K|\vec{m}|A\rangle\right)\\[2ex]+\displaystyle\sum_{K(K\neq A)}\frac{\langle K|\vec{\mu}|A\rangle}{\Delta E_{KA}}\bullet\left(\langle A|\vec{m}|J\rangle\times\langle J|\vec{m}|K\rangle\right)\end{array}\right]$$

$$\overline{C}_0 = -\frac{i}{3|d_g|}\sum\langle A|\vec{\mu}|A\rangle\bullet\left(\langle A|\vec{m}|J\rangle\times\langle J|\vec{m}|A\rangle\right)$$

Here d_g is the degeneracy of $|A\rangle$ and if only $|A\rangle$ is degenerate $\mathcal{A}_1 = C_0$. Note that these expressions are comprised of the dot product of a magnetic dipole vector $\left(\vec{\mu} = \vec{L} + 2\vec{S}\right)$ with the cross product between two electric dipole vectors.

Orbital degeneracy is a prerequisite for $\mathcal{A}$- and C-term MCD. These degenerate states are split by the application of an external magnetic field (Zeeman effect), and the magnitude of the splitting is a function of the electronic structure of the molecule or site. The combination of the ligand field and Jahn-Teller effect typically results in transition metal complexes displaying orbitally *nondegenerate* ground states. However, if they are paramagnetic they do possess *spin* degeneracy. The MCD selection rules are such that $\Delta M_L = \pm 1$ and $\Delta M_S = \pm 0$, and the *orbital degeneracy* necessary to observe C-term MCD results from out-of-state spin-orbit coupling between one or more excited states and the ground state. This spin orbit mixing is also the origin of g-tensor anisotropy, as well as single ion anisotropy (zero-field splitting, D) in $S>1/2$ centers. . Finally, the presence of $\mathcal{B}$-terms arrises from out-of-state Zeeman mixing involving an intermediate state $|K\rangle$. The electronic origin of the $\mathcal{A}$-, $\mathcal{B}$-, and C-terms are depicted graphically in Figure 1.

The C-terms deserve special mention, as these are the MCD spectral features most often observed for paramagnetic active sites studied at high magnetic fields and low-temperatures (i.e. T~4K, H~7T). Expanding the expression for $\overline{C}_o$ yields *(10)*:

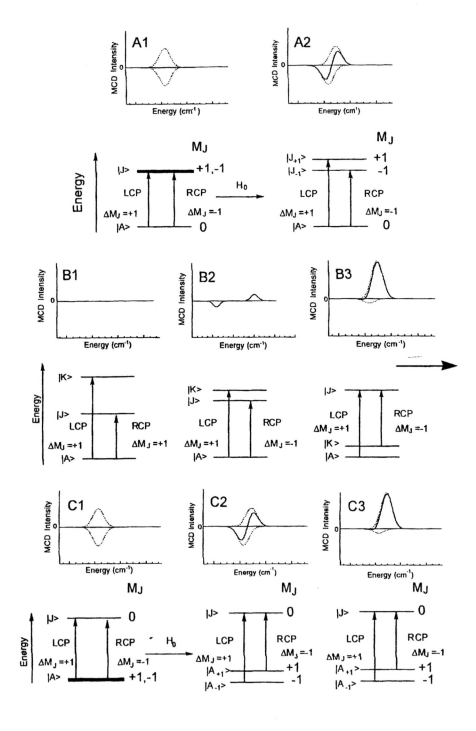

Figure 1. Origin of MCD 𝒜- (AX), ℬ- (BX) and C-terms (CX). Total angular momentum states are given as $|m_j>$. In A1, only the excited state $|J>$ is degenerate. $|J>$ is split by the applied magnetic field H_0 in A2 to yield a derivative-shaped MCD spectrum resulting from equal and oppositely signed components. Provided non-degenerate states $|J>$ and $|K>$ are greatly seperated in energy or do not mix by the Zeeman operator, the magnitude of the ℬ-term is zero (B1). However, out-of-state Zeeman mixing between $|J>$ and $|K>$ yields equal and oppositely signed ℬ-terms (B2). The intensity is proportional to the magnitude of the Zeeman matrix element that connects $|J>$ and $|K>$ as well as inversely proportional to the energy difference, ΔE, between these states. In C1, $|A>$ is degenerate and $|E>$ is non-degenerate. Upon application of a very small applied field in the high-temperature limit $|A>$ does not split appreciably, resulting in essentially equal and oppositely signed components that sum to zero. Increasing the field in C2 results in very weak derivative shaped MCD. When the field is increased further and the temperature reduced, the MCD takes on an absorptive appearance, as shown in C3.

$$C_o = -\frac{i}{3|d_g|} \text{Im} \sum [\langle A|\mu_x|A\rangle(\langle A|m_y|J\rangle \times \langle J|m_z|A\rangle - \langle A|m_z|J\rangle \times \langle J|m_y|A\rangle)$$

$$+ \langle A|\mu_y|A\rangle(\langle A|m_x|J\rangle \times \langle J|m_z|A\rangle - \langle A|m_z|J\rangle \times \langle J|m_x|A\rangle)$$

$$+ \langle A|\mu_z|A\rangle(\langle A|m_x|J\rangle \times \langle J|m_y|A\rangle - \langle A|m_y|J\rangle \times \langle J|m_x|A\rangle)]$$

Here it is clearly seen that for C-term intensity one must have two mutually orthogonal transition dipoles which themselves are mutually orthogonal to the magnetic field direction, and this results in the proportionality, $C_o \propto g_z M_x M_y + g_y M_x M_z + g_x M_y M_z$. Neese and Solomon (8) have recently expanded on this and have presented elegant expressions for the C-term MCD intensity in cases where the ground state is not orbitally degenerate in first-order, but derives MCD intensity through spin-orbit coupling between ground and excited states. Their approach is a general one which only requires the determination of ground state spin expectation values, $\sum_i \langle S_{x,y,z}\rangle_i$. Furthermore, this formalism makes it very easy to interpret variable-temperature, variable-field (VTVH) MCD data in the non-linear (saturation) limit. They show that there are 3 contributions to the C-term intensity which are given below and depicted graphically in Figure 2.

$$\frac{\Delta\varepsilon^{(1)}}{E} = \frac{\gamma}{S}\vec{l}(\bar{D}^{AJ} \times \Delta\bar{D}^{JA})\Delta_{JA}^{-1}\sum_i N_i\left(\bar{L}_x^{AJ}\langle S_x\rangle_i + \bar{L}_y^{AJ}\langle S_y\rangle_i + \bar{L}_z^{AJ}\langle S_z\rangle_i\right)$$

$$\frac{\Delta\varepsilon^{(2)}}{E} = \frac{\gamma}{S}\sum_{K \neq A,J}\vec{l}(\bar{D}^{KA} \times \bar{D}^{AJ})\Delta_{KJ}^{-1}\sum_i N_i\left(\bar{L}_x^{KJ}\langle S_x\rangle_i + \bar{L}_y^{KJ}\langle S_y\rangle_i + \bar{L}_z^{KJ}\langle S_z\rangle_i\right)$$

$$\frac{\Delta\varepsilon^{(3)}}{E} = \frac{\gamma}{S}\sum_{K \neq A,J}\vec{l}(\bar{D}^{AJ} \times \bar{D}^{JK})\Delta_{KA}^{-1}\sum_i N_i\left(\bar{L}_x^{KA}\langle S_x\rangle_i + \bar{L}_y^{KA}\langle S_y\rangle_i + \bar{L}_z^{KA}\langle S_z\rangle_i\right)$$

Here S is the ground state spin, $\vec{l} = (\sin\theta\sin\phi, \sin\theta\cos\phi, \cos\theta)$, $\bar{D}^{nm}$ are transition dipole matrix elements, N_i is the Boltzmann weighted population of states, and $\bar{L}_{x,y,z}^{nm}$ are spin-orbit matrix elements that connect states n and m. The first contribution is generally quite small since the change in dipole moment $\Delta\bar{D}^{JA}$ is often colinear with the transition dipole moment. The second expression usually dominates, and arises from out-of-state spin orbit coupling between an an excited state $|J>$ and an intermediate state $|K>$. This leads to the MCD "sum rule", requiring the integrated intensity of the MCD over the entire energy spectrum to be zero. Finally, the third contribution to the C-term

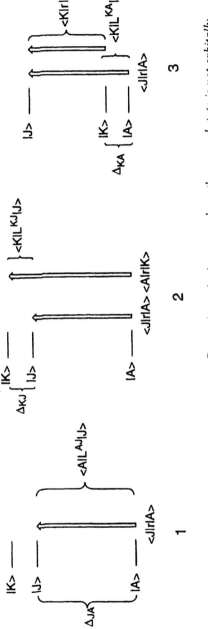

Figure 2. Origin of the three contributions to C-term intensity in cases where the ground state is not orbitally degenerate in first-order, but derives MCD intensity through spin-orbit coupling between ground $|A>$, excited $|J>$, and intermediate states $|K>$. Electric dipole matrix elements are given in bracket notation as $<n|r|m> = \bar{D}^{nm}$. Similarly, the spin-orbit matrix elements in the figure are given as $<m| L_{x,y,z}^{nm} |n>$ which are related to
$$\bar{L}_{x,y,z}^{nm} = \mathrm{Im}\langle nSS|\sum_i h_{x,y,z}(i)s_{o(i)}|mSS\rangle_0 .$$
See text for additional details.

intensity derives from out-of-state spin-orbit mixing between an intermediate state |K> and the ground state |A>. This spin-orbit interaction is responsible for violations of the "sum-rule" particularly when the energy of |K> is near that of the ground state.

Information Content of MCD

To obtain the full information content of the technique, MCD spectra are usually interpreted in the context of other available spectroscopic data such as electronic absorption, resonance Raman, EPR, etc. As the MCD dispersion possesses sign, overlapping bands in absorption spectra are often resolved in the MCD. This is of particular importance in highly covalent systems, as they often display multiple, low-lying charge transfer (CT) transitions that overlap with ligand field (LF) bands. The MCD intensity is often expressed in terms of $C_o/\mathcal{D}_o$ ratios, where $\mathcal{D}_o$ is the dipole strength. Since $\mathcal{D}_o$ is small for LF transitions and the metal spin-orbit coupling constant usually dominates over that of the ligand, $C_o/\mathcal{D}_o$ ratios are typically largest for LF bands aiding in there spectral assignment.

The temperature and field dependence of the MCD spectrum (VTVH MCD) contains considerable information regarding the nature of the ground state wavefunction and is therefore complementary to techniques such as EPR and magnetic susceptibility. VTVH MCD allows for the determination of zero-field splitting parameters (D, E), g-values, and for multimetallic systems, exchange coupling constants (J). It is also possible to obtain polarization information regarding optical transitions in randomly oriented samples from an analysis of the VTVH MCD data *(8)*. Finally, MCD is particularly sensitive to paramagnetic chromophores due to the temperature dependence of the C-term.

As a result, weakly absorbing paramagnetic centers present in metalloproteins that also possess strongly absorbing diamagnetic chromophores may be probed site-selectively by temperature difference MCD methods *(11,12)*. Below we detail how MCD spectroscopy has been used to gain additional insight into the electronic structure of two pyranopterin Mo enzymes, xanthine oxidase and sulfite oxidase.

Freeze-Quench MCD of the Xanthine Oxidase "Very-Rapid" Intermediate.

Absorption spectra of xanthine oxidase are dominated by contributions from the two 2Fe-2S clusters and FAD, masking the much weaker spectral contributions arising from the Mo active site *(12)*. The aerobic xanthine oxidase - 2-hydroxymethlypurine (HMP) reaction kinetics of Hille *(13)* demonstrated that redox equivalents do not accumulate in the 2Fe-2S clusters or FAD upon reaction with HMP. The Mo active site of the "very-rapid" intermediate is in the paramagnetic Mo(V) oxidation state while the 2Fe-2S clusters and FAD are anticipated to be fully oxidized. This results from the facile removal of reducing equivalents from these endogenous redox cofactors due to efficient electron transfer to dioxygen under these conditions. Therefore, MCD spectroscopy is an ideal technique for probing the active site electronic structure of this important catalytic intermediate. However, the kinetic work *(13)* was performed in aqueous solution, while low-temperature MCD data must be collected on enzyme samples prepared in a way to ensure that a high-quality optical glass is formed. We often use PEG-400 as an effective glassing agent/cryoprotectant in low-temperature MCD studies *(11,12)*. However, the high viscosity of these solutions may retard the mass transfer rates during catalysis which could result in additional redox equivalents being present in the Mo(V)-P intermediate sample. Of the reduced metal sites, only the 2Fe-2S clusters are paramagnetic. Both the Mo(IV) center and $FADH_2$ are diamagnetic, and the reduced 2Fe-2S cluster $S_T = 1/2$ paramagnetism results from the strong antiferromagnetic exchange coupling between Fe(II) (S = 2) and Fe(III) (S = 5/2) centers within each dimer *(14)*. Therefore, it is essential to determine the C-term MCD spectral signature of the paramagnetic reduced 2Fe-2S clusters, *a priori*, and this is shown in Figure 3. The intense temperature dependent C-term spectrum of XO_{red} is characteristic of reduced 2Fe-2S clusters with ground state spin $S_T = 1/2$.

The temperature difference (4.7K - 80K) MCD spectra of a typical xanthine oxidase "very rapid" intermediate preparation is shown as the dashed line in

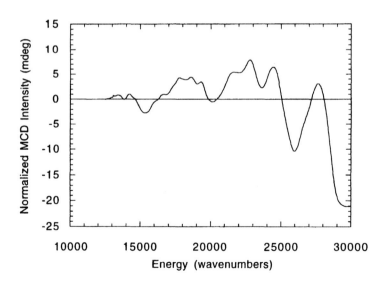

Figure 3. 4.7K minus 80K difference spectrum yielding the C-term MCD of reduced 2Fe-2S clusters in XO_{red}. Adapted from reference 11.

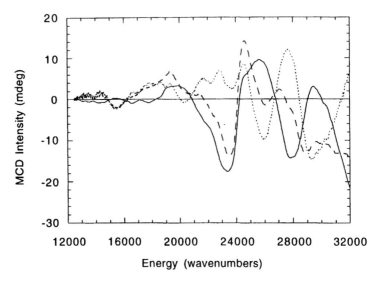

Figure 4. Temperature-dependent MCD spectrum of the "very rapid" Mo(V)-
$P_{product}$ center of XO (solid) obtained by subtraction of the temperature-
dependent XO_{ox} spectrum (dotted) from the temperature-dependent spectrum of
the xanthine oxidase intermediate (dashed). Adapted from Reference 11.

Figure 4. The difference spectrum represents the pure C-term contribution to the total MCD dispersion, and includes contributions from all paramagnetic chromophores present in the sample. The Mo(V)-$P_{product}$ intermediate giving rise to the "very rapid" EPR signal is an S=1/2 system, however, reduced 2Fe-2S clusters and FAD• semiquinone radical are also possible contributors to the C-term MCD spectrum. We observe that the MCD of oxidized and dithionite reduced XO preparations are essentially identical, differing only in relative intensity. This indicates that the oxidized sample contains a small, but measurable, quantity of paramagnetic reduced 2Fe-2S clusters.

Since the reduced enzyme component present in the oxidized XO sample is catalytically unreactive toward the HMP substrate it is also present at the same concentration in the "very-rapid" intermediate sample as well (Figure 4, dotted). Therefore, the contribution of reduced enzyme to the C-term MCD spectrum of the intermediate can be removed by spectral subtraction, and this is shown as the solid line in Figure 4. The C-term difference spectrum originates solely from electronic transitions arising from the Mo(V) center. A comparison of the MCD spectrum of the "very rapid" intermediate with that of the model compound $(LN_3)MoO(bdt)$ is given in Figure 5. There exists a remarkably close spectral correspondence over a relatively large energy range. This is a surprising result, since the X-ray structure of oxidized XO related aldehyde oxidoreductase from

D. gigas has been interpreted as possessing a terminal sulfido ligand oriented *cis* to the ene-1,2-dithiolate chelate and *cis* to the terminal oxo. We argue that the similarity of spectral features over such a large energy range can only be the result of a nearly identical manifold of excited states *(11)*. This is important, since the energy and electronic description of each individual excited state

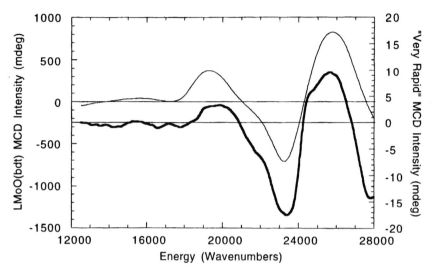

Figure 5. Comparison of the MCD spectrum from the Mo(V) product bound xanthine oxidase intermediate (bold line, right) at 4.7K/7T with the intensity scaled solid-state mull MCD spectrum of (LN₃)MoO(bdt) in poly(dimethylsiloxane) (light line, left) at 4.86K/7T. Adapted from Reference 11.

derives from the nature of the metal-ligand bonding as well as the stereochemical arrangement of the donor ligands at the molybdenum site.

We have recently undertaken a detailed electronic absorption, resonance Raman, and MCD spectroscopic study on a series of (LN_3)MoO(dithiolate) complexes including (LN_3)MoO(bdt) *(15-19)*. This work shows that all of the observed transitions below ~30,000 cm^{-1} in (LN_3)MoO(bdt) derive from ligand-to-metal charge transfer (LMCT) transitions originating from the coordinated dithiolate ligand. The assignment of the 19,000 cm^{-1} band as an in-plane dithiolate $S_{ip} \rightarrow$ Mo d_{xy} LMCT transition is based on the resonance enhancement of highly coupled in-plane Mo-S stretching and bending modes *(15)*. The higher energy LMCT transitions at 22,000 cm^{-1} and 24,000 cm^{-1} have been assigned as out-of-plane dithiolate $S_{op} \rightarrow$ Mo $d_{xz,yz}$ transitions consistent with their pseudo-$\mathcal{A}$ term MCD bandshapes and resonance enhancement of the high frequency Mo≡O stretch accompanying excitation into these bands *(15)*. Given the strong correlation in MCD spectral features of $(LN_3)MoO(bdt)$ and the enzyme, the same spectral assignments have been applied to the "very rapid" intermediate.

The MCD spectral similarity between $(LN_3)MoO(bdt)$ and the "very rapid" intermediate strongly suggests that the stereochemical relationship between the $Mo\equiv O$ bond and the coordinated ene-1,2-dithiolate portion of the pyranopterin in XO is the same as in $(LN_3)MoO(bdt)$. Since the $Mo\equiv O$ bond is oriented *cis* to the dithiolate chelate in $(LN_3)MoO(bdt)$, while the crystallographic interpretation indicates that the terminal sulfido ligand is oriented *cis* to the ene-1,2-dithiolate plane in oxidized XO related aldehyde oxidoreductase, we propose that a change in Mo coordination geometry occurs during catalysis. This initial catalytic event may be envisioned as either a covalent or non-covalent association of the substrate at the active site which triggers a conformational change at the Mo center and drives the oxo ligand out of the ene-1,2-dithiolate plane and into a position normal to it yielding the MoO(dithiolate) stereochemistry observed in *(LN₃)MoO(bdt)*. Covalent interactions between the redox active Mo d_{xy} orbital and the S_{ip} orbitals of the dithiolate are maximized when the oxo ligand is oriented *cis* to the dithiolate plane *(15)*, underlying the importance of the ene-1,2-dithiolate portion of the cofactor in catalysis. Furthermore, there is mounting evidence that efficient coupling of the reduced Mo site into protein mediated superexchange pathways *(15)* is facilitated by the pyranopterin, as the structure of *D. gigas* aldehyde oxidoreductase shows the amino group of the pterin is hydrogen bonded to a cysteinyl sulfur of a 2Fe-2S cluster *(3,4)*. We hypothesize that the presence of substrate/product at the active site maximizes the magnitude of the electron transfer coupling matrix element (H_{DA}) between donor and acceptor sites by virtue of a conformationally induced increase in covalent interactions between the Mo d_{xy} redox orbital and the S_{ip} orbitals of the ene-1,2-dithiolate chelate portion of the pyranopterin.

Finally, a brief discussion of the catalytically essential sulfido ligand is warranted, since the interpretation of the MCD spectrum of the "very rapid" intermediate has been restricted to the energy region for which corresponding spectral features were observed in $(LN_3)MoO(bdt)$. The MCD spectra of the enzyme and model do diverge at energies $>28,000$ cm . This is likely due, in part, to the lack of a terminal sulfido ligand in $(LN_3)MoO(bdt)$. We therefore anticipate $S_{sulfido} \rightarrow$ Mo LMCT transitions to occur at energies greater than $28,000$ cm^{-1}. EPR studies of xanthine oxidase enriched with [33]S at the sulfido position show strong and anisotropic coupling of the [33]S nuclear spin to the unpaired Mo(V) electron *(20)* and this is interpreted to result from a very covalent Mo d_{xy} - sulfido $S_{p\pi}$ bonding interaction. As such, one might expect the energy of the Mo d_{xy} orbital to be destabilized relative to the $S_{dithiolate}$ orbitals. Surprisingly, the MCD data for the intermediate suggests that the sulfido ligand increases the energy of the dithiolate $S_{ip} \rightarrow$ Mo d_{xy} LMCT transition by only ~ 500 cm^{-1} relative to that in $(LN_3)MoO(bdt)$. Currently, we are performing detailed spectroscopic studies on $[MoOS]^{2+,1+}$ model complexes which possess a terminal sulfido ligand in order to address this issue *(21)*.

MCD Spectroscopy of the Mo(V) Site in Chicken Sulfite Oxidase.

We have also used VT difference MCD to probe the electronic structure of the Mo active site in chicken SO *(12)*, as vertebrate SO possesses a strongly absorbing *b*-type heme. The enzyme may be poised in the catalytically relevant [Mo(V):Fe(II)] state by anaerobic reduction of the enzyme with sulfite, in the absence of the physiological oxidant cytochrome *c*. The [Mo(V):Fe(II)] state is a proposed catalytic intermediate in the oxidative half reaction, where SO is reoxidized to the resting [Mo(VI):Fe(III)] state by two sequential one-electron transfers to cytochrome *c via* an endogenous *b*-type heme. We have acquired low temperature MCD spectra of both the fully oxidized [Mo(VI):Fe(III)] and the 2e⁻ reduced SO in order to quantify the spectral contributions of the heme domain, and to ensure that no ferric contamination was present in the [Mo(V):Fe(II)] spectrum upon incubation with excess sulfite. As anticipated, the ferric and ferrous *b*-type heme MCD spectra are essentially identical to those reported for various cytochromes *b (22-24)*. The MCD spectrum of the [Mo(VI):Fe(III)] state is extremely temperature dependent (*C*-term behavior) in the B- and Q-band regions due to the S=1/2 paramagnetism of the oxidized *b*-type heme originating from the low spin ferric iron. However, the MCD spectrum of the [Mo(V):Fe(II)] state is dominated by temperature independent A- and B- terms, in addition to the weak *C*-terms arising from the paramagnetic Mo(V) active site. Contamination of the catalytically relevant [Mo(V):Fe(II)] spectrum with even small amounts of the ferric heme would be problematic as the weak C-term features anticipated to arise from the Mo(V) active site would be obscured by the intense *C*-term features originating from the ferric heme. Fortunately, we found that the addition of excess sulfite results in complete reduction of the heme to the ferrous state. The absence of ferric heme contributions in the MCD spectrum of sulfite-reduced SO allows the direct use of temperature difference MCD techniques to reveal the intrinsic *C*-term contributions arising solely from the Mo(V) active site in the [Mo(V):Fe(II)] state. Figure 6 shows the 1.8 K minus 40 K difference MCD spectrum of sulfite reduced SO. The [Mo(V):Fe(II)] EPR spectrum generated from this preparation is identical to that of *hpH* SO *(25)* and confirms that the *C*-term MCD signal arises from the Mo site. The intense 𝒜-term MCD and high oscillator strength of the ferrous heme B- and Q-bands preclude the observation of a [Mo(V):Fe(II)] MCD signal in certain regions of the spectrum. However, a substantial spectral window exists, and three MCD *C*-terms are observed at 22,250 cm⁻¹ (positive C-term), 26,500 cm⁻¹ (positive *C*-term), and 31,000 cm⁻¹ (negative *C*-term). Interestingly, the MCD spectra of both *hpH* and *lpH* SO are indistinguishable from one another, implying that the differences observed in the EPR spectra of *lpH* and *hpH* SO are not be discernable by MCD.

A key observation is that *no low energy CT bands occur below 17 000 cm^{-1} in the enzyme*, in marked contrast to the low-energy CT features observed in (LN$_3$)MoO(dithiolate) model compounds *(15-19)*. This observation was unexpected, and suggests significant electronic structure differences between

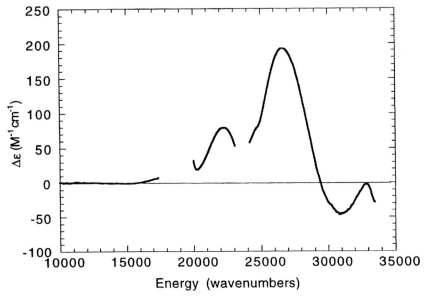

Figure 6. Temperature difference MCD of the [Mo(V):Fe(II)] state of SO. Adapted from Reference 12.

current models and the [Mo(V):Fe(II)] active site of SO. One possible explanation suggested for the lack of observable low energy CT features in the [Mo(V):Fe(II)] state is that there is poor orbital overlap between the dithiolate S_{dt}^{op} orbitals and the in-plane Mo d_{xy} redox orbital, and negligible mixing between ene-1,2-dithiolate S_{dt}^{ip} and S_{dt}^{op} orbitals *(15)*.

$S_{cys} \rightarrow$ Mo d_{xy} CT transitions are also anticipated to occur at energies below ~20 000 cm^{-1}, and this is supported by detailed spectroscopic studies on [PPh$_4$]MoO(SPh)$_4$ and (L-N_2S_2)MoO(SR) *(26-28)*. These compounds exhibit low energy, high oscillator strength $S_{cys} \rightarrow$ Mo d_{xy} CT transitions which are easily observed in their MCD spectra as temperature dependent *C*-terms. The intensity of the low energy $S_{cys}^{v} \rightarrow$ Mo d_{xy} CT transition should be proportional to the square of the $<S_{cys}^{v}|d_{xy}>$ overlap integral. Since no MCD bands are observed in the enzyme below ~17 000 cm^{-1}, a 90° O-Mo-S$_{cys}$-C dihedral angle with

$< S^v_{cys}|d_{xy}>^2 \sim 0$ is implied for the S_{cys} - Mo interaction in the [Mo(V):Fe(II)] state of SO. It is satisfying to note that the spectroscopically determined S_{cys}-Mo bonding description is consistent with the crystallographically determined O-Mo-S-C dihedral angle of $90\pm5°$ (6).

These studies allow for the tentative assignment of MCD bands originating from Mo(V) in the [Mo(V):Fe(II)] state of SO. We have assigned the 22,250 cm^{-1} positive C-term feature as a $S_{cys}\rightarrow$Mo CT transition. The absence of a positive C-term feature at $\sim$22,250 cm^{-1} in the MCD spectra of (L-N_3)MoO(dithiolate) models implies that this transition in SO does not derive from pyranopterin ene-1,2-dithiolate coordination in the enzyme. Although the S^σ_{cys} orbital dominantly interacts with $d_{x^2-y^2}$, the low symmetry environment of the Mo site in SO allows for a small degree of d_{xy} - $d_{x^2-y^2}$ orbital mixing. Therefore, the 22,250 cm^{-1} transition likely arises from a weak $S^\sigma_{cys}\rightarrow$Mo d_{xy} charge transfer. The two higher energy C-term features in SO are more difficult to assign since only recently have model compounds been synthesized that contain both dithiolate *and* thiolate donors in the equatorial plane *cis* to the Mo=O bond (29,30). An MCD feature, similar to that observed at 26,500 cm^{-1} in SO, has been assigned in (L-N_3)MoO(bdt) as the high energy component of an ene-1,2-dithiolate $S^{op}_{dt} \rightarrow$ Mo $d_{xz,yz}$ CT transition (15,31). This implies that the corresponding feature in SO may also be the high energy component of a S^{op}_{dt} $\rightarrow$ Mo $d_{xz,yz}$ transition possessing a positive pseudo-$\mathcal{A}$ term, provided that the lower energy negative component is obscured by the positive C-term at 22,250 cm^{-1}. Finally, we anticipate a positive C-term feature at $\sim$19 000 cm^{-1} arising from the Mo site, which has been assigned as a dithiolate $S^{ip}_{dt}\rightarrow$Mo d_{xy} CT (11,18) in the MCD of (L-N_3)MoO(dithiolate) models. Unfortunately, this band is most likely buried under the strong Q-band MCD arising from the ferrous heme. Ongoing MCD studies in our laboratories on new (L-N_2S)MoO(dithiolate) models should help address this issue.

The spectroscopic results support our hypothesis that the S^{ip}_{dt} orbitals of the ene-1,2-dithiolate couple the molybdenum center into efficient ET pathways involving the σ-orbitals of the pyranopterin. The absence of low energy CT transitions in the MCD spectrum of SO suggests that the geometry of the active site maximizes the dithiolate S^{ip}_{dt}-Mo d_{xy} orbital overlap while significantly reducing the out-of plane S^v_{cys}-Mo d_{xy} bonding interaction. The X-ray structure of chicken liver sulfite oxidase suggests a O-Mo-S_{cys}-C dihedral angle of $\sim$90° in oxidized enzyme, and this result is fully consistent with the lack of observable low energy CT transitions in the MCD spectrum of the [Mo(V):Fe(II)] state. The obvious implication is that the S_{cys} does not play a direct role in ET regeneration of the active site, since this geometry precludes a significant S_{cys} - d_{xy} bonding interaction. However, the S_{cys} should still be an effective σ-donor and indirectly affect the reduction potential of the active site by σ-mediated charge donation to the empty d_x2_-y2 orbital. This would result in a reduction of the effective nuclear charge on Mo, and poise the active site at more negative reduction potentials. Thus, the S_{cys}-Mo d_{xy} bonding description in SO appears to be radically different from the S_{cys}-Cu d_x2_-y2 bonding scheme in blue copper proteins *(32)* and underscores the very different function of S_{cys} in these metalloproteins.

Conclusions

MCD spectroscopy is a powerful spectroscopic method for selectively probing paramagnetic active sites in metalloenzymes, particularly in the presence of additional, highly chromophoric endogenous redox centers. The MCD data presented here represent initial MCD studies of XO and SO. The interpretation of the enzyme MCD is aided by parallel studies on small molecule analogues, in conjunction with the application of additional complementary electronic structure probes such a EPR/ENDOR, electronic absorption, XAS, resonance Raman, and photoelectron spectroscopies.

Acknowledgements

The author would like to thank Prof. John H. Enemark and Prof. Russ Hille for wonderful ongoing collaborations and numerous stimulating conversations. Also, a special thank-you is extended to all of the students and postdoctoral researches who have been involved in our collaborative studies on SO and XO. Finally, M.L.K. would like to acknowledge the National Institutes of Health (GM-057378) for generous support of the work detailed in this contribution.

References

1. Hille, R. *Chem. Rev.* **1996**, *96*, 2757-2816.
2. Hille, R. *JBIC* **1996**, *1*, 397-404.
3. Huber, R.; Hof, P.; Duarte, R.O.; Moura, J.J.G.; Moura, I.; Liu, M.-Y.; LeGall, J.; Hille, R.; Archer, M.; Romao, M.J. *Proc. Natl. Acad. Sci., USA* **1996**, *93*, 8846-8851.
4. Romao, M.J.; Archer, M.; Moura, I.; Moura, J.J.G.; LeGall, J.; Engh, R.; Schneider, M.; Hof, P.; Huber, R: *Science* **1995**, *270*, 1170-1176.
5. Turner, N.A.; Bray, R.C.; Diakun, G,P. *Biochem. J.* **1989**, *260*.563-571.
6. Kisker, C.; Schindelin, H.; Pacheco, A.; Wehbi, W.A.; Garrett, R.M.; Rajagopalan, K.V.; Enemark, J.H.; Rees, D.C. *Cell* **1997** *91*, 973-983.
7. Solomon, E. I.; Pavel, E. G.; Loeb, K. E.; Campochiaro, C. *Coord. Chem. Rev.* **1995**, *144*, 369.

8. Neese, R.; Solomon, E. I. *Inorg. Chem.* **1999**, *38*, 1847-1865.
9. Gerstman, B. S.; Brill, A. S. *J. Chem. Phys.* **1985**, *82*, 1212.
10. *Group Theory in Spectroscopy with Applications to Magnetic Circular Dichroism;* Sean P. McGlynn, Ed.; Wiley-Interscience Monographs in Chemical Physics: New York, 1983.
11. Jones, R.M.; Hille, R.; Kirk, M. L. *Inorg. Chem.*, **1999**, *38*, 4963-4970.
12. Helton, M.E.; Pacheco, A.; McMaster, J.; Enemark, J.H.; Kirk, M.L. *J. Inorg. Biochem.* **2000**, *80*, 227-233.
13. McWhirter, R. B.; Hille, R. *J. Biol. Chem.* **1991**, *266*, 23724.
14. Spencer, J.T., *Coord. Chem Rev.*, **1983**, *48*, 59-82.
15. Inscore, F. E.; McNaughton, R.; Westcott, B. L.; Helton, M. E.; Jones, R.; Dhawan, I. K.; Enemark, J. H.; Kirk, M. L. *Inorg. Chem.* **1999**, *38*, 1401-1410.
16. Helton, M. E.; Gebhart, N. L.; Davies, E. S.; McMaster, J.; Garner, C. D.; M. L. Kirk, *J. Am. Chem. Soc.*, **2001**, *123*, 10389-10390.
17. Helton, M. E.; Gruhn, N.; McNaughton, R. L.; Kirk, M. L. *Inorg. Chem.* **2000**, *39*, 2273-2278.
18. Helton, M. E.; Kirk, M. L. *Inorg. Chem.* **1999**, *38*, 4384-4385.
19. Inscore, F. E.; Rubie, N.; Joshi, H. K.; Kirk, M. L.; Enemark, J. H. *submitted for publication.*
20. Malthouse, J. P.; George, G. N.; Lowe, D. J.; Bray, R. C. *Biochem. J.* **1981**, *199*, 629-637.
21. Rubie, N.; Doonan, C.; George, G. N.; Young, C. G.; Kirk, M. L. *Manuscript in preparation.*
22. Fridén, H.; Cheesman, M. R.; Hederstedt, L.; Andersson, K. K.; Thomson, A. J. *Biochim. et Biophys. Acta* **1990**, *1041*, 207-215.
23. Moore, G. R.; Williams, R. J. P.; Peterson, J.; Thomson, A. J.; Mathews, F. S. *Biochim. et Biophys. Acta* **1985**, *829*, 83-96.
24. Goldbeck, R. A.; Einarsdottir, O.; Dawes, T. D.; O'Connor, D. B.; Surerus, K. K.; Fee, J. A..; Kliger, D. S. *Biochemistry* **1992**, *31*, 9376-9387.
25. Dhawan, I. K.; Enemark, J. H. *Inorg. Chem.* **1996**, *35*, 4873-4882.
26. McMaster, J.; Carducci, M. D.; Yang, Y.-S.; Enemark, J. H.; Solomon, E. I. *Inorg. Chem.* **2000**, *39*, 5697-5706.
27. R. L. McNaughton, R. L.; Tipton, A. A.; Conry, R. R.; Kirk, M. L. *Inorg. Chem.* **2000**, *39*, 5697-5706.
28. McNaughton, R. L.; Cosper, M. M.; McMaster, J.; Enemark, J. H.; Kirk, M. L. *Submitted.*
29. Peariso, K.; Chohan, B. S.; Carrano, C. J.; Kirk, M. L. *Submitted for publication.*
30. Lim, B. S.; Willer, M. W.; Miao, M. M.; Holm, R. H. *J. Am. Chem. Soc.* **2001**, *123*, 8343-8349.
31. Carducci, M. D.; Brown, C.; Solomon, E. I.; Enemark, J. H. *J. Am. Chem. Soc.* **1994**, *116*, 11856-11868.
32. Holm, R. H.; Kennepohl, P.; Solomon, E. I. *Chem. Rev.* **1996**, *96*, 2239-2314.

Chapter 20

X-ray Magnetic Circular Dichroism: A Primer for Chemists

Stephen P. Cramer

Department of Applied Science, EU 111, University of California, Davis, CA 95616

X-ray magnetic circular dichroism – XMCD – is the difference in absorption of left- and right-circularly polarized x-rays by a magnetized sample. Although MCD with x-rays is only about 15 years old, the physics is essentially the same as for UV-visible MCD that has been known since 1897. For (bio)inorganic chemists and materials scientists, XMCD has the advantage of elemental specificity that comes with all core electron spectroscopies. Thanks to simple sum rules, XMCD can provide quantitative information about the distribution of spin and orbital angular momentum. Other strengths include the capacity to determine spin orientations from the sign of the XMCD signal, to infer spin states from magnetization curves, and the ability to separate magnetic and non-magnetic components in heterogeneous samples. With new synchrotron radiation sources and improved end stations, XMCD measurements on biological samples are, if not routine, at least no longer heroic. One goal of this review is to encourage chemists, materials scientists, and biologists to consider XMCD as an approach to understanding the electronic and magnetic structure of their samples.

Introduction

X-ray magnetic circular dichroism, XMCD [1], measurements compare the relative absorption of left- and right-circularly polarized [2] x-rays by magnetized samples (**Figure 1**). Circularly polarized x-rays have oscillating electric and magnetic fields that are 90 degrees out of phase (**Equation 1**). We use the convention of Born and Wolf [1], in which the instantaneous electric field E_{rcp} for a right circularly polarized photon propagating in the z direction resembles a right-handed screw (**Figure 1**).

$$\vec{E}_{rcp} = E_0 \{ \sin[\omega t - kz + \phi_0] \mathbf{i} + \cos[\omega t - kz + \phi_0] \mathbf{j} \} \qquad (1)$$

In this equation, ω is the angular frequency, $\omega = 2\pi\nu$, k is the wave number, $k = 2\pi/\lambda$ where λ is the wavelength, ϕ_0 is an arbitrary phase shift, and $\mathbf{i}$ and $\mathbf{j}$ are unit vectors along the x and y axes respectively. With the above definition, it turns out that left circularly polarized photons carry $\hbar$ angular momentum.

Although the Born and Wolf convention is standard for optics and chemistry literature, most physics literature uses the opposite definition, and one should check how the polarization is defined if the sign of the MCD effect is to be meaningful. The papers of deGroot and Brouder generally use the Born and Wolf convention, while those of Thole, van der Laan, and Carra use the physics or 'Feynman' definition [2]. The pitfalls of describing circular polarization have been cogently described by Kliger *et al.* [3].

In 1690, Huygens discovered that either of the two light rays refracted by a calcite crystal could be extinguished by rotation of a second 'analyzer' crystal [4]. One evening, more than a century later (1808), Malus observed that sunlight reflected from a window pane had similar properties, and by analogy with magnetic bodies, he called this light 'polarized' [5]. Although these early observations were first interpreted in terms of Newton's 'corpuscular' theory [6], by the 1820's Fresnel had developed the mathematical foundation for polarized light in terms of two perpendicular transverse waves [7]. In 1846, Faraday demonstrated rotation of the plane of polarization induced by a magnetic field [8,9]. This 'Faraday effect' or 'magnetic optical rotation' is the result of circular birefringence – which is a difference in the real part of the index of refraction for left and right circular polarization. Prompted in part by these results, the theoretical work of Maxwell completed the picture of light as an electromagnetic wave [10].

The first demonstration of a magnetically induced difference in absorption (the imaginary part of the complex index of refraction) came with the Nobel prize-winning work of Zeeman [2]. After first observing both linear and magnetic circular dichroism in the sodium D emission lines in a magnetic field

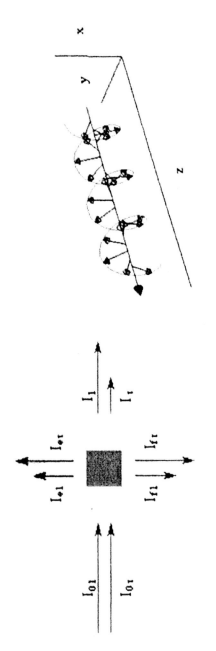

Figure 1 – (Left) Schematic of XMCD experiment. I_{0r} and I_{0l} are respectively the incident beam intensities for rcp and lcp x-rays, and I_r and I_l are the transmitted intensities. The remaining arrows correspond to rcp and lcp electron or fluorescence yields. (Right) Illustration of the electric field direction along the propagation axis for right circularly polarized light.

[11], he reversed the experiment and observed MCD effects on the D absorption lines of Na vapor [12]. After the discovery of x-rays by Röntgen in 1895 [13], attempts were made to observe magnetic effects on x-ray spectra [14,15]. However, a successful experiment would have to wait another 80 years. The modern history of XMCD begins with Erskine and Stern predicting a magneto-optical Kerr effect (MOKE) for ferromagnetic Ni at the $M_{2,3}$ edge ($3p \rightarrow 3d$) [16]. Subsequent attempts to see XMCD in a GdFe alloy at the Gd L_3 edge were unsuccessful [17]. A year later, Thole, van der Laan, and Sawatzky predicted strong XMCD and x-ray magnetic linear dichroism (XMLD) in the $M_{4,5}$ ($3d \rightarrow 4f$) edges of rare earths [18], and the latter was reported in 1986 [19]. The XMCD effect was finally observed at the K-edge of metallic Fe by Schütz and coworkers in 1987 [20]. A much stronger soft x-ray MCD at the Ni $L_{2,3}$ edge was reported in 1990 [21] (**Figure 2**). Our group reported the first XMCD for a paramagnetic metalloprotein in 1993 [22]. Since then, the growth of XMCD for materials science applications has been explosive [23], leading to more than 1000 papers over the past decade.

Experimental Considerations

"In principle therefore all polarization experiments which are
possible with visible light can be performed with x-rays.
... it is in practice difficult to obtain sufficient intensity "
Skalicky and Malgrange, 1972 [24]

The key ingredients for an XMCD measurement include (1) a source of circularly polarized x-rays, (2) a monochromator and optics (a 'beamline'), (3) a means for producing a magnetized sample, and (4) an x-ray absorption detection system. Items (3) and (4) are considered the 'endstation'.

Sources of Circularly Polarized X-Rays

Before the introduction of synchrotron radiation beamlines, sources of circularly polarized high energy photons were exotic, such as magnetically oriented radioactive nuclei [25,26] and astronomical synchrotron radiation [27]. Although the concepts behind x-ray circular polarizers had been demonstrated with Cu Kα radiation [28], the resultant beams were not bright enough (~15 photons/sec) for practical applications.

Bend Magnets. The bend magnets associated with particle storage rings are the simplest sources of circular polarization. By viewing the electron (or positron) beam off-axis, one observes a charge accelerated in an elliptical orbit, while viewing the beam on-axis reveals only a horizontal component to the acceleration (**Figure 3**). Thus, the synchrotron radiation emitted from bend magnets is highly polarized – ranging from pure linear polarization in the plane

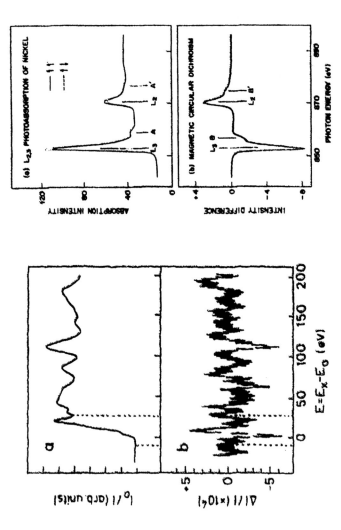

Figure 2 – The first Fe K-edge XANES and EXAFS (different scales) XMCD spectrum, from Schütz et al. [20]. (Right) The first soft x-ray MCD spectrum, reported for Ni metal by Chen and coworkers [21].

of the orbit to nearly circular far out of the plane. Borrowing from Kim [29], the relative amplitudes of the horizontal and vertical electric field components (E_x and E_y) are given by **Equation 2**:

$$\begin{pmatrix} E_x \\ E_y \end{pmatrix} = \begin{pmatrix} K_{2/3}(\eta) \\ \dfrac{i\gamma\psi \; K_{1/3}(\eta)}{\sqrt{1+(\gamma\psi)^2}} \end{pmatrix} \tag{2}$$

where $K_{1/3}$ and $K_{2/3}$ are modified Bessel functions, y is the ratio of photon energy to the critical energy, γ is the ratio of the electron energy to its rest mass energy, and $\eta = (y/2) \, [1+ (\gamma y)^2]^{3/2}$. Defining r as the ratio of the minor to major axes of the polarization ellipse, given by $r = E_y \, / \, iE_x$, yields the degree of circular polarization P_c (defined as P_3 by Kim): $P_c = 2r \, /(1+r^2)$ [29].

Although bend magnets can provide any desired degree of polarization, this comes at a price – the flux falls dramatically as ψ increases. The strongest XMCD is obtained with pure circular polarization, but as $P_c \rightarrow 1$, $I \rightarrow 0$. One therefore has to make a tradeoff between flux and polarization. A figure of merit for most XMCD experiments is P^2I, and the angle for optimal P^2I depends on the photon energy and the critical energy of the ring (**Figure 3**). Apart from limited P_c, other drawbacks of bend magnets are modest brightness and the emission of lcp and rcp in different directions. Better XMCD measurements can be done with insertion device beamlines.

Insertion devices. Insertion devices are magnetic structures 'inserted' into straight sections of the storage ring lattice to produce synchrotron radiation with special characteristics. Although other insertion devices, such as asymmetric wigglers, crossed undulators, and helical undulators [30,31] are sometimes used, the elliptical undulator (EPU) is the most successful device for the production of circularly polarized synchrotron radiation. In an EPU, the magnetic field vector rotates as a particle passes through the device, causing the particle to spiral about a central axis. Both electromagnetic and permanent magnet versions have been developed. Permanent magnet EPUs consist of 4 banks of magnets – two on top and two below (**Figure 4**).

The peak energy of the undulator output is changed by varying the vertical separation between the magnet assemblies, a so-called 'gap scan', while the polarization is varied by changing the relative positions (phases) of adjacent rows of magnets – a 'row scan'. In the case of the EPU on ALS beamline 4.0.1, the polarization can be changed from left to right circular polarization in a few seconds, and the peak energy can be varied as quickly as the monochromator scans [32].

Soft X-Ray Beamlines. In the soft x-ray region, beamlines use grazing incidence mirrors and gratings. In these geometries, the source polarization is almost completely preserved as it passes through the optics. For example, ALS beamline 4.0.1 employs an initial horizontally deflecting toroid (M1) at 2°, a plane pre-mirror (M3) at ~3°, a plane grating at variable glancing angle, a

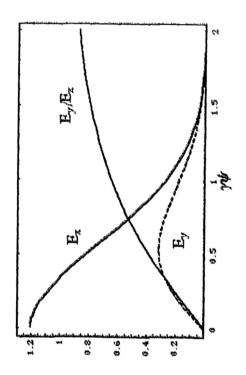

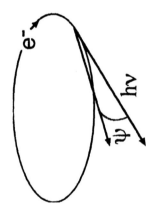

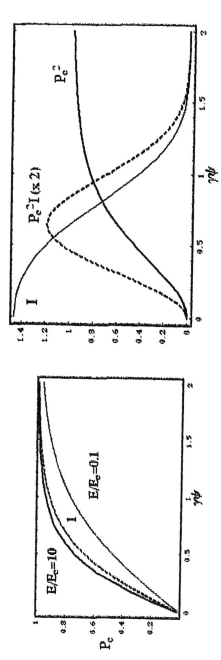

Figure 3 – (Top left) Definition of the observation angle ψ for synchrotron radiation vs. the orbital plane. (Top right) Amplitudes and ratio of x- and y-components of the electric field vs. vertical viewing angle ($\gamma\psi$). (Bottom left) Degree of circular polarization (P_c) as function of $\gamma\psi$ □□□ ratio of photon energy to critical energy (E/E_0). (Bottom right) Tradeoff between flux (I) and polarization (P_c), and figure of merit ($P_c^2 I$).

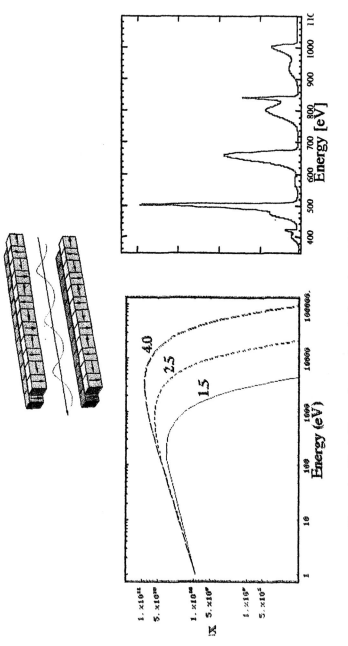

Figure 4 – (Top) Charged particle path through an EPU. Arrows represent field direction of permanent magnets. (Bottom left) Calculated spectra for different electron beam energies and bend magnet with 12.7 m bend radius (SSRL). Flux is in units of photons sec^{-1} mrad^{-1} mA^{-1} 0.1% bandwidth^{-1}. (Right) Measured output of ALS EPU.

cylindrical magnifying mirror (M4) at ~1°, and a final refocusing mirror (M5) at ~1° (**Figure 5**). This beamline provides ~10^{12} photons/sec with $\Delta E/E \sim 10,000$ from 50-2000 eV [32].

Hard X-Ray Beamlines and Quarter Wave Plates. Hard x-ray applications employ crystal monochromators, and the effect of the crystal optics on the beam polarization can be considerable. A well-known phenomenon with visible light reflection is the so-called Brewster angle – the angle of incidence at which the electric field in the plane of incidence (π component) of a reflected beam is totally suppressed. This results in pure linear polarization along the out-of-plane or σ direction. The same phenomenon can occur with Bragg reflection from crystals. After all the hard work of producing perfect circular polarization from a hard x-ray EPU, the crystal optics can degrade the degree of circular polarization and in some cases produce pure horizontal polarization!

The solution is the same approach used with UV-visible MCD experiments: start with linear polarization and convert to circular polarization with a retarder, commonly a quarter wave plate. X-ray quarter wave plates exploit the birefringence of crystals for σ and π electric field components for geometries on or close to diffraction conditions. Although early work emphasized on-reflection Bragg or Laue geometries [24,28,33-35], some of the most popular x-ray quarter wave plates now operate in the off-Bragg transmission geometry. In this case, the crystal is adjusted to one of the wings of the Bragg reflection, at an angular deviation $\Delta\Theta$ from the center of the rocking curve for Bragg angle Θ_B. The crystal planes are placed at a 45-degree angle to the incident polarization direction, to allow equal intensity for σ and π components of the electric field. The phase shift $\Delta\phi$, for a sufficiently large $\Delta\Theta$, depends on the difference in indices of refraction ($n_\sigma - n_\pi$), as well as the beam path t and wavelength λ [36]:

$$\Delta\phi = \frac{2\pi}{\lambda}\left(n_\sigma - n_\pi\right)t = -\left|\frac{r_e^2 F_h F_{\bar{h}}}{2\pi V^2}\frac{\lambda^3 \sin 2\theta_B}{\Delta\Theta}\right|t = A\frac{t}{\Delta\Theta}. \tag{3}$$

where F_h and $F_{\bar{h}}$ are structure factors for h and $\bar{h}$ reflections, r_e is the classical electron radius, and V is the unit cell volume. By rocking a low-Z crystal such as diamond [37], Be [36], or LiF [38], from one side of the Bragg reflection to another, the polarization can be switched from lcp to rcp, with a $P_c > 90\%$. APS 4-ID-D is a specific beamline that uses such quarter wave plates (**Figure 6**) [39].

Magnetic Field and Temperature Control

The simplest samples for XMCD experiments are permanent magnets, such as the domains in disk storage devices. These samples can be magnetized separately from the XMCD measurement. For example, Stöhr and coworkers have used the XMCD effect to image magnetic domains in disk storage devices

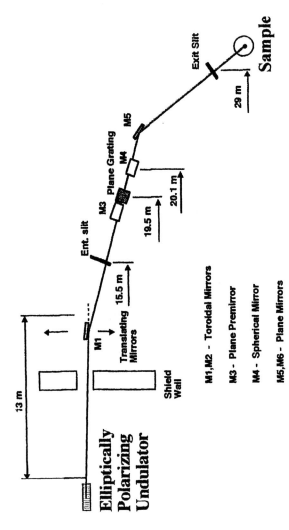

Figure 5 – ALS EPU beamline. M2 (mirror for 2nd undulator) is not yet installed.

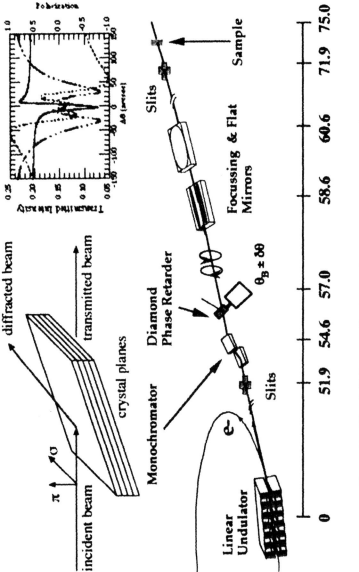

Figure 6 – (Top left) Bragg reflection and transmission geometries for x-ray quarter wave plates. (Top right) Phase shift difference for σ and π components as function of offset (Δθ) from Bragg condition. (Bottom) APS 4-ID-D, a hard x-ray line with quarter wave plates to provide circularly polarized x-rays [39].

[40]. Other groups have magnetized thin film samples *in situ* with a pulsed magnetic field, which is turned off during the experiment [41,42]. Ferromagnets are also easy to study, since only a small field need be applied. For example, the early measurements of Chen and coworkers used an external permanent magnet in close proximity to the Ni foil. The field was switched by manually reversing the external magnet [21]. NdFeB alloy permanent magnets are still used in some experiments to polarize the sample. Electromagnets make this process easier to automate, and several groups have used them in a variety of ways [43].

Superconducting magnets are required to achieve the highest magnetic fields. Our group has employed two different split coil designs. In one instrument, a 6 Tesla split coil is used, with a transverse gap between the coils [22,44]. This device has the advantage that full magnetization of the sample is achieved at 2-3 K, which is readily achieved with pumped ^{4}He cryostats. A temperature of ~0.5 K can be reached by using a ^{3}He inset with this device [45]. It has two limitations. First, the small gap limits the size of the Ge detector that can be introduced between the coils, hence we lose some solid angle of fluorescence collection. More important, the relatively large (31 Henry) inductance limits the rate at which the magnetic field can be reversed.

Our most recent instrument employs a ^{3}He-^{4}He dilution refrigerator, a split-coil 2-Tesla superconducting magnet system, and a 30-element windowless Ge fluorescence detector [46]. This device has a larger gap (8.25 cm), allowing insertion of the 30-element Ge detector close to the sample. It also has a low inductance winding (1.3 Henry), allowing field sweeps from +2T to –2 T in ~10 sec. Several layers of thermal shielding allow temperatures below 0.5 K to be reached routinely (**Figure 7**); in principle, 0.1 K should be possible.

Detection Methods

XMCD is essentially a measurement of relative absorption coefficients, hence all the detection methods used for conventional XAS can in principle be used. In practice, the three most important modes are transmission, fluorescence, and electron yield. As discussed long ago by Lee *et al.* [47], the optimum mode depends on the concentration and spatial distribution of the element under investigation, as well as fluorescence yields and matrix absorption coefficients. Since the factors involved in optimizing hard x-ray measurements are well known, we concentrate on the special requirements for MCD with soft x-rays.

Transmission is the simplest measurement, and it should be used whenever possible. The incident beam intensity (I_0) is often measured using the electron yield or photocurrent from a partially transmitting metal grid. The intensity after the sample (I) can be measured using a second grid, a solid metal plate, or a Si photodiode. Chen and coworkers have done careful transmission measurements of the XMCD of thin metal foils to check the accuracy of the sum rules [48], and with such samples transmission works beautifully. Unfortunately, in the soft x-

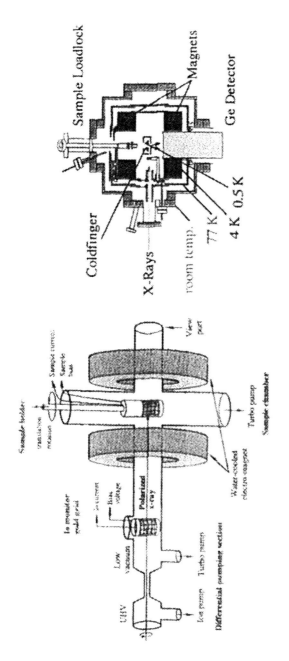

Figure 7 – (Left) A water-cooled electromagnet XMCD system for electron yield detection employed by Stöhr and coworkers [51]. (Right) Sketch of the dilution refrigerator XMCD instrument.

ray region, it is difficult to prepare metal complexes as homogeneous, pinhole-free samples with the required sub-micron thickness. As discussed by Stern [49] and Goulon [50], thick or porous samples will have suppressed absorption features due to 'leakage' effects, and consequently diminished XMCD amplitudes, especially if harmonics are present in the incident beam. Since the particle size of most samples is greater than the $1/e$ ($<1\mu$) path-length, diluting the sample in a low Z powder or mull is of no use.

For concentrated inorganic samples, electron-based detection methods are preferred. One can either measure the electron yield directly with a channeltron electron multiplier (CEM), or indirectly as the photocurrent flowing to the sample from ground. Since electrons are emitted only from approximately the first 25-50 Å of sample, these approaches are very sensitive to oxidation or other surface reactivity. For very high cross sections, they can also suffer from saturation effects [51]. A problem for inorganic chemists is that many coordination complexes are poor conductors at the ~ 4 K temperature required for XMCD of paramagnets. To some extent, this can be overcome by (1) making very thin samples, (2) using a high collection voltage, (3) embedding the sample in a metallic grid, (4) pressing the sample particles into an indium foil, or (5) mixing the sample with a good conductor (Ag or graphite dust). The one generalization we can make from experience is that every sample is different.

Apart from S/N issues, electron methods may suffer from artifacts if magnetic field switching is used to measure the XMCD effect. The trajectories of emitted photoelectrons depend not only on their initial velocity, the applied voltage, and the geometry of collector placement, but also on the magnetic field. The apparent absorption cross section will vary if changing the field affects the fraction of photoelectrons that are accepted. Thus, electron-based XMCD measurements are best done with variable photon polarization.

Fluorescence yield might seem immune from magnetic field artifacts, but most detectors convert x-rays into electrons, and the resolution or gain can be influenced by a strong field. The detector sensitivity needs to be checked before varying the field for XMCD measurements or magnetization curves (**Figure 8**). Furthermore, as discussed by deGroot [52,53] and others [54], the fluorescence-detected excitation spectrum is not necessarily the same as the absorption spectrum, because the fluorescence yield can vary for different excited states. In extreme cases, a line can even be missing from the excitation spectrum (**Figure 8**) [54]! However, as noted by van Veenendaal and others, the effect is not usually that severe for XMCD [55,56]. They point out that, 'although in principle fluorescence yield is unequal to x-ray absorption, in the presence of a crystal field or of strong core-hole spin-orbit coupling fluorescence yield can be used to obtain ground state expectation values of L_z and S_z.' (*vide infra*) [55]. From an experimentalist's point of view, a fluorescence-detected spectrum with known limitations is better than no spectrum at all.

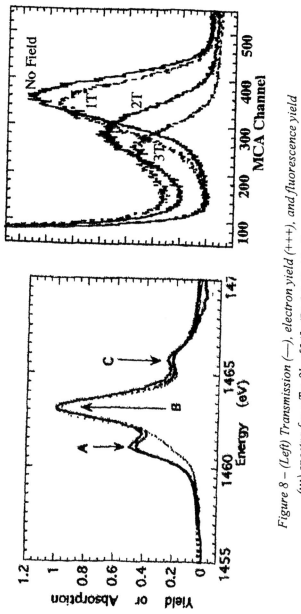

Figure 8 – (Left) Transmission (—), electron yield (+++), and fluorescence yield (···) spectra for a Tm film [54]. (Right) Field effects on Ge detector.

Finally, we should mention that even polarization-switched measurements suffer from potential artifacts. If the effective source point for left- and right-circularly polarized beams is slightly different, this can transform into an energy difference between the two beams at a given monochromator position. The slight mismatch will result in a derivative shaped contribution to the spectrum that is stronger for sharper features. Since this effect is independent of the applied field, one should check that there is indeed no XMCD effect in the absence of sample magnetization. (This presumes there is no 'natural' CD effect, but that is another story [57,58]).

Simplified Theory

"as simple as possible, but not simpler"
Albert Einstein

There are two common pictures used to describe the origin of the XMCD effect – Stöhr and Wu have described these as the '1-electron picture' and the 'configuration picture' [59]. A better name for the latter is the multiplet approach, because XMCD falls out naturally from ligand field multiplet theory. Below we compare the predictions of both pictures.

One-Electron Theory

A large fraction of the XMCD literature, especially papers involving magnetic thin film and metallic samples, uses the 1-electron model along with a 2-step approach to XMCD [59]. In this picture, the first step is to write the initial

Table 1 – Wave functions used in 1-electron model		
1-Electron Label	$|l\,s\,j\,m_j\rangle$ basis: m_l	$Y_l^{m_l}\,\Phi_{m_s}$
$^2p_{1/2}$	1/2	$\frac{1}{\sqrt{3}}(Y_1^0\alpha - \sqrt{2}Y_1^1\beta)$
	-1/2	$\frac{1}{\sqrt{3}}(\sqrt{2}Y_1^{-1}\alpha - Y_1^0\beta)$
$^2p_{3/2}$	3/2	$Y_1^1\alpha$
	1/2	$\frac{1}{\sqrt{3}}(\sqrt{2}\,Y_1^0\alpha + Y_1^1\beta)$
	-1/2	$\frac{1}{\sqrt{3}}(Y_1^{-1}\alpha + \sqrt{2}Y_1^0\beta)$
	-3/2	$Y_1^{-1}\beta$

orbitals for first transition metal L-edges as spin-orbit split $2p$ wave functions, as summarized below in **Table 1** [59].

The next step is to find expressions for the cross section matrix elements for transitions to various states with d symmetry. Assuming a constant radial matrix element R, Stöhr and Wu write these, citing Bethe and Salpeter [60], as:

$$\langle n', l+1, m_l + 1 | P_1^{(1)} | n, l, m_l \rangle = \sqrt{\frac{(l + m_l + 2)(l + m_l + 1)}{2(2l + 3)(2l + 1)}} R \quad (4)$$

and

$$\langle n', l+1, m_l - 1 | P_{-1}^{(1)} | n, l, m_l \rangle = \sqrt{\frac{(l - m_l + 2)(l - m_l + 1)}{2(2l + 3)(2l + 1)}} R, \quad (5)$$

where

$$P_1^{(1)} = \frac{1}{\sqrt{2}} (x + iy) = r\sqrt{\frac{4\pi}{3}} Y_1^1 \quad \text{and}$$

$$P_{-1}^{(1)} = \frac{1}{\sqrt{2}} (x - iy) = r\sqrt{\frac{4\pi}{3}} Y_1^{-1} . \quad (6)$$

These matrix elements are then evaluated for specific d-orbitals in spherical symmetry (Y_2^{ml}, β), and the 'oscillator strength' for different edge and polarization combinations is calculated by summing over $m_l = -2, -1, 0, 1,$ and 2:

$$I_{L_3}^+ = \sum_{i,f} |\langle f | P_1^{(1)} | i \rangle|^2 = \frac{1}{3} R^2 , \quad I_{L_3}^- = \frac{5}{9} R^2 , \quad I_{L_2}^+ = \frac{1}{3} R^2 , \text{ and } I_{L_2}^- = \frac{1}{9} R^2 . \quad (7)$$

Finally, the XMCD effect, for transitions solely to spin-down orbitals, is given by $\Delta I = I^+ - I^-$. For the L_3 edge, $\Delta I_{L3} = -(2/9)R^2$, while at the L_2 edge, $\Delta I_{L2} = +(2/9)R^2$. The key result from these calculations is that with lcp (using our 'optical' definition), *at the L_3 edge* the atom preferentially (5/8 of the time) emits *spin-up* electrons, while *at the L_2 edge* the atom preferentially (3/4 of the time) emits *spin-down* electrons.

The 1-electron model can also be used to explain the spin polarization of K-edges. In these cases, spin-orbit coupling of the final state p-electron is invoked [20,61]. However, as noted by Brouder and Hikam [62], at K-edges the relative amounts of spin-up and spin down states depend on the absorbing atom, the neighboring atoms, and the energy above threshold. Because spin-orbit coupling is much weaker, the degree of spin-polarization is typically 1% or less.

Band Theory and XANES XMCD

Once the spin polarization of the 'emitted' electron is established, the 1-electron model can be used to explain XMCD effects in both the XANES and EXAFS regions. In the 2-step model, the $p_{1/2}$ and $p_{3/2}$ shells are viewed as spin-polarized sources, and vacant spin-up and spin-down $3d$ bands are viewed as spin-sensitive 'detectors' [63]. For example, if only spin down states are available, then the asymmetry, $(\sigma^+ - \sigma^-)/(\sigma^+ + \sigma^-)$, at the L_2 edge should be a 50% effect, twice as large as at the L_3 edge (and opposite in sign). The simplest case, the rigid band 'Stoner model', is illustrated in **Figure 9**.

In a more sophisticated analysis, multiple scattering calculations using FEFF8 source code reproduced most of the Ni $L_{2,3}$ XMCD, including a controversial satellite 6 eV above the main resonance [64]. The latter peak was missing using a 13-atom cluster, but appeared in the 50-atom calculation. A lower energy feature at 3 eV was assigned to many body effects. The authors point out that no approach yet captures all the physics in the XMCD effect.

Scattering Theory and EXAFS XMCD

The EXAFS region is also sensitive to the spin polarization of emitted electrons – scattering by magnetic neighbors depends on the photoelectron polarization. When the neighboring atom is spin polarized, there will be an exchange contribution in addition to the Coulomb scattering potential. This will modify both the amplitude $f_0(\pi,k)$ and phase shift $\phi_0(k)$ for electron backscattering, so that the traditional formula for EXAFS for a single absorber-scatterer interaction requires modification. Schütz and coworkers proposed adding terms $f_c(\pi,k)$ and $\phi_c(k)$ for the magnetic backscattering amplitude and phase shift, scaled by the degree of photoelectron polarization σ_z [61]:

$$\chi(k) = \frac{e^{-2\sigma^2 k^2} e^{-R/\lambda}}{k R^2} |f(\pi,k)| \sin[2kR + \phi(k)] \qquad (8)$$

where:

$$f(\pi,k) = f_0(\pi,k) \pm \langle \sigma_z \rangle f_c(\pi,k) \quad \text{and} \quad \phi(k) = \phi_0(k) \pm \sigma_z \phi_c(k) . \qquad (9)$$

The other terms have their conventional meaning: R is the absorber-scatterer distance, σ is the rms variation in R, k is the magnitude of the photoelectron wave vector, and λ is the photoelectron mean free path.

Ligand Field Multiplet Theory

Ligand field multiplet theory (LFMT) is a multi-electron viewpoint that describes the initial and final states as multiplets that are mixed and split by the

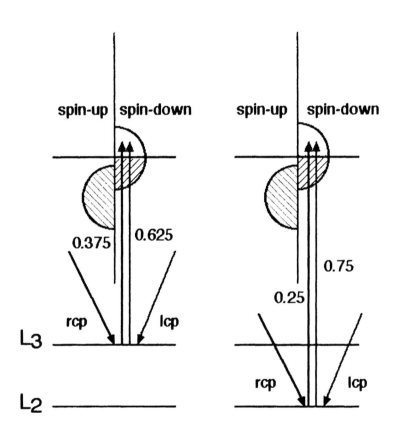

Figure 9 – Origin of L-edge XMCD in the 2-step model.

symmetry of the ligand field [65-69]. In this approach, the XMCD effect emerges naturally as a consequence of angular momentum selection rules. Spectra have been calculated for $3d^1$ through $3d^9$ systems with a range of crystal field and spin-orbital coupling strengths [70,71], so for many inorganic systems one can check beforehand to see the expected XMCD. Before discussing LFMT in detail, we illustrate the differences between 1-electron and multiplet approaches using the same $2p^63d^0 \rightarrow 2p^53d^1$ transition discussed previously.

In the L-S coupling scheme, the closed shell ground state for a d^0 system such as Ti^{4+} has zero spin and orbital angular momentum, $S = L = 0$, hence this is a ^1S term and the only level is $^1S_0$. For the $2p^53d^1$ final state configuration, there are 12 possible levels: the triplets – $^3P_{0,1,2}$, $^3D_{1,2,3}$, $^3F_{2,3,4}$, and the singlets – $^1P_1$, $^1D_2$, $^1F_3$. In the absence of any final state coupling, the ΔJ selection rule ($\Delta J = 0$, ±1, no 0 → 0) allows only a single transition: $^1S_0 \rightarrow {}^1P_1$. Turning on the $2p$ spin-orbit interaction mixes the different L-S levels, and produces two accessible levels, a 'triplet' at $-(1/2)\xi_p$ and a 'singlet' at $+\xi_p$. The relative strengths are the familiar 2:1 ratio. So far, the results are the same as for the 1-electron picture.

LFMT predictions diverge from the 1-electron picture when interactions between partially filled $2p^5$ and $3d^N$ shells are included. These Coulomb and exchange interactions, described by the Slater-Condon parameters F^2, G^1, and G^3 [72], cause additional mixing of terms, so that the lowest energy (mostly triplet) level acquires $^1P_1$ character. Thus, even in spherical symmetry, LFMT predicts additional features that cannot be explained by 1-electron theory (**Figure**).

If one next turns on the ligand field portion of the theory, then J is no longer a 'good quantum number', and further mixing of levels occurs. In an O_h field, the symmetry of the initial state is A_1, the dipole operator is T_1, and final states must also have T_1 symmetry. It turns out there are 7 such levels – 4 derived from the $10Dq$ splitting of the 2 main peaks, and an additional 3 transitions not explained in 1-electron theory (**Figure 10**) [68]. (Branching from $O_3 \rightarrow O_h$ is explained in Butler [73]). These features are observed in d^0 systems such as KF, CaF_2, $FeTiO_3$ and ScF_3 [68], and their prediction and confirmation was one of the initial successes that helped confirm the utility of the LFMT approach.

Of course, for d^0 systems, there are no magnetic effects. What does LFMT have to say about the paramagnetic systems of interest in materials science and (bio)inorganic chemistry? For transition metal XMCD, the simplest case is the $2p^63d^9 \rightarrow 2p^53d^{10}$ transition seen with Cu(II) and Ni(I) complexes, as explicated in a classic paper by van der Laan and Thole [71]. They begin by writing the initial and final states of an atom in a magnetic field as $|\alpha JM>$ and $|\alpha'J'M'>$, where J and M are the total angular momentum and magnetic moment respectively, and α designates all other quantum numbers. They note that the temperature-dependent line strength is given by:

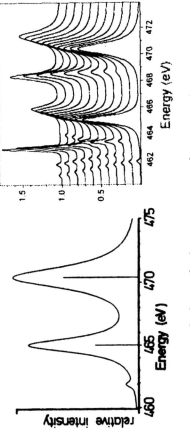

Figure 10 – $2p^6 3d^0 \rightarrow 2p^5 3d^1$ transitions for d^0 system (Ti^{4+}) in (left) O_3 symmetry with p-d coupling [68], and (right) O_h ligand field and p-d coupling. 10Dq is 0 for lowest curve and increases to 3.0 eV for top curve.

$$\left\langle S^q_{\alpha J,\alpha' J'}\right\rangle=\left\langle A^q_{JJ'}\right\rangle\left|\left\langle \alpha J\right|C^{(1)}\left|\alpha' J'\right\rangle\right|^2, \tag{10}$$

where the last factor is the line strength of the $\alpha J \to \alpha' J'$ transition, and the geometric factor $\left\langle A^q_{JJ'}\right\rangle$ distributes this intensity over the different $M \to M'$ transitions:

$$\left\langle A^q_{JJ'}\right\rangle=\left|\sum_M \begin{bmatrix} J & 1 & J' \\ M & q & M \end{bmatrix}^2 e^{-M/\theta}\right| / \sum_M e^{-M/\theta} \tag{11}$$

In the above equation θ is the reduced temperature, $\theta = kT/|g\mu_B|H$, and the squared term in the summation is a '3j symbol' [71]. Once the wave function is described in terms of M and J, the XMCD intensities at T = 0 derive from the angular momentum algebra contained in the 3j symbol.

Life is simple in spherical symmetry. The initial d^9 ^{2}D term is split by spin-orbit coupling into $^2D_{3/2}$ and $^2D_{5/2}$ levels by the 3d spin-orbit interaction, and in a magnetic field; the latter is split by the Zeeman effect into 6 distinct states. At T = 0, there is only one allowed transition, $^2D_{5/2}$ ($M_J = -5/2$) $\to$ $^2D_{5/2}$ ($M_J = -3/2$), thus $\Delta M_J = q = +1$. This corresponds with absorption of lcp x-rays with our optical definition (**Figure 11**).

Of more interest to chemists is the effect of a ligand field on the energies and intensities of different transitions. For example, for Cu^{2+} in D_4 symmetry, the wave function can have B_1, A_1, B_2, and E irreducible representations [74]. Splitting by spin-orbit coupling and a magnetic field along the z-axis yields a Γ_8 ground state that is a mixture of $|5/2,3/2\rangle$, $|5/2,-5/2\rangle$, and $|3/2,3/2\rangle$ levels [71]. When the spin-orbit splitting is small compared to crystal field splittings, first order perturbation theory gives the ground state wave function as [71]:

$$|\Gamma_8\rangle = |b_1\rangle + \zeta_d/\Delta(b_2)|b_2\rangle + (1/\sqrt{2})\,\zeta_d/\Delta(e)|e\rangle \tag{12}$$

The results of the van der Laan and Thole analysis are reproduced in **Table 2**. Note the major difference between LFMT and '2-step' predictions. In the absence of ground state spin-orbit coupling, LFMT predicts a branching ratio of 1 for $q = -1$ (a 100% XMCD effect at the L_2 edge). In contrast, 1-electron theory predicts at most a 50% asymmetry.

The complexity of XMCD spectra increases rapidly as (a) the number of d-electrons nears a half-filled shell, (b) charge-transfer effects require inclusion of multiple configurations, (c) lower symmetry requires inclusion of orientation dependence, and (d) zero field splittings complicate the magnetic field dependence. Since this is a primer and not an encyclopedia, we merely mention potential complications and point the interested reader toward the relevant literature.

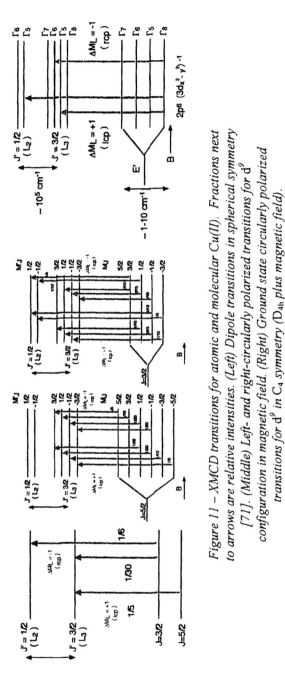

Figure 11 – XMCD transitions for atomic and molecular Cu(II). Fractions next to arrows are relative intensities. (Left) Dipole transitions in spherical symmetry [71]. (Middle) Left- and right-circularly polarized transitions for d^9 configuration in magnetic field. (Right) Ground state circularly polarized transitions for d^9 in C_4 symmetry (D_{4h} plus magnetic field).

Table 2 – Line Strength and Branching Ratio for d^9 System with Γ_8 Character in C_4 Symmetry.		
Dipole Excitation	Line Strength (P)	Branching Ratio (B)
isotropic	1/5	$[2+2\zeta_d/\Delta(b_2) + \zeta_d/\Delta(e)] / 3$
$q = +1$	$[1-2\zeta_d/\Delta(b_2)]/10$	$[1-2\zeta_d/\Delta(b_2) + 2\zeta_d/\Delta(e)] / 3$
$q = -1$	$[1+2\zeta_d/\Delta(b_2)]/10$	1
$q = 0$	$[\zeta_d/\Delta(e)]^2/20$	1

First of all, the calculations become more complex towards the middle of the transition series. As noted by deGroot [75], there are 1512 possible final states for a $2p^5 3d^5$ final state (for example, a Mn^{3+} initial state)! Furthermore, in lower symmetry structures, additional parameters (Ds, Dt) are required to describe the ligand field, and the potential for artificially good simulations rises proportionally. Van der Laan has pointed out that in C_1 symmetry, the XMCD is a sum over three fundamental spectra, and that measurements in four different geometries are required [76]. Searle and van Elp have discussed how the XMCD can vary dramatically for different molecular orientations, especially when there is a zero field splitting comparable or larger than the applied Zeeman splitting [77]. Finally, if charge transfer is significant between the metal ion and the ligands (L), then two or more configurations (e.g. $3d^N$, $3d^{N+1}\underline{L}$) may be required to describe the electronic structure of both the initial and final states. Configuration interaction will add additional free parameters to the spectroscopic model. Given the potential complications to spectral simulation, it is fortunate that an alternate approach exists that requires far fewer assumptions – sum rule analysis.

Sum Rule Analysis

Sum rules are equations based on integrated spectra, and they allow derivation of valuable information without resort to simulation or fitting techniques. For example, most chemists are familiar with the Kuhn-Thomas sum rule, which states that the sum of oscillator strengths f_{no} is equal to the number of electrons N_e [78]:

$$\sum_n f_{no} = N_e \qquad (13)$$

The first x-ray sum rule to consider states that the integrated intensity over particular absorption edges reflects the number of empty states with the appropriate symmetry for the transition [79]:

$$\int_{J_++J_-} (\mu_0 + \mu_1 + \mu_{-1})\, d\omega \propto \frac{4l+2-n}{2l+1} P_{cll}^2 \qquad (14)$$

For $L_{2,3}$ edges, which have an initial p level, the transitions are primarily to states of d-symmetry, since the $2p \rightarrow 4s$ transitions are ~20-fold weaker than the $2p \rightarrow 3d$ transitions [80]. The decrease in the number of d-vacancies across the first transition series is nicely illustrated by comparing the white line intensities of the pure metals [51]. We have used this sum rule to quantify the amount of electron density transferred from copper to its ligands in blue copper proteins [81] and the Cu_A site, as well as the number of $3d$ vacancies in compounds with different Ni oxidation states [82].

For XMCD, the most important sum rules involve projections of the spin $<S_z>$ and orbital $<L_z>$ angular momentum of the absorbing species. In general terms:

$$\delta = \frac{\int_{J_-} (\mu_1 - \mu_{-1})\, d\omega - \frac{c+1}{c} \int_{J_-} (\mu_1 - \mu_{-1})\, d\omega}{\int_{J_++J_-} (\mu_0 + \mu_1 + \mu_{-1})\, d\omega} = \frac{l(l+1) - 2 - c(c+1)}{3c(4l+2-n)} < S_z > \hbar + \alpha < T_z > \hbar \qquad (15)$$

and

$$\rho = \frac{\int_{J_++J_-} (\mu_1 - \mu_{-1})\, d\omega}{\int_{J_++J_-} (\mu_0 + \mu_1 + \mu_{-1})\, d\omega} = \frac{1}{2} \frac{l(l+1) + 2 - c(c+1)}{l(l+1)(4l+2-n)} < L_z > \hbar \qquad (16).$$

Stöhr and König have shown that the $<T_z>$ angular term averages to zero in 'powder' samples [83], so this term in **Equation 15** can often be omitted. The quantities involved in sum rule analysis are illustrated graphically in **Figure 12**. Using these A, B, and C terms to represent the appropriate integrals, neglecting the $<T_z>$ term, and assuming that $\mu_0 = (\mu_1 + \mu_{-1})/2$ [84], yields the following simple expressions for the sum rules:

$$\langle L_z \rangle = \frac{2(A+B)}{3C} n_h \qquad \langle S_z \rangle = \frac{A - 2B}{2C} n_h. \qquad (17)$$

The sum rules have been tested by comparison with experimental measurements [48,51] and theoretical calculations [85,86]; they are generally

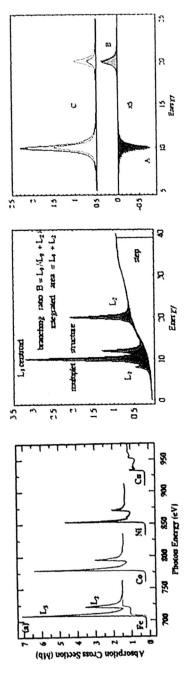

*Figure 12 – (Left) White-line intensities of some first transition metals.
Important quantities in hypothetical L-edge (middle) and XMCD (right) spectra.
C is the average area under the absorption curves.*

thought to be accurate within about 10%. Sainctavit, Arrio, and Brouder have done analytical calculations for Cu(II) in an octahedral ligand field that address the sum rule assumptions [87]. They found that at low temperature, the $<T_z>$ term makes a large contribution to the spin sum rule and cannot be ignored. Others have shown that $<T_z>$ can also be significant for other first transition metals, especially at lower symmetry surface sites [88].

Chemical Applications of XMCD

Deciphering Mixtures

Real world samples are often inhomogeneous. In some spectroscopic techniques, such as EPR, this is not a major problem − non-magnetic components do not give a signal, and overlapping spectra can be separated by exploiting different power saturation curves or by other methods. Inhomogeneity is more a problem for conventional x-ray spectroscopy, where chemical shifts are relatively small compared to natural line widths. XMCD provides an extra tool for separating magnetic and non-magnetic components in an x-ray spectrum, as well as for distinguishing components with different magnetic moments.

For example, in collaboration with David Grahame, we have studied the β*-subunit 'A-cluster' Ni site of *M. thermophila* ACDS protein [89]. This is presumably related to the active site for acetyl-CoA synthesis in the α subunit of carbon monoxide dehydrogenase/acetyl-CoA synthase (CODH/ACS) from *C. thermoaceticum*, where a recent crystal structure has revealed a unique Ni-Cu-Fe_4S_4 cluster (**Figure 13**) [90].

Both high spin (paramagnetic) and low-spin (diamagnetic) Ni(II) have been proposed as constituents of various Ni enzymes [91,92]. When we examined the Ni L-edge spectrum of one particular sample (that might have had some O_2 exposure), we observed a complex spectrum with at least three bands (**Figure 13**). The XMCD spectrum showed that the low and high energy features were magnetic, and most likely represented two components of a single high-spin Ni(II) spectrum. The central peak did not show an XMCD effect and most likely represented low-spin Ni(II). One hypothesis is that Ni occupies both 'external' sites in this particular protein. Our working hypothesis is that the low-spin Ni occupies the more square planar 'M_b' site, while the high-spin Ni occupies the more tetrahedral 'M_a site'. The observed heterogeneity would be hard to infer from the K-edge XANES or EXAFS alone.

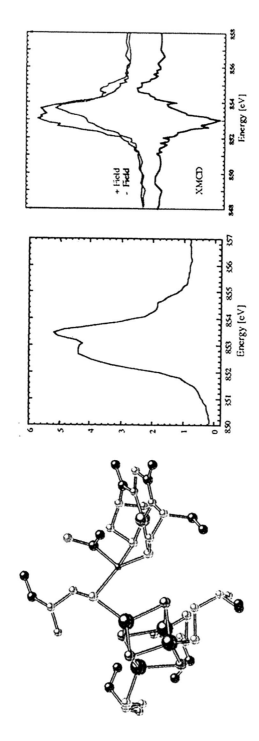

Figure 13 – (Left) Proposed structure for the 'A-cluster' of M. thermophila ACDS protein [90]. (Middle) Ni L-edge absorption spectrum for the β*-subunit A-cluster. (Right) XMCD spectrum for the same sample.

Magnetic Coupling

Since the sign of the XMCD effect reveals the net spin orientation for a given element or oxidation state, this technique can probe the interaction between different species in magnetically coupled systems. The easiest cases to study are interactions between different elements, because the edges are usually well separated in energy. The first such application was a temperature dependent XMCD study of $Fe_3Gd_3O_{12}$ [93]. At room temperature, the primary Fe XMCD signal was negative, while the Gd signal was positive. This indicated that the bulk magnetic moment was dominated by the contribution from the Fe spins, and that the Gd was antiferromagnetically coupled to the Fe. At low temperature, the Gd M_3 edge showed a strong negative XMCD, indicating that the Gd moment became the dominant factor, while positive Fe L_3 edge XMCD again indicated antiferromagnetic coupling.

In collaboration with Edward Solomon and Kenneth Karlin and their co-workers, we have used XMCD to observe antiferromagnetic coupling between Fe and Cu in Karlin's cytochrome oxidase model, $[(F_8\text{-TPP})Fe^{III}\text{-}(O^{2-})\text{-}Cu^{II}(TMPA)]^+$, **Figure 14**. This is a total spin $S_t = 2$ system resulting from antiferromagnetic coupling between $S = 5/2$ high-spin Fe(III) and $S = 1/2$ Cu(II). As expected, Fe and Cu have opposite sign XMCD.

XMCD analysis can also be used to study the magnetic coupling in mixed valence of the same element, provided there is a useful chemical shift between different oxidation states. Our long-term goal is to use XMCD for interpreting the spectra of complex clusters, such as the M center in nitrogenase [94] and the oxygen-evolving complex (OEC) of photosystem II [95]. As a model for the latter problem, we studied the C_{15}-carboxylate derivative of a 'single-molecule magnet' $Mn_{12}O_{12}(O_2CR)_{16}(H_2O)_4$ cluster system from George Christou's lab [96], **Figure 15**. The spectrum shows a strong bipolar signal at both L_3 and L_2 edges. The negative XMCD at the L_3 edge is assigned primarily to the set of 8 Mn(III) ions that are ferromagnetically coupled, while the positive signal at higher energy corresponds mostly to the central cube of 4 Mn(IV) ions whose magnetic moments are predominantly opposite to the net magnetization. A more detailed analysis with LFMT simulations is still in progress.

More complex XMCD is seen for $[Fe_2(II,III)(bpmp)(\mu\text{-}O_2CC_2H_5)_2][BPh_4]_2$ [97]. In a 6 T field, both signals are negative, indicating that the spins are mostly parallel and that the Zeeman interaction overwhelms both the zero field splittings D and the exchange interaction J_{AB}. The observed XMCD was ~34% of the effect expected for two independent and totally oriented Fe(II) and Fe(III) ions, indicating that the temperature and field were not sufficient to achieve total spin alignment. In a weaker field (1 T), the Fe(II) XMCD became quite weak.

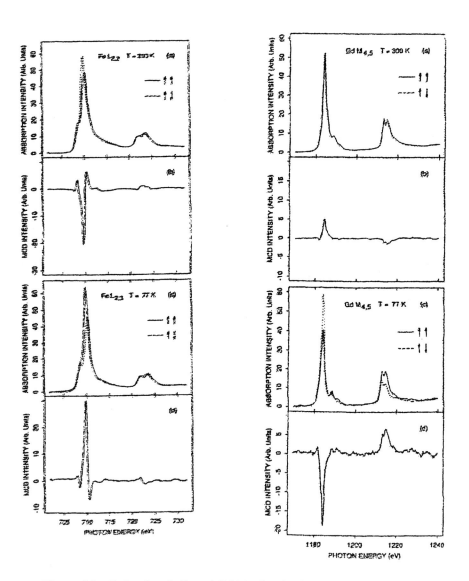

Figure 14 – Fe L-edge (left) and Gd M-edge (middle) XMCD spectra of
Fe₃Gd₃O₁₂ reported by Rudolf et al. [93]. (Right) XMCD for (top) Cu and

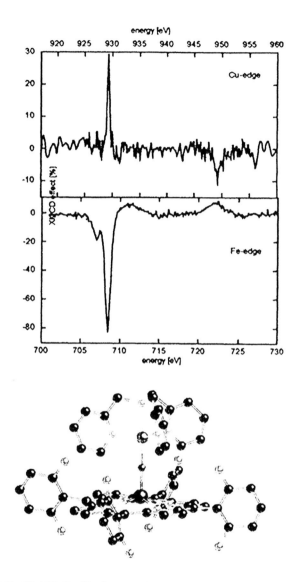

(bottom) Fe XMCD for Karlin's FeCu complex (structure shown below).

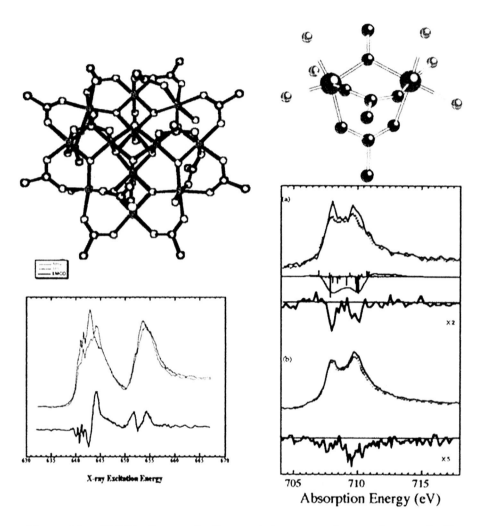

Figure 15 – XMCD of magnetically coupled systems of same element. (Left) XMCD for [Mn₁₂(III,IV] complex in 6 T field. Spectra with lcp (- - -) and rcp (—) and XMCD. (Middle) XMCD spectra for [Fe₂(III,II)(bpmp)(μ-O₂CC₂H₅)₂][BPh₄]₂ in different fields. (a) Top: spectra with lcp (- - -) and rcp

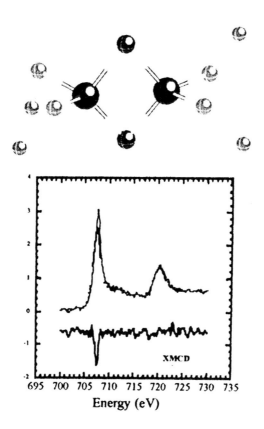

(—); middle: sum of calculated XMCD for Fe(II) and Fe(III); bottom: experimental XMCD (b) XMCD at 1 T: spectra with lcp (- - -) and rcp (—) and XMCD spectrum [97]. (Right) Fe(II)Fe(III) 2Fe ferredoxin and XMCD at 6 T.

Neglecting zero field splittings, the antiferromagnetically coupled Fe(II) should eventually have a positive XMCD at a sufficiently weak magnetic field.

One of the most difficult systems we have encountered is the 'simple' $S = 1/2$ Fe(II)Fe(III) site encountered in reduced 2Fe ferredoxins. Subtraction of a reduced ($S = 2$) Fe(II) rubredoxin (Rd) XMCD signal from the oxidized $S = 5/2$ Fe(III) Rd spectrum yields a bipolar spectrum distinctly different from the experimental spectrum (**Figure 15**). At least two factors complicate this naïve analysis. First, Mössbauer data show that the covalency of Fe in the binuclear site is higher than in mononuclear Rd centers, and assuming a proportionality between isomer shift and L-edge shift, smaller x-ray shifts are expected. Second, one must consider the spectra from all the possible M_S states that give rise to the total $S = 1/2$. Not only is there a contribution from $M_{Fe(III)} = 5/2 + M_{Fe(II)} = -2$, but there are 4 other combinations (3/2, -1; 1/2,0; -1/2,1; -3/2,2) weighted in proportion to their Clebsch-Gordan coefficients [98].

Magnetic Moments from Sum Rule Analysis

Element specific magnetic spin moments have been one of the major applications of XMCD. Thanks to the 'charge', 'spin', and 'orbital sum rules', simple integration of properly normalized spectra can reveal the number of vacancies and the magnetic moments. The first bioinorganic application of XMCD sum rule analysis involved the 'blue Cu' site in plastocyanin. Application of the orbital sum rule yielded a Cu $3d$ specific $<L_z>$ of ~0.07 $\hbar$ and $<S_z>$ of ~0.18 $\hbar$ per Cu, both within 15% of values derived from SCF-Xα-SW calculations (**Figure 16**) [99]. Similar analyses are underway on Ni complexes (**Figure 16**).

Element-Specific Magnetization Curves

The field and temperature dependence of the XMCD effect yield information about the magnetization of a sample that can be interpreted independently of 1-electron or LFMT models. The advantage of x-rays over optical techniques such as the Kerr effect is that each element can be probed separately. An important technological application has been the study of the magnetization of different elements in magnetic multilayers, such as the 'spin-valve' heads used in modern high-density read heads [100]. These devices employ the giant magnetoresistance (GMR) effect to produce a large change in electrical current from a small change in applied field, thus allowing higher density information storage. We remind the reader that the magnetization of ferromagnetic samples can depend not only on temperature and the current applied field, but also on its previous values; in other words, samples can exhibit 'hysteresis' [101].

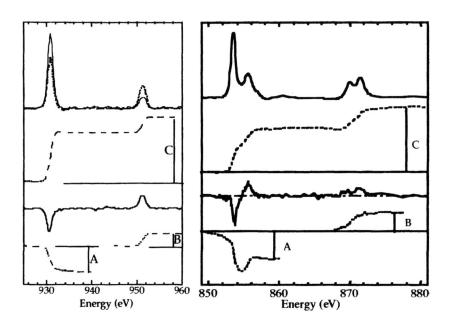

Figure 16 – (Left) XMCD spectra and sum rule integrations for the blue Cu site in plastocyanin. (Right) Absorption and XMCD spectra and sum rule integrations for the Ni(II) site in Ni-doped MgO.

For example, scientists at ESRF have studied a model trilayer system consisting of a (soft) 50 Å $Ni_{80}Fe_{50}$ layer, a variable thickness metallic Cu spacer, and a (hard) 50 Å Co layer, using fluorescence detected XMCD at the Ni and Co L_3 edges. With a thin (60 Å) Cu spacer and a slowly changing applied field, both Ni and Co reverse magnetization with the same coercive field, showing that the layers are strongly ferromagnetically coupled. With a thicker (100 Å) Cu spacer, the $Ni_{80}Fe_{50}$ layer requires a much smaller coercive field – the magnetic layers have become decoupled. By employing a pump-probe technique, the authors were able to observe the dynamics of the magnetization process on a nanosecond time scale [102].

For the paramagnets of interest in (bio)inorganic chemistry, the simplest model for the magnetization of N interacting atoms in volume V is given by $M = M_0 B(x,J)$, where M_0 is the saturation magnetization, $x = (g\mu_B H)/(kT)$ and $B(x,J)$ is the Brillouin function:

$$B(x,J) = \frac{2J+1}{2J}\coth\left(\frac{2J+1}{2J}x\right) - \frac{1}{2J}\coth\left(\frac{1}{2J}x\right) \quad (18)$$

which reduces to XMCD ~ $\tanh[(g\mu_B H)/(kT)]$ for S = 1/2 systems with no orbital moment [101]. The curves in **Figure 17** illustrate the potential of XMCD magnetization curves as a characterization tool separate from sum rule analysis and multiplet simulations. Magnetization curves should be especially useful for the analysis of mixtures, where different uncoupled species might magnetize at different rates. Of course, as noted by Pavel and Solomon [103], systems with zero field splittings such as Fe(II) can exhibit far more complex magnetization curves. Only recently has the quality of magnetization curves improved enough to warrant a more sophisticated analysis.

Electronic Structure

In UV-visible spectroscopy, MCD is often used to bring out detail in absorption spectra that are otherwise broad and featureless. L-edge XMCD can be used in the same manner. For example, the L-edge spectra of V(III) complexes are relatively uninformative, and can be modeled by a wide set of ligand field parameters. In contrast, the XMCD spectra contain a wealth of structure that puts additional constraints on any LFMT simulation (**Figure 18**).

XMCD can also help to distinguish $p \rightarrow d$ transitions from those that are primarily charge-transfer. In **Figure 18**, we show recent XMCD spectra for a Ni(I) complex from Charles Riordan's lab. As expected for a d^9 system, there is a strong signal at the L_3 edge, and the XMCD approaches a 100% effect at the L_2 edge. In contrast, a feature near 856 eV has almost no dichroism. We attribute the latter to a charge-transfer transition to an empty π^* orbital. DFT calculations are being done with Jorge Rodriguez to quantify these assignments.

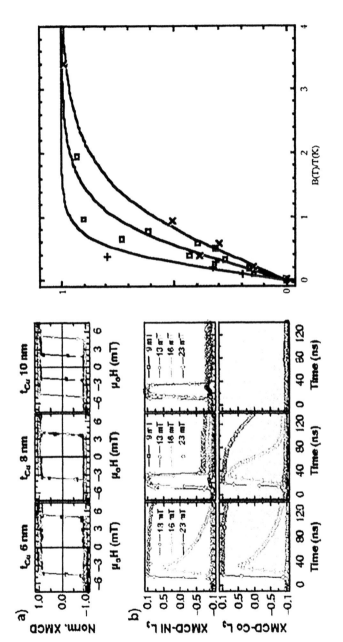

Figure 17 – (Left, top) XMCD magnetization curves obtained on the $Ni_{80}Fe_{20}/Cu/Co$ trilayer system. Ni XMCD - filled circles, Co XMCD - open circles. (Left, bottom) Time dependence of Co and Ni magnetization. (Right) XMCD magnetization curves for Cu(II) in plastocyanin (x x x), compared with Ni(II) ($\Box$ $\Box$ $\Box$) and Mn(II) (+ + +) doped into MgO.

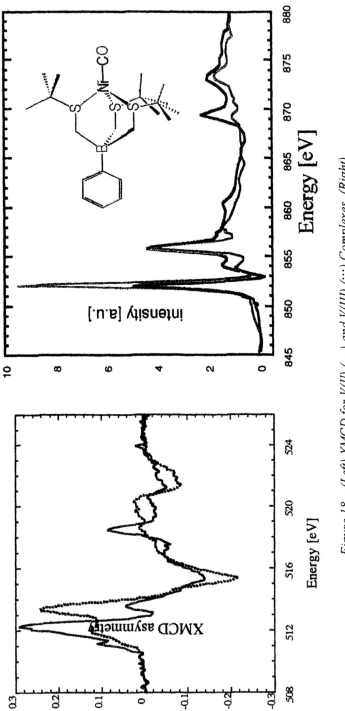

Figure 18 – (Left) XMCD for V(II) (—) and V(III) (···) Complexes. (Right)
XMCD for Riordan's Ni(I)CO complex (structure shown in inset).

Summary – A Dose of Reality

'caveat emptor'

Few would argue that XMCD spectra, in principle, contain a wealth of information. As circularly polarized beamlines proliferate worldwide, access to and conduct of these measurements should become significantly easier. In ideal cases, XMCD should be able to reveal (a) the distribution of spin and orbital angular momentum in transition metal complexes (from sum rule analysis), (b) the strength of magnetic coupling between different centers (from the field dependence), and (c) the total magnetic moment (from the magnetization curve). In practice, intelligent use of the technique requires some caution.

For example, for systems more complicated than d^p ions, the XMCD effect at a given edge is often bipolar or even more complex (see the calculations of van der Laan and Thole [70]). For the analysis of coupled systems, one is often faced with the question – is the XMCD bipolar because of antiferromagnetically coupled ions, or because of side lobes in a single component spectrum? In the extreme cases of Fe-S clusters, it seems that individual features can completely overlap, leaving a relatively featureless XMCD with multiple interpretations.

Experimental artifacts are also a concern, particularly radiation damage. Many samples are photoreduced in seconds on modern beamlines, and precautions such as sample motion or rapid scanning are often essential. Other potential problems include surface oxidation of reactive samples, beam heating at cryogenic temperatures, and surface anisotropy of magnetic properties. As discussed above, each detection scheme has the potential for mistakes as well.

Despite the potential difficulties, XMCD opens a window into electronic and magnetic structure and provides information that is often difficult to obtain by other techniques. For well-chosen problems, it should become a significant tool for the inorganic and bioinorganic communities.

Acknowledgements

'it takes a village ...'
African proverb

Our XMCD work has been made possible by generous support from the NSF (BIR-9015323, BIR-9317942, and SMB-9107312), NIH (GM-44380), and the DOE Office of Biological and Environmental Research. The support staff at NSLS, SSRL, and the ALS were essential for successful measurements, as were the beamline scientists – George Meigs at NSLS U4-B, Jeff Moore at SSRL 8-2, and Elke Arenholz and Tony Young at ALS 4.0.1. The hard work of postdocs

(Simon George, Jan van Elp, Jie Chen, Gang Peng, Marie-Anne Arrio, Lisa Miller, Tobias Funk, and Hongxin Wang) and students (Jason Christiansen, Xin Wang, Craig Bryant, and Wei-wei Gu) is deeply appreciated. Finally, I thank C. T. Chen and Francesco Sette for help getting XMCD started at the NSLS 'Dragon' beamline, George Sawatzky for initial encouragement to try paramagnetic XMCD, and Frank deGroot for a long and patient collaboration.

References

[1.] Born, M.; Wolf, E. Principles of Optics; Fifth ed.; Macmillan: New York, 1974.

[2.] Feynman, R. P.; Leighton, R. B.; Sands, M. The Feynman Lectures on Physics; Addison-Wesley: Reading, 1963; Vol. 1.

[3.] Kliger, D. S.; Lewis, J. W.; Randall, C. E. Polarized Light in Optics and Spectroscopy; Academic Press: San Diego, 1990.

[4.] Huygens, C. Traité de la lumière Leiden, 1690.

[5.] Maulus, É. L. Bul. Sc. Soc. Phil., **1809**, 1.

[6.] Newton, I. Opticks: or, a treatise of the reflexions, refractions, inflexions and colours of light.; S. Smith and B. Walford: London, 1704.

[7.] Fresnel, A. Oeuvres complètes; Imprimerie Imperiale: Paris, 1866.

[8.] Faraday, M. Experimental Research London, 1855; Vol. 3.

[9.] Faraday, M. Experimental Researches in Electricity; Green Lion Press, 2000.

[10.] Maxwell, J. C. A Treatise on Electricity and Magnetism Oxford, 1873.

[11.] Zeeman, P. Nature, **1897**, 55, 347.

[12.] Zeeman, P. Nobel Lecture, **1902**.

[13.] Röntgen, W. C. Nature, **1896**, 53, 274-276.

[14.] Forman, A. H. Phys. Rev., **1914**, 3, 306-313.

[15.] Forman, A. H. Phys. Rev., **1916**, 7, 119-124.

[16.] Erskine, J. L.; Stern, E. A. Phys. Rev. B, **1975**, 12, 5016.

[17.] Keller, E.; Stern, E. A., in EXAFS and Near Edge Structure III; Hodgson, K. O., Hedman, B. and Penner-Hahn, J. E., Ed.; Springer: Berlin, 1984, pp. 507-508.

[18.] Thole, B. T.; van der Laan, G.; Swatazky, G. A. Phys. Rev. Lett., **1985**, 55, 2086-2088.

[19.] van der Laan, G.; Thole, B. T.; Sawatzky, G. A.; Goedkoop, J. B.; Fuggle, J. C.; Esteva, J.-M.; Karnatak, R.; Remeika, J. P.; Dabkowska, H. A. Phys. Rev., **1986**, 34, 6529-6531.

[20.] Schütz, G.; Wagner, W.; Wilhelm, W.; Kienle, P.; Zeller, R.; Frahm, R.; Materlik, G. Phys. Rev. Lett., **1987**, 58, 737.

[21.] Chen, C. T.; Sette, F.; Ma, Y.-J.; Modesti, S. Phys. Rev. B, **1990**, 42, 7262-7265.

[22.] van Elp, J.; George, S. J.; Chen, J.; Peng, G.; Chen, C. T.; Tjeng, L. H.; Meigs, G.; Lin, H. J.; Zhou, Z. H.; Adams, M. W. W.; Searle, B. G.; Cramer, S. P. Proc. Nat. Acad. Sci. U. S. A., **1993**, 90, 9664-9667.

[23.] Goering, E.; Will, J.; Geissler, J.; Justen, M.; Weigand, F.; Schuetz, G. J. Alloys Compounds, **2001**, 328, 14-19.

[24.] Scalicky, P.; Malgrange, C. Acta Crys. A, **1972**, 28, 501-507.

[25.] Cox, J. A. M., Ph. D. Thesis, University of Leiden, 1954.

[26.] Huiskamp, W. J., Ph. D. Thesis, University of Leiden, 1958.

[27.] Reynolds, S. P. Space Science Rev., **2001**, 99:, 177-186.

[28.] Hart, M. Phil. Mag. B, **1978**, 38, 41-56.

[29.] Kim, K.-J. SPIE Proc., **1991**, 1548, 73-79.

[30.] Kim, K.-J. SPIE Proc., **1991**, 1345, 116.

[31.] Kitamura, H. Syn. Rad. News, **1992**, 5, 14-20.

[32.] Young, A. T.; Martynov, V.; Padmore, H. J. Elect. Spec. Rel. Phen., **1999**, 103, 885-889.

[33.] Golovchenko, J. A.; Kincaid, B. M.; Levesque, R. A.; Meixner, A. E.; Kaplan, D. R. Phys. Rev. Lett., **1986**, 57, 202-205.

[34.] Mills, D. M. Nucl. Inst. Meth. A, **1988**, 266, 531.

[35.] Mills, D. M., in Third Generation Hard X-Ray Synchrotron Radiation Sources: Source Properties, Optics, and Experimental Techniques; Mills, D. M., Ed.; John Wiley & Sons: New York, 2002, pp. 41-99.

[36.] Giles, C.; Malgrange, C.; Goulon, J.; Debergevin, F.; Vettier, C.; Fontaine, A.; Dartyge, E.; Pizzini, S.; Baudelet, F.; Freund, A. Rev. Sci. Inst., **1995**, 66, 1549-1553.

[37.] Varga, L.; C., G.; Zheng, Y. L.; Pizzini, S.; de Bergevin, F.; Fontaine, A.; Malgrange, C. J. Syn. Rad., **1999**, 6, 1125-1132.

[38.] Leitenberger, W.; Eisenschmidt, C.; Höche, H.-R. J. App. Crys., **1997**, 30, 164-170.

[39.] Freeland, J. W.; Lang, J. C.; Srajer, G.; Winarksi, R.; Shu, D.; Mills, D. M. Rev. Sci. Inst., **2002**, 73, 1408.

[40.] Stöhr, J.; Wu, Y.; Hermsmeier, B. D.; Samant, M. G.; Harp, G. R.; Koranda, S.; Dunham, D.; Tonner, B. P. Science, **1993**, 259, 658-661.

[41.] Tobin, J. G.; Waddill, G. D.; Jankowski, A. F.; Sterne, P. A.; Pappas, D. P. Phys. Rev. B, **1995**, 52, 6530-6541.

[42.] Dürr, H. A.; van der Laan, G.; Spanke, D.; Hillebrecht, F. U.; Brookes, N. B. Phys. Rev. B, **1997**, 56, 8156-8162.

[43.] Nakajima, R., Ph. D. Thesis, Stanford University, 1997.

[44.] George, S. J.; van Elp, J.; Chen, J.; Mitra-Kirtley, S.; Mullins, O. C.; Cramer, S. P., in Synchrotron Radiation in Biosciences; Chance, B., Ed.; Oxford University Press: Oxford, 1994.

[45.] Christiansen, J.; Peng, G.; A. T. Young, A. T.; LaCroix, L. B.; Solomon, E. I.; Cramer, S. P. Inorg. Chim. Acta, **1996**, 118, 229-232.

[46.] Bryant, C., M. S. Thesis, University of California, Davis, 1998.

[47.] Lee, P. A.; Citrin, P. H.; Eisenberger, P.; Kincaid, B. M. Rev. Mod. Phys., **1981**, 53, 769-806.

[48.] Chen, C. T.; Idzerda, Y. U.; Lin, H. J.; Smith, N. V.; Meigs, G.; Chaban, E.; Ho, G.; Pellegrin, E.; Sette, F. Phys. Rev. Lett., **1995**, 75, 152-155.

[49.] Stern, E. A.; Kim, K. Phys. Rev. B, **1981**, 23, 3781-3787.

[50.] Goulon, J.; Goulon-Ginet, C.; Cortes, R.; Dubois, J. M. J. Physique, **1982**, 43, 539-548.

[51.] Stöhr, J.; Nakajima, R. IBM J. Res. Develop., **1998**, 42, 73-88.

[52.] de Groot, F. M. F.; Arrio, M.-A.; Sainctavit, P.; Cartier, C.; Chen, C. T. Sol. State Comm., **1994**, 92, 991-995.

[53.] de Groot, F. M. F.; Arrio, M.-A.; Sainctavit, P.; C., C.; Chen, C. T. Phys. B, **1995**, 208-209, 84-86.

[54.] Pompa, M.; Flank, A. M.; Lagarde, P.; Rife, J. C.; Stekhin, I.; Nakazawa, M.; Ogasawara, H.; Kotani, A. Phys. Rev. B, **1997**, 56, 2267-2272.

[55.] van Veenendaal, M.; M. Goedkoop, J. B.; Thole, B. T. Phys. Rev. Lett., **1996**, 77, 1508-1511.

[56.] Goedkoop, J. B.; Brookes, N. B.; van Veenendaal, M.; Thole, B. T. J. Elec. Spect. Rel. Phenom., **1997**, 86, 143-150.

[57.] Carra, P.; Benoist, R. Phys. Rev. B, **2000**, 62, R7703-R7706.

[58.] Peacock, R. D.; Stewart, B. J. Phys. Chem. B, **2001**, 105, 351-360.

[59.] Stöhr, J.; Wu, Y., in New Directions in Research with Third-Generation Soft X-Ray Synchrotron Radiation Sources; Schlacter, A. S. and Wuilleumier, F. J., Ed.; Kluwer: Amsterdam, 1994, pp. 221-250.

[60.] Bethe, H. A.; Salpeter, E. E. Quantum Mechanics of One- and Two-Electron Atoms; Plenum: New York, 1977.

[61.] Schütz, G.; Fischer, P.; Goering, E.; Attenkofer, K.; Ahlers, D.; Rößl, W. Syn. Rad. News, **1997**, 10, 13-26.

[62.] Brouder, C.; Hikam, M. Phys. Rev. B, **1991**, 43, 3809-3820.

[63.] Since this review was written at World Series time, consider a baseball analogy. The L_2 edge can be likened to a pitcher is either left- or right-handed. The opposing batter can also be left- or right-handed ...

[64.] Nesvizhskii, A. I.; Ankudinov, A. L.; Rehr, J. J.; Baberschke, K. Phys. Rev. B, **2000**, 62, 15295-15298.

[65.] Yamaguchi, T.; Shibuya, S.; Suga, S.; Shin, S. J. Phys. C, **1982**, 15, 2641.

[66.] Thole, B. T.; van der Laan, G.; Butler, P. H. Chem. Phys. Lett., **1988**, 149, 295-299.

[67.] van der Laan, G.; Thole, B. T.; Sawatzky, G. A.; Verdaguer, M. Phys. Rev. B, **1988**, 37, 6587-6589.

[68.] de Groot, F. M. F.; Fuggle, J. C.; Thole, B. T.; Sawatzky, G. A. Phys. Rev. B, **1990**, 41, 928-937.

[69.] de Groot, F. M. F.; Fuggle, J. C.; Thole, B. T.; Sawatzky, G. A. Phys. Rev. B, **1990**, 42, 5459-5468.

[70.] van der Laan, G.; Thole, B. T. Phys. Rev. B, **1991**, 43, 13401-13411.

[71.] van der Laan, G.; Thole, B. T. Phys. Rev. B, **1990**, 42, 6670-6674.

[72.] Weissbluth, M. Atoms and Molecules; Plenum Press: New York, 1978.

[73.] Butler, P. H. Point Group Symmetry. Applications, Methods and Tables; Plenum: New York, 1981.

[74.] Cotton, F. A. Chemical Applications of Group Theory; 3rd ed.; Wiley-Interscience: New York, 1990.

[75.] de Groot, F. M. F.; Vogel, J., in, 2002, pp. in press.

[76.] van der Laan, G. J. Phys. Soc. Jap., **1994**, 63, 2393-2400.

[77.] van Elp, J.; Searle, B. G. J. Elect. Spec. Rel. Phenom., **1997**.

[78.] Atkins, P. W.; Friedman, R. S. Molecular Quantum Mechanics; Third ed.; Oxford University Press: Oxford, 1997.

[79.] Starace, A. F. Phys. Rev. B, **1972**, 5, 1773-1784.

[80.] Stöhr, J. J. Electron Spec. Rel. Phenom., **1995**, 75, 253-272.

[81.] George, S. J.; Lowery, M. D.; Solomon, E. I.; Cramer, S. P. J. Am. Chem. Soc., **1993**, 115, 2968-2969.

[82.] Wang, H.; Ge, P.; Riordan, C. G.; Brooker, S.; Woomer, C. G.; Collins, T.; Melendres, C.; Graudejus, O.; Bartlett, N.; Cramer, S. P. J. Phys. Chem. B, **1998**, 102, 8343-8346.

[83.] Stöhr, J.; Konig, H. Phys. Rev. Lett., **1995**, 75, 3748-3751.

[84.] Carra, P. Syn. Rad. News, **1992**, 5, 21-24.

[85.] Wu, R. Q.; Wang, D.; Freeman, A. J. Phys. Rev. Lett., **1993**, 71, 3581–3584.

[86.] Wu, R. Q.; Freeman, A. J. Phys. Rev. Lett., **1994**, 73, 1994–1997.

[87.] Sainctavit, P.; Arrio, M. A.; Brouder, C. Phys. Rev. B, **1995**, 52, 12766-12769.

[88.] Crocombette, J. T.; Thole, B. T.; Jollett, F. J. Phys.: Condens. Matter, **1996**, 8, 4095-4105.

[89.] Grahame, D. A.; Demoll, E. J. Biol. Chem., **1996**, 271, 8352-8358.

[90.] Doukov, T. I.; Iverson, T. M.; Seravelli, J.; Ragsdale, S. W.; Drennan, C. L. Science, **2002**, 298, 567-572.

[91.] Wang, H.; Ralston, C. Y.; Patil, D. S.; Jones, R. M.; Gu, W.; Verhagen, M.; Adams, M. W. W.; Ge, P.; Riordan, C.; Marganian, C. A.; Mascharak, P.; Kovacs, J.; Miller, C. G.; Collins, T. J.; Brooker, S.; Croucher, P. D.; Wang, K.; Stiefel, E. I.; Cramer, S. P. J. Am. Chem. Soc., **2000**, 122, 10544-10552.

[92.] Ralston, C. Y.; Wang, H.; Ragsdale, S. W.; Kumar, M.; Spangler, N. J.; Ludden, P. W.; Gu, W.; Jones, R. M.; Patil, D. S.; Cramer, S. P. J. Am. Chem. Soc., **2000**, 122, 10553-10560.

[93.] Rudolf, P.; Sette, F.; Tjeng, L. H.; Meigs, G.; Chen, C. T. J. Mag. Mag. Mat., **1992**, 109, 109-112.

[94.] Rees, D. C.; Howard, J. B. Curr. Op. Struct. Biol., **2000**, 4, 559-566.

[95.] Barber, J.; Anderson, J. M. Phil. Trans. Royal Soc. London B, **2002**, 357, 1325-1328.

[96.] Sessoli, R.; Tsai, H.-L.; Schake, A. R.; Wang, S.; Vincent, J. B.; Folting, K.; Gatteschi, D.; Christou, G.; Hendrickson, D. N. J. Am. Chem. Soc., **1993**, 115, 1804-1816.

[97.] Peng, G.; van Elp, J.; Jang, H.; Que, L.; Armstrong, W. H.; Cramer, S. P. J. Am. Chem. Soc., **1995**, 117, 2515-2519.

[98.] van Elp, J.; Peng, G.; Zhou, Z. H.; Mukund, S.; Adams, M. W. W. Phys. Rev. B, **1996**, 53, 2523-2527.

[99.] Wang, H.; Bryant, C.; Randall, D. W.; LaCroix, L. B.; Solomon, E. I.; LeGros, M.; Cramer, S. P. J. Phys. Chem. B, **1998**, 102, 8347-8349.

[100.] Tsang, C. H. F., R.E., Jr.; Lin, T.; Helm, D.E.; Gurney, B.A.; Williams, M.L. IBM J. Res. Dev., **1998**, 42, 103-116.

[101.] Cullity, B. D. Introduction to Magnetic Materials; Addison-Wesley: Reading, Mass., 1972.

[102.] Bonfim, M.; Ghiringhelli, G.; Montaigne, F.; Pizzini, S.; Brookes, N. B.; Petroff, F.; Vogel, J.; Camarero, J.; Fontaine, A. Phys. Rev. Lett., **2001**, 86, 3646-3649.

[103.] Pavel, E. G.; Solomon, E. I., in Spectroscopic Methods in Bioinorganic Chemistry; Solomon, E. I. and Hodgson, K. O., Ed.; American Chemical Society: Washington, D. C., 1998; Vol. 692, pp. 119-135.

[104.] Maruyama, H.; Suzuki, M.; Kawamura, N.; Ito, M.; Arakawa, E.; Kokubun, J.; Hirano, K.; Horie, K.; Uemura, S.; Hagiwara, K.; Mizumaki, M.; Goto, S.; Kitamura, H.; Namikawae, K.; Ishikawa, T. J. Syn. Rad., **1999**, 6, 1133-1137.

Indexes

Author Index

Subject Index

Printed in the United States
36312LVS00001BA/27